ENVIRONMENTAL PLANNING AND MANAGEMENT IN AUSTRALIA

ARTHUR CONACHER & JEANETTE CONACHER

OXFORD
UNIVERSITY PRESS

OXFORD

UNIVERSITY PRESS

253 Normanby Road, South Melbourne, Victoria, Australia 3205

Oxford University Press is a department of the University of Oxford.
It furthers the University's objective of excellence in research, scholarship,
and education by publishing worldwide in

Oxford New York

Athens Auckland Bangkok Bogotá Buenos Aires Calcutta
Cape Town Chennai Dar es Salaam Delhi Florence Hong Kong Istanbul
Karachi Kuala Lumpur Madrid Melbourne Mexico City Mumbai Nairobi
Paris Port Moresby São Paulo Singapore Taipei Tokyo Toronto Warsaw

with associated companies in Berlin Ibadan

OXFORD is a registered trade mark of Oxford University Press
in the UK and certain other countries

National Library of Australia
Cataloguing-in-Publication data:

Conacher, A.J. (Arthur J.), 1939– .
  Environmental planning and management in Australia.

  Bibliography.
  Includes index.
  ISBN 0 19 553819 6.

  1. Environmental management—Australia. 2. Land use, Rural—Australia. 3. Regional planning—
  Environmental aspects—Australia. 4. Environment impact analysis—Australia. 5. Regional planning—
  Citizen participation. 6. Environmental law—Australia. 7. Regional planning—Environmental
  aspects—New Zealand. 8. Natural resources—Law and legislation—Australia.
  I. Conacher, Jeanette. II. Title.

333.760994

Edited by Tim Fullerton
Cover design by Modern Art Production Group
Typeset by Desktop Concepts Pty Ltd, Melbourne
Printed through Bookpac Production Services, Singapore

CONTENTS

List of Figures

LIST OF PLATES

LIST OF TABLES

ABBREVIATIONS

ABS	Australian Bureau of Statistics
ACF	Australian Conservation Foundation
ACLEP	Australian Collaborative Land Evaluation Program (steering committee has representatives from State and Territory agricultural/natural resources agencies plus CSIRO and the Commonwealth AFFA, Bureau of Resource Science and AGSO)
ACRES	Australian Centre for Remote Sensing
ACT	Australian Capital Territory
AFFA	Agriculture, Fisheries and Forestry Australia (previously part of DPIE)
AGSO	Australian Geological Survey Organisation
AgWA	Agriculture Western Australia
ALES	Automated Land Evaluation System
ALGA	Australian Local Government Association
ALIC	Australian Land Information Council—now ANZLIC
ANAO	Australian National Audit Office
ANZECC	Australia and New Zealand Environment and Conservation Council
ANZFA	Australia and New Zealand Food Authority
ANZLIC	Australia and New Zealand Land Information Council
ARC	Australian Research Council
ARMCANZ	Agriculture and Resource Management Council of Australia and New Zealand
ASDD	Australian Spatial Data Directory
ASDI	Australian Spatial Data Infrastructure
ASTEC	Australian Science and Technology Council
AUSLIG	Australian Surveying and Land Information Group
BMP	Best Management Practices
BRS	Bureau of Rural Sciences (Commonwealth)
CALM	Department of Conservation and Land Management (WA)
CaLP	Catchment and Land Protection (Vic.)
CAMBA	China–Australia Migratory Birds Agreement
CAR	A 'comprehensive, adequate and representative' reserve system (under the terms of the 1992 National Forests Policy Statement)

CCC	Catchment Co-ordinating Committee (Qld)
CEARC	Canadian Environmental Assessment Research Council
CEPA	Commonwealth Environment Protection Agency (superseded by the Environment Protection Group in Environment Australia)
CEQ	Council on Environmental Quality (United States, Federal)
CER	Consultative Environmental Review (a low-level EIA—WA)
CFCs	Chlorofluorocarbons (in relation to depletion of ozone in the upper atmosphere)
CIA	In this context it means Cumulative Impact Assessment
CITES	Convention of International Trade in Endangered Species of Wild Fauna and Flora
CLPR	Centre for Land Protection Research (Vic.)
CMA	Catchment Management Authority (Vic.)
CMC	Catchment Management Committee (NSW)
COAG	Council of Australian Governments
CONCOM	Council of Nature Conservation Ministers (Commonwealth)
CRA	Comprehensive Regional Assessments, produced in relation to RFAs
CRC	Co-operative Research Centre
CSIRO	Commonwealth Scientific and Industrial Research Organisation
CSNF	Campaign to Save Native Forests (WA conservation group)
CWMB	Catchment Water Management Board (SA)
CYPLUS	Cape York Peninsula Land Use Study
DCP	Development Control Plans (NSW)
DEH	Department of the Environment and Heritage (Federal)
DEHAA	Department of Environment, Housing and Aboriginal Affairs (SA)
DELM	Department of Environment and Land Management (Tas.)—see DPIWE
DENR	Department of Environment and Natural Resources; now PIRSA (South Australia)
DEP	Department of Environmental Protection (WA)
DEST	Department of Environment, Sport and Territories (Federal) (now Department of the Environment and Heritage (DEH))
DHRD	Former Federal Department of Housing and Regional Development, now DTRD
DHUD	Department of Housing and Urban Development (SA)
DISR	Department of Industry, Science and Resources (Federal)
DLPE	Department of Lands, Planning and Environment (NT)
DLWC	Department of Land and Water Conservation (NSW)
DME	Department of Mines and Energy (SA)
DNR	Department of Natural Resources (Qld)
DNRE	Department of Natural Resources and Environment (Vic.)
DPIE	Former Department of Primary Industries and Energy (Federal), now AFFA
DPIWE	Department of Primary Industries, Water and Environment (Tas.)—incorporates DELM
DPLE	Department of Lands, Planning and Environment (NT)

DPUD	Department of Planning and Urban Development (WA): renamed the Ministry for Planning (MfP) in 1996
DTRD	Department of Transport and Regional Development (Federal)—now DTRS
DTRS	Department of Transport and Regional Services (Federal)
DTUPA	Department of Transport, Urban Planning and the Arts (SA)
DUAP	Department of Urban Affairs and Planning (NSW)
DURD	Former Department of Urban and Regional Development (Federal): later, DHRD and now DTRS
EARP	Environmental Assessment Review Process (Canada)
EC	Electrical conductivity (a measure of soluble salt concentration)
EDO	Environmental Defenders Office
EES	Environmental Effects Statement (Vic.)—equivalent to an EIS
EIA	Environmental Impact Assessment (a process)
EIS	Environmental Impact Statement (the report arising from an EIA)
EIU	Environmental Impact Unit (in relation to the Batelle environmental system)
EMA	Environment Management Authority (ACT)
EMS	Environmental Management Systems
EPA	Environment(al) Protection Authority or Agency
EPP	Environmental Protection Policy
ERDB	Eyre Regional Development Board (SA)
ERIN	Environmental Resource Information Network
ERMP	Environmental Review and Management Program (WA)—equivalent to an EIA
ESD	Ecologically Sustainable Development
ESRI	Environmental Systems Research Institute
EU	European Union
EWS	Engineering and Water Supply Department, South Australia (proposed to become Southern Water and Power)
FAO	Food and Agriculture Organisation of the United Nations, Rome
FEARO	Federal Environment Assessment Review Office (Canada)—now renamed
FOI	Freedom of Information (legislation)
GIS	Geographic(al) Information System
GNP	Gross National Product
HEP	Hydro-electric power
HMASPST	Hobart Metropolitan Area Strategy Plan Study Team (Tas.)
IAIA	International Association for Impact Assessment
ICESD	Intergovernmental Committee on Ecologically Sustainable Development
ICM	Integrated Catchment Management
IDAS	Integrated Development Assessment System—centrepiece of Queensland's *Integrated Planning Act 1998*
IGAE	Intergovernmental Agreement on the Environment
IIA	Integrated Impact Assessment
IPA	Integrated Planning Act (Qld)
IPM	Integrated Pest Management
IUCN	International Union for the Conservation of Nature and Natural Resources

JAMBA	Japan–Australia Migratory Birds Agreement
LCD	Land Conservation District
LCDC	Land Conservation District Committee
LCI	Land Condition Index (used in SA)
LD	Lethal Dose (in relation to the evaluation of the toxicity of substances)
LEP	Local Environment Plans (NSW)
LGA	Local Government Authority
LIST	Land Information System Tasmania
LRIS	Land and Resources Information System (Qld)
LUPIS	Land Use Planning and Information System (CSIRO)
LWRRDC	Land and Water Resources Research and Development Corporation
MAB	Man and the Biosphere program of UNESCO/UNEP
MDBC	Murray–Darling Basin Commission (or Council)
MEDALUS	Mediterranean Desertification and Land Use: a major research project initiated by the European Community under the direction of Professor John Thornes, King's College London
MfP	Ministry for Planning (WA)
MPA	Management Priority Area (a type of reserve in the State forests of WA)
NEPC	National Environment Protection Council
NEPA	National Environment Policy Act 1969 (USA)
NEPM	National Environment Protection Measure
NFPS	National Forests Policy Statement
NHMRC	National Health and Medical Research Council (Australia)
NHT	Natural Heritage Trust
NIMBY	Not in my back yard
NLP	National Landcare Program
NLWR	National Land and Water Resources audit
NOAA	National Oceanic and Atmospheric Administration
NOLG	National Office of Local Government (Federal)
NPI	National pollutant inventory, the setting up of which was a task of CEPA
NPWS	NSW National Parks and Wildlife Service
NRIC	National Resource Information Centre, within the DPIE
NRM	Natural resource management
NSCP	National Soil Conservation Programme
NSW	New South Wales
NT	Northern Territory
NWMPA	North West Master Planning Authority (Tas.)
OCM	Office of Catchment Management (WA)
OECD	Organisation for Economic Co-operation and Development, Paris
ORRCA	Organisation for the Rescue and Research of Cetaceans in Australia
ORSTOM	Office de la Recherche Scientifique et Technique Outre-Mer
PEP	Protection of the Environment Policy (NSW)
PER	Public Environmental Review (WA, Commonwealth: less than a full EIA)
PIRSA	Primary Industries and Resources SA
PIU	Parameter Importance Units (used in the Batelle weighting system)

PMP	Property Management Program (SA)
Qld	Queensland
REEP	Regional Environmental Employment Programs (Federal, under the former Department of Housing and Regional Development)
REP	Regional Environment Plans (NSW)
RFA	Regional Forest Agreements between the States and the Commonwealth, under the terms of the National Forests Policy Statement of 1992
RIRDC	Rural Industries Research and Development Corporation, Canberra
RMPS	Resource Management and Planning System (Tas.)
RPAC	Regional Planning Advisory Committee (Qld)
RPAG	Regional Planning Advisory Group (Qld)
SA	South Australia
SCARM	Standing Committee on Agriculture and Resource Management (Commonwealth)
SEA	Strategic Environmental Assessment
SEPP	State Environmental Planning Policies (NSW)
SFDF	South West Forests Defence Foundation (WA conservation group)
SMMPA	Southern Metropolitan Master Planning Authority (Tas.)
SMPA	Southern Metropolitan Planning Authority (Tas.)
SoE	State of the Environment
SPC	State Planning Commission (WA); renamed WA Planning Commission (WAPC) in 1996
SPOT	Satellite Pour l'Observation de la Terre
SPP	Statement of Planning Policy (WA)
SRDR	Sustainable Rural Development Regions (WA)
SROC	The Southern Region of Councils (SA)
Tas.	Tasmania
TRMPA	Tamar Regional Master Planning Authority (Tas.)
TSS	Total soluble salts
UNCED	United Nations Conference on Environment and Development
UNDP	United Nations Development Programme, New York
UNEP	United Nations Environment Programme, Nairobi
UNESCO	United Nations Economic, Social and Cultural Organisation
US(A)	United States (of America)
USDA	United States Department of Agriculture
USLE	Universal Soil Loss Equation
Vic.	Victoria
WA	Western Australia
WALIS	Western Australian Land Information System
WAPC	Western Australian Planning Commission
WHO	World Health Organisation of the United Nations, Geneva
WISALTS	Whittington Interceptor Salt Affected Land Treatment Society
WRC	Water and Rivers Commission (WA)

PREFACE

The purpose of this book is to examine planning and management responses and solutions to the range of problems and issues associated with resource allocation, land use and environmental degradation in Australia, with an emphasis on non-metropolitan areas. Pastoral and agricultural land uses are spatially dominant; other land uses include mineral exploration and extraction, tourism and recreation, forestry, preservation of flora and fauna, and Aboriginal occupation. Infrastructure, secondary industries, towns and hobby farms also occupy land and impinge on other land uses and the environment in non-metropolitan regions.

Problems and issues posed by the above activities include on- and off-site land degradation, and competition for land and resources amongst the various existing and/or potential users. This competition often leads to conflict amongst the user groups, conflicts which have been particularly intense in or associated with Australia's forests, National Parks, Aboriginal land, coastal zones and rural–urban fringes.

Several approaches are possible in writing a book on managing these problems of land degradation and competition for land and resources.

1   The book could tackle the problems sectorally, by dealing with each land use in turn. A sectoral approach was adopted in most of the discussion papers which led to the formulation of the Australian Commonwealth's *Ecologically Sustainable Development Strategy* (Commonwealth of Australia 1992c). But such an approach makes it very difficult to deal with *integrative* solutions, which we consider to be the key to resolving the problems.

2   The book could tackle the problems by dealing with biophysical *environments* or bioregions—for example, the tropical north, the coastal zone, temperate forests, the high country, the semi-arid and arid regions. This would make an integrative approach possible and relate the various problems directly to the nature of the biophysical environment. Such an approach has been favoured by natural scientists, for example, as discussed in *The National Strategy for the Conservation of Australia's Biodiversity* (DEST 1996; SCARM 1998). But a number of solutions, or management

methods, would be common to different environments, leading to repetition and unnecessary length, as well as other organisational difficulties.

3   In order to resolve the problems of land degradation and competition for land and resources, planning must precede management. It is also necessary to plan and manage 'land' (including natural resources) in an integrated way on a regional basis. In other words, what is needed is a book which shows how regional planning and management can integrate land-use planning (and allocation and management) with environmental protection.

That is the approach taken here; further justification will be provided in some detail in several places. While the approach uses and relies on Australian examples, the methods involved are not unique to this country. Nevertheless, integrative approaches to land and resource use management and planning in Australia are now receiving considerable attention at national, State and local levels.

The book is structured into five Parts, each containing several chapters. In order to clarify some important differences in concepts, **Part 1** first considers some fundamental definitions—of 'resource', 'environment' and 'environmental management', and the distinction between environmental 'problems' and 'issues'. An understanding of these is critical to defining the parameters for environmental policy, assessment, management and planning procedures. Two case studies are then presented, concerning intensive forestry practices (an environmental *issue*) and rural land degradation (an environmental *problem*). They are intended to demonstrate the need for integrated land-use and environmental planning and management on a regional basis. The case studies are also drawn on for illustration in several subsequent chapters.

Part 2 considers the Australian legislation which is designed to resolve environmental and land-use and resource allocation problems, and the agencies which have been set up to implement the legislation. It is concluded that these arrangements, usually with a narrow focus, are often ineffectual in dealing with problems of competing land/resource users and in protecting the Australian environment. Better methods are needed for the evaluation of land and resources and the effects on them of human actions. However, positive changes are taking place, and these are covered in Part 2 and again in Part 5.

The ready availability of sound environmental data is crucial to good environmental planning and management. **Part 3** deals with different methodologies for environmental and land assessment and evaluation, ending with environmental impact assessment (EIA). It is argued that EIA has considerable deficiencies—in particular, its focus on project-specific proposals—whereas other methods are generally more appropriate to integrated planning at regional (and other) scales. It is further maintained that effective planning requires not only appropriate legislation and methodologies, but also (on grounds of morality, equity and pragmatism) adequate involvement by the community in the decision-making processes. **Part 4**, then, looks at public participation in environmental decision-making.

The first four Parts highlight a range of deficiencies in the legislation, agencies, methods and decision-making relating to land management in Australia, and point to the need for environmental protection to be integrated with land-use planning and management at a hierarchy of national, State, regional and local scales, underpinned by sound, strategic planning policies and action plans. **Part 5** reviews current regional planning and management laws, practices, methods and implementation in Australia. It draws attention to their strengths and weaknesses and concludes with a discussion of future prospects.

The rapid and uneven rate of changes in legislation, policies, agencies and nomenclature across all jurisdictions, often associated with changes of government and reform processes, has been a significant problem in researching and writing this book. Partly for this reason there are undoubtedly factual errors in the book—hopefully not too many or too serious—and the authors would appreciate having necessary corrections brought to their attention.

ACKNOWLEDGMENTS

For Mark, Sheree, Andrew, Amy and Max

Many people have assisted us in the preparation of this book and we apologise if, through an oversight, anyone has been omitted from the following.

Over several years, the following people assisted materially with willing and unstinting provision of materials and comments. In no particular order they are: **Tasmania**: Elizabeth Fowler, Director of Planning, Planning Division, of the previous Tasmanian Department of Environment and Land Management; John Pretty, Manager, Planning Services, Department of Primary Industries, Water and Environment. **Queensland**: Peter Bek, Assistant Manager, SoE Unit, EPA; Janelle Allison, Queensland University of Technology; Rod Hewitt, Manager, Southwest Strategy, Charleville; Geoff Edwards, Principal Planning Officer, Department of Natural Resources. **New South Wales**: Dick Smyth, planning consultant; Glen Searle, University of Technology, Sydney. **Victoria**: Jack Krohn, Environment Assessment Officer with the Office of Planning and Heritage, Department of Planning and Development; Wayne A. Chamley, Director, Strategic Planning Division, Department of Conservation and Natural Resources; Kevin Love, Assistant Secretary, Resources and Infrastructure Branch, Department of Premier and Cabinet; Kim McGough, Manager Planning, South Western Region, Department of Infrastructure. **Northern Territory**: Peter Cardiff, then in the Planning Branch, Northern Territory Department of Lands, Housing and Local Government (now the Department of Lands, Planning and Environment); Barbara Singer, Graham Bailey and Randall Scott, in the Department of Lands, Planning and Environment. **South Australia**: Michael Lennon, Chief Executive Officer of the South Australian Department of Housing and Urban Development; Les Heathcote, School of Social Sciences, Flinders University; Andrew Johnson, Department of Primary Industries and Resources; Darryl Hockey, Chief of Staff to the Hon. Rob Kerin,

Minister for Primary Industries, Natural Resources and Regional Development; Nick Harvey, University of Adelaide. **Western Australia**: Sue Thomas, Department of Environmental Protection; Karl Knox-Robinson, University of Western Australia; Hugo Bekle, Edith Cowan University. **Canberra**: Peter Waterman, consultant; Ed Wensing, Policy Manager, Australian Local Government Association; Hayley Odgers, Environment ACT. **Commonwealth**: the Hon. Sharman Stone, Parliamentary Secretary to the Minister for the Environment and Heritage; Terry Ryan, Senior Adviser to the Hon. Mark Vaile, Minister for Agriculture, Fisheries and Forestry; Morag Brand, Environment Australia.

Many others have influenced us consciously or otherwise over the years through their personal example, writing and/or their teaching. Invidious as it may seem, we owe a particular debt of gratitude to Dr John Dalrymple (retired), Professor Tim O'Riordan and (the late) Professor Derrick Sewell.

We thank Julie Kennedy who drew the diagrams with care and patience, and Steve Randles of Oxford University Press for his professional advice on their design; Tim Fullerton, freelance editor for Oxford University Press, and Jill Henry, publishing editor of Oxford University Press, for her good-humoured patience as the deadline for the manuscript blew out by some 18 months. Research for the book was assisted materially by an ARC grant which is gratefully acknowledged. Two anonymous reviewers/readers of a draft of the manuscript made constructive suggestions.

Finally, we thank the generosity of the following for their permission to reproduce materials without charge: The Food and Agriculture Organisation of the United Nations for Tables 10.3, 10.4, 10.5, 10.6 and 10.7; the Ministry for Planning, Western Australia, for Fig. 17.3; and the Scientific Committee on Problems of the Environment for figs 11.4 and 11.5, and Tables 11.2 and 11.5.

Arthur and Jeanette Conacher
Darlington, Western Australia
September 1999

Part 1

The need for Integrated, Regional, Environmental and Land-use Planning: Some Definitions and Two Case Studies

PART 1

Part 1 first introduces some important definitions. The concepts of resource and environment are differentiated and hence resource management and environmental management. These terms have at times been used interchangeably, resulting in confusion or conflicting objectives within some government agencies (and in decision-makers' perceptions). In turn, this has been an important underlying cause of environmental degradation and the failure of politicians to devote adequate resources to preventing and resolving the problem. The distinction is then drawn between environmental issues and environmental problems. The latter is seen as being a relatively straightforward situation—first understanding the processes and mechanisms responsible for environmental degradation and then developing effective solutions. On the other hand, environmental issues are characterised by human conflicts, publicity and politicisation of the situation, and they require solutions which focus on people rather than technology.

Chapters 2 and 3 are case studies of an environmental issue (the conflict over wood chipping of native forests) and an environmental problem (secondary salinity, as one aspect of land degradation). The case studies are developed in some detail for the following reasons: to show the kind of detail required to understand a situation before attempting to devise solutions, since considerations of length prevent the development of other case studies; to provide information and insights which are drawn on in the remainder of the book; and to demonstrate the increasingly urgent need to develop more effective means of resolving environmental issues and problems in Australia.

CHAPTER 1

THE DIFFERENCES BETWEEN RESOURCES AND ENVIRONMENTAL MANAGEMENT AND BETWEEN ENVIRONMENTAL PROBLEMS AND ISSUES

Before introducing the case studies, it is necessary to recognise that there is often some confusion between the concepts and definitions of 'resource' and 'environment'. Any such confusion needs to be clarified at the outset, as the distinctions between the two concepts emerge repeatedly in this book. For similar reasons, it is also desirable to distinguish between an environmental *problem* and an environmental *issue*. Early discussions of these concepts appeared in Conacher (1978, 1980) and are referred to in this chapter, with the addition of more recent studies.

Johnson *et al.* (1997) defined and standardised the term 'environment' and its closely related lexicons (natural environment, non-natural environment, environmental changes, environmental degradation, environmental quality). These may be used as core or key terms in environmental science, ecology, agriculture/agronomy, forestry, archaeology and atmospheric/hydrologic/earth sciences; but they are also used by politicians, farmers and practitioners of the growing fields of conservation biology, environmental ethics, landscape ecology and sustainable agriculture. The meanings of the terms may vary widely both within and between the various disciplines. While definitions generally attempt to simplify, attempts to 'define' environment become complex when qualifications appear and when function (utility), process and human impact are alluded to.

RESOURCE

The important point about the term 'resource'—as is also true of 'environment'—is that it is a *concept* about things (or objects). The concept of 'resource' incorporates several aspects: a resource is utilitarian and anthropic—a thing is not a resource unless it can be used; a resource is used to fulfil human needs—for food, shelter, warmth, transportation and gratification; and it must be placed in a social, economic, cultural, political, administrative and technological context. Uranium, for example, is regarded as a resource by technologically developed societies, but has no significance for a primitive society.

The idea that something may not be a resource introduces Haggett's (1972) associated concept of *stock*. The sum total of all the material components of the earth, including both mass and energy, and physical and biological matter, can be described as *total stock*. Despite its abundance, the vast proportion of the earth's total stock is of little interest to people. Either it is wholly inaccessible under existing technology, or it consists of substances which people have not yet learned to use. Resources are therefore a cultural concept. A stock *becomes* a resource when it can be of some utility to people. It is worth noting that the transformation of stock to resource can work in reverse: with changing technology, or for other reasons such as economic, social or cultural collapse, a resource may lose its utility.

Characteristics of Resources

Despite Haggett's definition of stock, resources can be both physical (for example, mineral ores) and human. Human attributes of numbers, work capacity, intelligence, knowledge and vitality are just as fundamental a resource as are coal deposits, timber, land and water. Indeed, within the utilitarian definition of resource, one cannot exist without the other.

Traditionally, resources are classified as non-renewable (finite) or renewable (self-reproducing or cyclic), with the latter category further subdivided according to the degree to which the resource is depleted, sustained or increased according to the degree of human use or misuse.

Measurement of Resources

Methods used to measure resources depend on the nature of the resource. Measurement usually involves an attempt to delimit quantity, flow, renewability or quality. But the quality of the resource is narrowly assessed in terms of its utility, and a monetary parameter or value is often applied: namely, what is it worth in dollar terms? Measurement is part of the intention to satisfy the utilitarian needs and wants of people, and underpins modern economies—economic development and standards of living/quality of life.

Resource Management

The goal of resource management, in its strict sense, relates to the previous comments. Couched in traditional economic terms, resource management is devoted to ensuring a continuous supply of goods and services to satisfy peoples' needs and wants. Associated objectives are those of maximising profits, production and efficiency, and minimising costs—which may lead to over-exploitation of the resource. Thus, resource management may be defined as:

> the set of technical, economic and managerial practices which use resources for the purpose of satisfying peoples' utilitarian needs and wants under prevailing socio-economic and technological conditions (Conacher 1978:438).

The practices referred to in the definition include extraction, manufacturing, marketing, waste disposal and recycling. All of these processes can be included in the idea of a 'resource system' (Fig. 1.1).

The human needs and wants in the definition are often seen by resource managers in monetary terms. The objectives of the managers, therefore, are narrowly conceived in practice (but theoretically need not be, as later discussions will show). This characteristic of traditional resource management has often given rise to environmental problems. But there is a further important element: resource managers such as mining engineers, business managers, farmers and senior public servants, are generally specialised by training and hence in their outlook and attitudes. By virtue of their training and job focus, they often have strong links to powerful politico-economic processes. The narrow conception of resource management is hence perpetuated.

Further, it would be unrealistic to expect a company manufacturing a product to sell competitively on the open market to give equal consideration in its endeavours to other, environmental objectives, which may conflict with the company's prime function, or at best not enhance it. This is not to say that the company may behave irresponsibly, but to point out that the objectives and nature of resource management differ from those of environmental management. For some additional readings on the nature of economics and environmental economics, refer to Simmons (1993), O'Riordan (1995:Chapters 2 and 3), Beder (1996) and Owen and Unwin (1997).

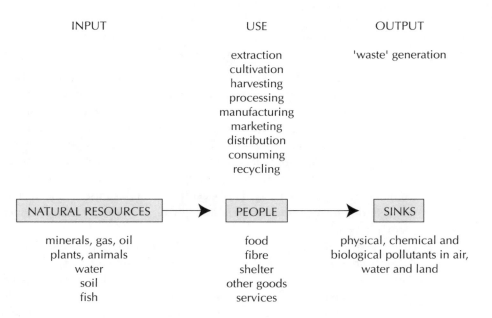

Fig. 1.1 Diagrammatic illustration of a resource system.

ENVIRONMENT

In order to define 'environment', it is helpful to turn to general systems theory—a framework which became very popular in the 1960s and which has become absorbed (or redefined) in the structure of many disciplines. Terms such as 'system' and 'equilibrium' are used widely, often without an appreciation of their origin.

It is not appropriate here to enter into a detailed discussion of the origins and meanings of general system theory. Suffice it to say that it is not strictly a theory at all, but an *approach* to or framework for research and understanding, which was popularised by the biologist, Ludwig Von Bertalanffy (1968).

The development of general systems theory was based on the observation by a number of researchers that the things which are the object of study of widely disparate disciplines, stand in a relationship with one another in ways that are common to many disciplines. 'Systems' themselves are defined mathematically by simultaneous differential equations; in non-technical terms, a 'system' simply refers to a set of objects which have some interaction with one another. Systems *analysis* is seen as an aid in understanding the complexities of ecosystem (and other systems) behaviour, focusing on the dynamic rather than the static. Systems can be used in attempts to simplify reality, to measure change, to set up different assumptions and predict outcomes (Grant *et al.* 1997).

The relationship between a 'system' and its environment is central to general systems 'theory'. It has led to considerable scientific debate, and it is particularly germane to the present discussion.

As defined by Hall and Fagen (1956):

> a system's environment is the set of all objects whose behaviour is influenced
> by the behaviour of the primary system, and those objects whose behaviour
> influences the behaviour of the primary system.

Since many environmental problems relate to *resource systems* as discussed in Fig. 1.1, the above definition of 'environment' can be rephrased by replacing Hall and Fagen's term 'primary system' with 'resource system', as illustrated in Fig. 1.2. It should be noted that the 'objects' in one resource system's environment will include other resource systems. Regardless of the actual wording of the definition, the point that needs to be stressed is that 'environment' must be defined *in relation to something*: it cannot exist in abstract, it must be the environment *of* something. This essential point has also been emphasised by referring to the primary system as a 'pivot', or 'node', in relation to which 'environment' is defined.

One such node is people, or humankind. For example, the Australian Commonwealth's *Environment Protection (Impact of Proposals) Act 1974*, defines 'environment' quite simply as being 'the surroundings of man' (this was before the days when political correctness made the use of such a gender-specific definition unacceptable, even though it is clearly used in a generic sense). According to Simmons (1993), humans construct a world with a set of objects and rela-

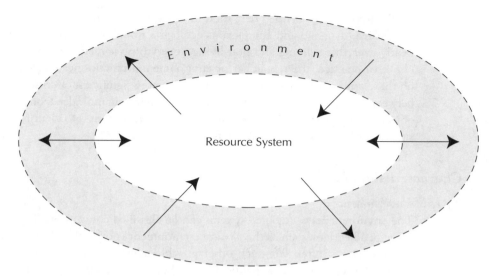

Fig. 1.2 Diagrammatic illustration of a resource system and its environment, indicating the definitive links. It is stressed that this two-dimensional and spatially-limited diagram is a grossly inadequate depiction of the concept of environment.

tionships amongst which they live, rather in the manner of a bubble beyond their skin. This envelope may be quite large (given today's technology and communications), overlap with others, and constantly change in size. The term for the sum total of a person's involvement in the cosmos within which he or she resides is the 'lifeworld'. Simmons (1993:8), citing Berger and Luckman (1967), refers to the human mind as a 'node' in the interplay of individuals, societies and their environments. He also argues that natural sciences overlook the cognitive links which join phenomena with their backgrounds and their history, and so are always limited in their understanding (refer also to Pritchard *et al.* 1998).

Other things can also be the node around which to hang the idea of environment. Animals, insects and plants individually or in groups have environments (which include people) which influence them and which are influenced by them. In the case of animals these environments are more usually termed habitats. In the case of plants we refer to the edaphic factors which affect their growth, health and survival. Both instances comprise parts of the study of ecology, which is simply defined as 'the science of the relations of living organisms with each other and their non-living environments'. In modern terms, this includes the *dynamics* of such interactions in terms, for example, of population ecology, population energetics and cycling of nutrients at a variety of scales (Simmons 1993:5).

The above discussion focuses attention not only on the 'objects' which comprise the environment and the pivotal or nodal resource system, but also on the flows or *links* between resource systems or pivots and their environments. These links are important not only from the point of view of defining environment, but also its measurement, function and behaviour, as discussed further below.

'Environment' as defined here (and in general systems theory) is very comprehensive, dynamic and complex: as ecologists would argue, it encompasses nearly everything, living and non-living ('everything is related to everything else'). In practice, the researcher or environmental practitioner concentrates on those aspects which are judged to have the more significant or critical links between the resource system (or pivot) and its environment. Indeed, one of the objectives or challenges of environmental impact assessment is to identify these significant linkages correctly (Chapter 11).

Characteristics of Environment

Types of Environment

The environments of resource systems can be classified conventionally with a range of adjectives, such as the *physical* environment (note—the all-important *of* something is often left out); other possible adjectives include biological, chemical, social, economic, political and cultural. This range of possible types of 'environment' follows clearly and inevitably from the definition. It also highlights the inadequacy of some of the earlier environment protection Acts in some of the Australian States, which restricted the term to aspects of the physical and biological environments (of what is often not stated) or, worse, left it open to interpretation (Chapter 7).

Linkages, Dynamism

Environments may also be classified by the nature of the definitive linkages between them and their pivots or resource systems (Fig. 1.2):
- by the direction of the link (is the resource system affecting the object in the environment, or *vice versa*, or both?);
- by whether the link is direct or indirect (does the interaction between the resource system and the objects in its environment occur directly or through other objects?)—this is a major problem in environmental impact assessment;
- by whether the interactions between the resource system and the objects in its environment are harmful (–ve) or beneficial (+ve) to either the resource system or the environmental object;
- by the strength of the link, and
- by the duration of the link (incorporating the factor of time).

Such an approach stresses the functional relationships between resource systems and their environments. In principle, it is much to be preferred to the static and more facile, conventional approach.

Equilibrium? Stability?

Environmental managers have also been concerned with the stability of the environment in question—in particular, the degree to which it exhibits a *tendency* to attain a state of dynamic equilibrium, or steady state, or homeostasis with its resource system. These terms mean much the same things although they are not always used interchangeably. All, however, indicate a *dynamic* and not a static relationship between a system and its environment. Environmental managers have been particularly interested in the potential for links between resource systems and their environments to break down, leading to disequilibrium conditions or possibly to a new, different equilibrium. The degree to which environments can adapt to future uncertainties is, therefore, an important characteristic.

More recently, it has been argued that in reality a 'stable' environment, or one in 'equilibrium' with its resource system (or *vice versa*) as defined in classical systems theory, does not exist. The idea of a self-regulating set of changes moving in the general direction of increasing complexity of organisation (in ecosystems) has been challenged by empirical work in the USA (cited in Simmons 1993). It was suggested that there is no determinable direction of change and that a system's interactions with its environment continue without reaching a condition of stability. There were none of the expected features predicted by ecosystem theories such as an increase in biomass or diversification of species. A forest is a shifting mosaic of trees and other species without an emergent, holistic collectivity except in the physiognomic sense: in fact, the systems are in a state of disequilibrium—but the *concept* of equilibrium still has validity. Similar arguments have long been raised in biogeography in relation to the now-discredited concept of a *climax* vegetation community (that is, vegetation in some kind of stable equilibrium with climate particularly, but also landform and soil). And in pedology, Bockheim's (1980) review and modelling of data from published research into chronosequences has shown that most soil properties continue to change even after more than a million years. More recently, Huggett (1998) showed that soils 'evolve' through continual creation and destruction at all scales, and may progress, stay the same, or retrogress, depending on the environmental circumstances. Thus disturbance and readjustment of natural systems are continual and the outcomes of change are poorly predictable. In this respect, ecological views look forward to chaos theory. They also anticipate the need for the precautionary principle in natural resource management and planning.

Simmons (1993) argued that the foundations of ecology as an over-arching narrative for human-environment relations might therefore be questioned. If ecology is the study of disequilibrium, how can it be determined what is worth measuring?

Chaos Theory

It appears that even simple systems have inherent uncertainties (aperiodicities, asymmetries) or 'errors' which generate motion so complex and sensitive that it appears random. Change may be constant as well as unpredictable (Gleick 1987; Worster 1990). Thus, this discussion has led quickly to the now-extensive literature relating to chaos theory, the class of behaviour which deals with mathematically *non-linear* relationships between objects. However, these relationships are not 'chaotic' in the layperson's sense of the term: curiously, there is an element of predictability about, or there are boundaries to, the ways in which chaotic systems behave. It is interesting that the vast majority of relationships in the real, natural world (as distinct from the carefully-controlled, experimental conditions created in laboratories) appear to be non-linear.

Measurement of Environment

Conventional ways of classifying 'environment' lead to traditional parameters of measurement—of soils, water, air and biota; or of population, buildings, land uses and transportation. Being essentially descriptive, lists of species, inventories or statements concerning the 'condition' of an environment, tell us nothing about the dynamic interactions which environmental managers are required to identify, predict, influence and manage. Nevertheless, traditional measurements are necessary and important; for example, to provide a basis for identifying and quantifying change. They are particularly important for developing environmental indicators and benchmarks (Chapter 8).

Measurements of environmental properties have generally not been made in economic terms—unlike measurements of resources. However, some economists moved into the area of environmental management (Department of Science and the Environment 1979) and, with others, have attempted to place a monetary value on various intrinsic environmental parameters (Daly and Cobb 1989; Costanza 1992; Lutz 1993). What, however, is the monetary value of wilderness? of clean air? of scenery? Various approaches have been taken in attempts to answer these kinds of questions, such as the amount of money people are prepared to spend to reach a wilderness, or the income which could be generated from the area if used in some other way. Some, such as Pimentel *et al.* (1980), have used cost/benefit analyses which also incorporate environmental and social factors (or 'externalities'). In one extraordinary attempt, ecologist Robert Costanza and co-authors valued world ecosystem services and natural capital at US$33 trillion per year, an exercise that was described as either foolhardy or heroic (Masood and Garwin 1998). Others have employed a different approach to cost or value the environment through an auditing or national accounting process akin to that used to calculate gross national product (GNP). The Australian Bureau of Statistics (ABS 1997b), in developing its environmental expenditure data, was guided by several international systems evolved by the OECD, United Nations and the World Bank. In general, these approaches gather data on expenditures covering goods and services provided by the environmental management 'industry' (to protect air, water, soils and biodiversity, and to control pollution and waste disposal).

Economic instruments such as cost/benefit analysis and measurements of so-called externalities, or contingent valuation, can be used, for example, to bring about changes in the way the environment is used (examples: Tietenberg 1993; Beder 1996). The 1996 Australian *State of the Environment* report sets out examples of the various economic mechanisms which might be used to conserve biodiversity (State of the Environment Advisory Council 1996:4.45).

Rather cleverly, these approaches have enabled economists to usurp the debate over 'sustainability', arguing that 'development' and 'environmental management' are compatible, although some (Beder 1996) would disagree. This well illustrates the point made at the beginning of this chapter concerning the confusion over the terms 'resource' and 'environment' and the tendency in some quarters to use them interchangeably. In the foregoing examples, nature is thus treated as a commodity (or a resource supplying goods and services for human needs and wants), subject to various market forces or policies.

None of these and other approaches provides a satisfactory outcome; though economists argue that by using a financial unit, otherwise contrasted land uses can be compared directly. How sobering that 50 years ago, biologist Aldo Leopold (1949) was arguing that the core of his 'land ethic' was that land was not a commodity to be bought and sold but a community of which humans form a part. Encouragingly, this community emphasis is still reflected today in, for example, the Western Australian Department of Environmental Protection's program, which is seen as being

> to ensure, with the Western Australian community, that our environment, with the life it supports, is protected for now and into the future (Government of WA 1995:49).

The program is underpinned by the basic philosophy that in environmental protection there is a threshold which distinguishes an acceptable from an unacceptable environment. Protection is seen largely as working with people to protect the environment *to meet aspirations based on human values* (emphasis added). Similar objectives are expressed in different ways by environment protection agencies in all Australian jurisdictions.

Thus, in marked contrast to the measurement of resources—and the views of some economists—*environmental* measurements are often made in terms of quality, increasingly in relation to some standard (as for noise, air and water quality, and safety). There are now internationally recognised (and Australia-wide) standards for these environmental attributes. A good example is the standard set by the World Health Organisation of 500 mg/L total soluble salts concentration in water used for human consumption. However, standard setting for economic, social, political and cultural attributes of the environment, by comparison, have lagged or are non-existent; although some are being addressed (Chapter 8). What levels of traffic congestion, housing density, crime or urban ugliness, for example, as expressed in terms of tolerance of critical stress levels, are 'acceptable' and to whom? The OECD (1998) has listed some key environmental data for its 29 member nations, covering matters such as threatened species, forested areas, air pollutant emissions, waste generated, noise levels, population densities, energy use and traffic volumes. The data are useful in a comparative sense and as performance measures, and the development of environmental indicators embraces similar approaches.

Additionally, measurements can be made of the linkages between environments and their resource systems (or pivots). Statistics can be used to measure the strength, direction and the nature of the linkages, using simple and multiple correlation and regression between two or more phenomena. An environmental impact matrix (Chapter 11) can be used to indicate the magnitude and importance of the effects of resource system activities on sets of environmental attributes, and assess whether those effects are positive (beneficial) or negative (harmful). Interviews of populations can be used to measure peoples' reactions to and perceptions of predicted changes in the relationships between resource systems and their environments: gaming simulation can be a useful tool in this context. Time series analysis can be used to measure or describe changes through time, to predict the responses of environments to change, and to identify their stability or degree of adaptability to (usually human-induced) change. Mathematical modelling can also be used in this context, although it has severe limitations in the number of parameters that can be included and in the restricting assumptions that have to be made.

And finally, there is the question of assessing the *significance* of any changes. The desirability, acceptability or otherwise of changes to the environments of resource systems, or of humankind, can only be assessed by the community (this is not the case when dealing solely with the environments of other living creatures). Although qualitative, this is an important part of evaluation and of the environmental management process (Part 4). The need to involve the community in identifying and measuring significance is particularly important when dealing with the 'commons' (Hardin 1968; O'Riordan 1995:Chapter 19; Ostrom *et al.* 1999)—oceans, rivers, groundwater, air, State-owned land and resources, and the condition of the community itself. Here it is worth stressing that knowledge is crucial if the community is to be able to assess accurately or reliably the desirability of change. Not only is measurement important, therefore, but so too is widespread and effective communication of the findings.

Environmental Management

It is now possible to approach a definition. From the discussion so far, it could be suggested that the *objectives* of environmental management are to enhance beneficial links and minimise adverse links between resource systems (pivots) and their environments. 'Beneficial' and 'adverse' can only be determined by the community (unless dealing with questions requiring scientific expertise, such as the requirements for maintaining the population of a particular species). Thus the objectives of environmental management need to incorporate community perceptions and desires. Similarly, another objective may be to attain environmental system states which are deemed to be 'desirable' by the community. Although both objectives are clearly anthropocentric and would therefore raise the ire of deep ecologists, it needs to be recognised that environmental management can only be carried out by people (at least, at present) and that the desirability (or otherwise) of maintaining biological diversity, for example, is a decision made by people. Thus, environmental management can be defined as:

those activities which enhance beneficial links and minimise adverse links between resource systems (or pivots) and their environments, and which seek to attain desirable environmental system states, in response to community perceptions and desires (Conacher 1978:439).

Human needs and desires appear in this definition and that of resource management. However, resource management is usually narrowly focused whereas environmental management is much more broadly conceived. The community is much more important in environmental management. Confusion is also engendered by the fact that the *objects* which comprise the environment may also be resources, and often are. The differences between resources and environments, and their management, therefore hinge on the ways in which they are perceived and the purposes for which (and the means by which) they are managed.

DIFFERENCES BETWEEN RESOURCES AND ENVIRONMENTS AND THEIR MANAGEMENT

The preceding discussion is now summarised and brought into sharper focus, and some further points are emphasised.

The characteristics of resources and environments, the ways of measuring them, and the objectives of the two sets of management, are quite distinct. There is often a clash of management objectives between the two management groups.

Resource management objectives are usually more specific, more immediate and more clearly defined—as a result of which, environmental management tends to be neglected.

The objectives of resource managers are often single-purpose whereas those of environmental managers are invariably multi-purpose. Resource managers focus on the resource system: environmental managers have to look at both the resource system and its environments *and* consider the total needs of the heterogeneous community.

Environmental managers are concerned with the adaptability of complex environments to future uncertainties and constant change; resource managers seek single or 'simple' solutions to problems within the resource system only, often embedded in engineering and economic terms. These solutions tend to include short-term planning based on minimal maintenance, or exploitation of raw materials, and have limited options for flexibility or adaptation to future uncertainties.

Recognition of community desires and needs is fundamental to the setting of objectives in environmental management. Because communities are heterogeneous, unanimity of desires and values is highly unlikely. Therefore, multiple-purpose objectives within a mediation process are inevitable or at least desirable. In contrast, resource managers consider community needs only insofar as they are needed for, or threaten to adversely affect, their specific resource system objectives, the need to make a profit and operational efficiency.

The skills and expertise of the two groups of managers are very different. Resource managers have relatively narrow skill bases focused on the specific objectives of their resource system. They are engineers, accountants, farmers or foresters. In contrast, environmental managers need broad-based skills and considerable flexibility, so that they can turn their attention from one set of problems to another (and juggle them) at short notice. They need to be knowledgeable about the functioning and complexities of the biophysical, social, economic and political environments.

The final comments refer to the two groups of managers' own perceptions of their roles. The nineteenth-century viewpoint of resource managers saw management as a rational process carried out by 'economic man' operating under conditions of certainty, complete knowledge and predictability, in the context of rapidly evolving, new industrial systems. Their objectives were to maximise profits and minimise costs. In 1978, Conacher wrote that: 'This viewpoint is now recognised as being wholly unrealistic. The important point, nevertheless, is that there is a clear tendency for resource managers to operate *as though* these conditions can be fulfilled and *as though* the objectives are still valid.' Nothing much has changed.

In contrast, environmental managers operate explicitly under conditions of uncertainty, incomplete knowledge and unpredictability. Their mode of operation is not couched in predominantly economic terms; instead, they talk of maintaining and improving environmental quality. They seek multiple objectives, which have to be flexible to accommodate future uncertainties, and they try to satisfy (if necessary, through mediation and compromise) the ideals, values and desires of heterogeneous communities.

In conclusion, what has been attempted here is to draw a clear distinction between environment, resource, environmental management and resource management. But it needs to be acknowledged that some reconciliation is taking place between the two groups of managers. The ecologically sustainable development (ESD) process, in particular, recognises the need to take account of environmental/social objectives but also to provide economic benefits to society over time (the inter-generational perspective). But that comment in turn raises a host of new issues concerning the multiple definitions of ESD, not discussed here.

THE NATURE OF ENVIRONMENTAL PROBLEMS AND ISSUES

The situations which lead to environmental problems were alluded to a number of times in the foregoing discussion. Sources of problems arise from differences in people's perceptions of objects as resources or parts of the environment, from different management objectives and methods for dealing with resources and environments, from the heterogeneity of human communities, and especially from the ways in which people respond to change. The discussion now focuses more specifically on the conditions which lead to environmental problems and the important differences between environmental problems and environmental issues.

The Initiation of Environmental Problems

The first point is that an environmental problem is perceived as something which has an adverse effect on people, directly or indirectly. Direct effects are clear; what is argued here is that even some adverse change in the relationship between an animal or plant and its environment is not considered to be a 'problem' unless and until it is recognised as such by a human being. This has been criticised as being an anthropocentric (human-centred) viewpoint, which indeed it is. 'Deep ecologists', for example, argue that 'nature' is entitled to exist in its own right, without any 'value' (other than an intrinsic one) being ascribed to it (Devall and Sessions 1985; Tobias 1985; Naess 1989). The point, however, is that in management terms, unless humans recognise a problem or feel threatened or pressured in some way, then no decisions or actions will be taken to rectify it.

Lovelock's (1979) Gaia hypothesis disagrees with this argument, in that he considers that the earth's systems are in some way self-maintaining or self-repairing (a sophisticated equilibrium theory). But even that view is not totally at odds with the above argument if it is conceded that humans are an integral part of an holistic earth system and not somehow separate from it; or, perhaps more problematically, that only humans have the ability to make *conscious* decisions to alter environmental system states.

Continuing with the suggestion that environmental problems arise as a result of some adverse effect on people, the next point is that they occur after there has been some *change* in the interaction between people and their environment. Manifestations of change are numerous: increasing air pollution; increasing social breakdown—crime, political instability, armed conflict; increasing congestion; increasing degradation of soils and water; loss of native plants and animals.

But change in itself is insufficient. The argument is that people are inherently conservative: most people do not welcome change, or feel uncomfortable about it. Thus change in turn *threatens* or challenges individuals or groups of people: it may threaten their economic activities, their social and cultural activities, their physical well-being, including comfort and health, or their psychological well-being (including their aesthetic sensibilities).

Threats set up stresses in people, which in turn activate them to become involved in some way. But people (as already noted) are not all the same, and they will respond differently to the same set of circumstances. Some (a small minority: the preservationists) will take direct action in an attempt to restore equilibrium (or to reverse the change—to return things to their original condition). Another group (the vast majority) will take no action, because they are lazy, indifferent, ineffectual (or unaware of the change). And a third group (another small minority) will approve the change.

Environmental Problems may become Issues

The fact that different groups of people may respond differently to an environmental problem (defined as something which has an adverse effect on humans resulting from some change in the interaction between people and their environment) may lead to *conflict*. The conflicts may be minor—such as the differences between a ratepayers' association and its local authority over the disposal of rubbish; or major—such as the pitched battles that took place between protesters and police over the new Tokyo airport, or between anti-nuclear protesters and police at Greenham Common in the UK. In Australia, some of the major environmental conflicts of recent decades have arisen over the proposal to dam the Gordon below Franklin River in Tasmania, the attempt to mine mineral sands on Fraser Island (Queensland), and the wood chipping of native forests in New South Wales, Victoria, Tasmania and Western Australia. It is these major conflicts, characterised by politicisation, intense publicity and sometimes arrests, that are here termed environmental *issues*.

A fundamental cause of the conflicts, again, is the heterogeneous nature of people: different people have different views on what constitutes 'quality of life'. Some people like living in cities; others feel claustrophobic in them and can't wait to get back to the farm. Some people can't bear to be without loud music; others find it unbearable.

The question then arises: how can such conflicts be resolved? Does one set out to change people's attitudes to conform with one's own? Of course, many people attempt to do just that, or to simply force their preferences on other, less powerful or less influential members of the community. But there is no *a priori* basis in logic (on first principles) for such a solution. It therefore follows that a solution which will be acceptable to the community as a whole and not solely to a dominant minority must at least make the attempt to incorporate equitably all value systems held by the community, difficult as that may be.

Another important point, particularly in the context of this book (and as will be seen in the following chapter), is that most, if not all, environmental conflicts have a *spatial* component. The conflicts do not occur in some kind of splendid isolation, or vacuum. Instead, they involve competition or conflict over the ways in which land or resources are to be used or allocated. Examples include disagreement over whether a particular piece of land should be quarried or protected as a flora and fauna reserve, or the conflict over whaling (still not entirely resolved in the case of Norway and Japan), or the range of practices which emit greenhouse gases and the possible nature, extent and implications of global warming as a consequence.

Such conflicts again draw attention to the attitudes of society concerning the use to which land and resources should be put, and questions concerning quality of life: what kind of quality? and for whom?

THE OBJECTIVES OF ENVIRONMENTAL MANAGEMENT REVISITED

The above discussion raises some points additional to those made previously concerning the objectives of environmental management.

1   Environmental management involves the resolution of conflicts between groups of people. It is not *just* a matter of repairing damage to the biophysical environment.

2   Environmental management should be proactive and not reactive. Emphasis needs to be placed on *preventing* (or at least minimising) conflicts from arising. Once a fully blown environmental issue has developed, it is very difficult to get the conflicting groups around a table and to persuade them to back down on highly public stances. It is much easier to mediate and seek consensus before opinions and viewpoints have hardened.

3   Environmental management involves equity in fully recognising and accommodating different value systems held by various groups of people, and their attitudes towards changes in the interactions (links) between people and their environments.

4   In order to resolve these differences, the community must be involved early, openly and directly in a consultative environmental management process. Solutions to conflicts underlain by deeply held, differing beliefs, attitudes and values cannot be imposed from 'above'.

5   The fact that environmental problems and conflicts involve differences concerning the use to which land or resources should be put, means that environmental management must be integrated with broad-based, land-use planning (or *vice versa*): which, fundamentally, is what this book is about.

CHAPTER 2

INTENSIVE FORESTRY PRACTICES, THE ENVIRONMENT AND PUBLIC CONFLICT: A CASE STUDY OF AN ENVIRONMENTAL ISSUE

This chapter discusses an example of an environmental *issue* as defined in the previous chapter; that is, a situation where disagreement over how a resource (or area of land) should be used, escalated to a highly public and politicised state. Amongst the arguments which are presented in this book for integrated regional planning is the need to prevent such conflicts before they arise. It will be shown that such conflicts or issues, although labelled *environmental* issues, in fact revolve around disagreements over land use. Thus, environmental issues not only have a *spatial* component, but they involve decisions over the allocation of land (or resources); that is, land-use planning. Environmental protection and the more classic idea of land-use planning therefore come together, and are seen as being essentially the same activity.

The example chosen is the conflict over wood chipping in the forests of southwestern Australia. This selection is not because this particular issue is any more (or less) important than any other, but because it is one with which the authors have some familiarity, it is topical, and it illustrates some of the more important themes to be highlighted in this book. Readers are referred to Mercer (1995) for some other examples.

THE ISSUE

The issue concerns the aims and objectives of managing the southwestern forests, especially the unique karri forests (*Eucalyptus diversicolor*). Should the forests be converted to plantations and managed for wood production, profit and jobs, or should they be managed to preserve the natural forest ecosystem, to protect native flora and fauna, and to give pleasure to people? Posing the question in this way perhaps creates too sharp a distinction, since forest preservation also provides income and jobs. Nevertheless, the distinction is very definitely perceived as being real by some commercial timber interests and some conservationists; and the dichotomy has been the basis of a serious and continuing land use and political conflict since 1973. The dichotomy challenges the achievement of ecologically sustainable development (ESD) objectives, and it was also well exemplified in the Regional Forest Agreement (RFA) process (see below).

Closely related questions are whether the two aims and therefore management objectives are, or can be, reconciled (or, whether economic imperatives can sit comfortably in partnership with environmental objectives) and whether it is even possible to 'manage' a *natural* ecosystem. However, this latter question, whilst of considerable scientific and possibly practical interest, is not what the primary conflict over wood chipping has been about, although it certainly impinges on the issue.

This chapter considers first, briefly, the nature of the woodchip industry, and then discusses the 'actors' who were and are involved in the conflict, the attitudes of and arguments used by the actors, a brief sequence of events, and finally the lessons to be learnt from the conflict.

WHAT IS THE WOODCHIP INDUSTRY?

The woodchip industry in southwestern Australia was initially located in a 490 000 ha 'woodchip licence area' in the State forests of the far southwest (Fig. 2.1) but abandoned in the 1990s, so that wood chipping can now take place in more extensive areas of the State forests and on private land. The licence was allocated to a private company, the WA Chip & Pulp Co., now owned by Bunnings, a subsidiary of Wesfarmers. The locational context is particularly important. In a predominantly arid and semi-arid State, the licence area occupied some of the prime areas of the State's cool, humid temperate climate zone, including 27% of the State's forests and 86% (about 163 000 ha) of the world's natural occurrence of the magnificent karri (Plate 2.1) (all the remaining 14% also occurs in the southwest of the State). Thus only about one-third of the woodchip licence area is karri forest, although that is the prized timber and a major tourist attraction. Much of the karri favours more fertile, deep red earths and it often grows in association with marri (*Eucalyptus calophylla*, the white flowering red gum—recently revised to *Corymbia calophylla*, a bloodwood). In addition, there are extensive areas of jarrah (*Eucalyptus marginata*), usually in association with marri, growing in less fertile soils derived from deep-weathered lateritic duricrusts. The jarrah/marri forests are particularly rich in understorey plant species. Indeed, the southwestern botanical province is considered Australia's richest: it contains approximately 27 forest ecosystems and 9000 species of vascular plants, of which more than 70% are endemic to the region (Commonwealth Government and Western Australian Government 1998a:2; Department of Environmental Protection 1997, Section 7:5–6). A rich fauna is associated with this vast plant community, possibly as many as 15 000 to 20 000 insect and closely allied species alone (Abbott 1995). Trayler *et al.* (1996) have described the uniqueness and richness of aquatic species in one southwest bioregion. It is also important that most of the State's few, perennial streams are

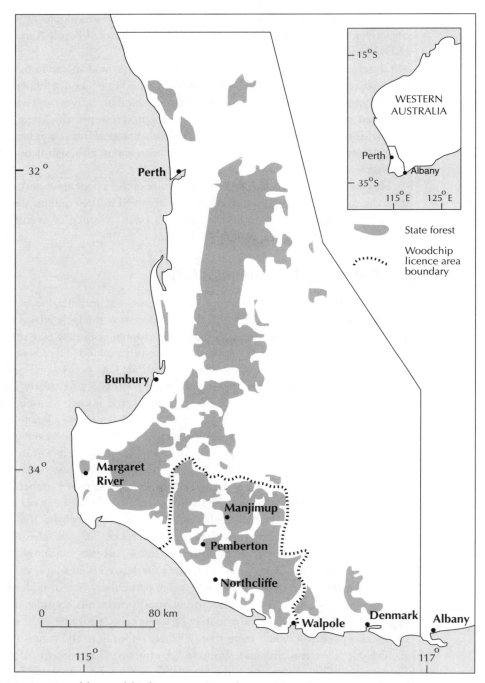

Fig. 2.1 Location of the woodchip licence area in southwestern Australia. From Conacher (1983, Fig. 5.1).

located in the region, which has implications for fauna diversity as well as for future sources of fresh water for the State's burgeoning population, most of which is located in the greater southwestern region.

Plate 2.1   Karri forest (all photographs were taken by Arthur Conacher unless otherwise acknowledged).

Thus, any disturbance to these rich and unique ecosystems will not be without a significant impact, as later discussion will illustrate.

The industry involves clear felling the native forests (Plate 2.2). Felling takes place in areas termed *coupes*, which in karri forest were not supposed to exceed 200 ha in area; in jarrah forest they could be as large as 800 ha (Forests Department, n.d.). The trees are felled and sawlog-quality timber is identified and sent to sawmills. Logs regarded as unsuitable for saw milling are sent to the main chipper, the Diamond chip mill, located south of Manjimup (figs 2.1 and 2.4; Plate 2.3). This mill currently chips about 700 000 tonnes each year. Another chipper is located at Whittakers mill at Greenbushes, chipping mainly sawmill waste, and some other sawmills also have small chippers dealing with sawmill offcuts. Under the original plans, 'seed trees' of karri were left (Plate 2.4) with the intention of

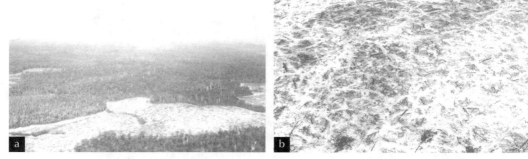

Plate 2.2   a) Aerial view of clear-felled coupe; b) detail, showing the extent of ground disturbance by vehicles.

providing seed for the next 'crop' of karri. The remaining vegetation is then rolled flat and, after drying, burnt in a high-intensity 'summer regeneration burn'. The objectives are first to remove competition, and second to provide an ashbed on which seeds from the seed trees would fall and germinate.

Logs at the Diamond chip mill are chipped in a giant 'pencil sharpener', transported to the port of Bunbury by rail and then shipped to Japan. The original Liberal Government agreement (in the form of the *Woodchip Agreement Act 1969*) was for an annual tonnage of 500 000 tonnes to be exported; that figure was increased to 680 000 tonnes under a Labor Government in 1973 (*Woodchip Agreement Amendment Act 1973*). The woodchip licence has been renewed on several occasions, most recently in 1997. By 1998 annual produc-

Plate 2.3   Aerial view, Diamond chip mill. The sizes of the logs waiting to be chipped can be assessed by comparing them with the 75-tonne logging truck. The conveyor loads the woodchips on to specially constructed railway carriages for transportation to the woodchip berth at Bunbury.

Plate 2.4   Karri seed tree, Beavis 1 coupe. A very small proportion of karri forest is regenerated by the seed tree method: most is clear felled as in Plates 2.2 and 2.10 and then seedlings are planted.

tion approached 1 million tonnes (Table 2.8)—almost one-third of Australia's hardwood woodchip exports—with an export value in 1997 of more than $76 million, some two-thirds the value of all timber-based exports from WA. Nearly 80% of chip logs from Crown land are obtained from mature forest, with approximately one-quarter of the logs coming from karri trees and the majority from marri. A similar quantity of chip material as that obtained from mature karri logs is obtained from regrowth thinnings, mostly karri, and an additional tonnage (73 000 in 1996) is obtained from sawmill residues. Timber from private land accounts for up to 15% of total woodchip production, and plantations on both public and private lands account for about 6% of woodchip production (CALM 1994, 1996; Commonwealth Government and Western Australian Government 1998b).

THE SEQUENCE OF EVENTS

The sequence of events associated with the wood chipping issue is outlined in Table 2.1. Although considerable protest activity understandably took place around the time of the industry's introduction, conflict has continued to the present day, escalating during the RFA process in the late 1990s. From time to time, increases in tonnage are announced, or proposals for a pulp mill are resurrected: indeed the *Woodchip Agreement Amendment Act 1973* required the WA Chip & Pulp Co. to investigate the feasibility of such an industry. In addition, people living in the region, especially around Pemberton, Northcliffe and

Table 2.1	Sequence of events associated with the Western Australian wood chipping issue 1969–99.
1969	State Liberal–Country Coalition Government passes *Woodchip Agreement Act* to export 500 000 tonnes of woodchips per year to Japan. WA Chip & Pulp Co. required to contribute towards the cost of constructing the new inner harbour at Bunbury. Export approval blocked by the Federal Liberal Government on the grounds that the industry is uneconomic.
1973	State Labor Government passes *Woodchip Agreement Amendment Act*, increases tonnage to 680 000 tonnes per year. WA Chip & Pulp Co. no longer required to contribute to the cost of the Bunbury inner harbour, already under construction. Alcoa, the only other user of the harbour, is required to contribute. Export approval obtained from the Federal Labor Government. Forests Department's EIS on the woodchip industry not tabled in State Parliament (Forests Department n.d.). Environmental Protection Authority (EPA) report expresses reservations, with particular reference to secondary salinity (EPA 1973). Not tabled in State Parliament until second reading of the Amendment Bill in the Upper House. Country Party the only political party to express reservations, on environmental grounds.
1973–75	Campaign to Save Native Forests opposes wood chipping industry. South-West Forests Defence Foundation formed in 1974 with express objective of taking legal action to halt the woodchip industry. The Routleys' book, *The Fight for the Forests*, is published in Canberra; quickly goes to three editions (Routley and Routley 1973). Also published: the Tasmanian Law Reform Group book on the woodchip industry (Jones 1975); and the UWA Geography Department Working Paper (Conacher 1975a). Considerable publicity in newspapers, TV and radio. Trades and Labor Council moves for a Green Ban on wood chipping: supported by the power, railways and harbour unions, but not by the Timber Workers Union. Woodchip berth at Bunbury bombed. Second EPA report in 1975 (EPA 1975) clears the woodchip industry, based on somewhat dubious scientific evidence—one report by CSIRO (Mulcahy *et al.* 1975), and another by the Department of Agriculture (Trotman 1974). In 1974, the Conservation Through Reserves Committee (CTRC) of the EPA recommends setting aside the Shannon Drainage Basin, in the woodchip licence area, as a national park (the 'Green Book') (Conservation Through Reserves Committee 1974). Applications by the SFDF for incorporation, and for legal aid, blocked repeatedly. Convener of SFDF and newspaper sued for libel by WA Chip & Pulp Co. Action withdrawn after six months. SFDF application for southwest forests to be listed on the Register of the National Estate ignored.
1976	Wood chipping starts. Acts of sabotage. Senate Standing Committee on Science and the Environment (1976) publishes its report, *Woodchips and the Environment*. Review Committee (1976) rejects the CRTC recommendations for the Shannon; recommends Forests Parks instead (the 'Brown Book'). Review Committee recommendations accepted by EPA (1976) (the 'Red Book') and by State Cabinet. Forests Department required to adopt the recommendations in its next Working Plan, but that Plan, in 1977, lists only 62% of the recommended areas, and renames the Forest Parks 'Management Priority Areas'.
1977	South-West Forests Defence Foundation (SFDF) publishes a report demonstrating non-compliance by the Forests Department of its management undertakings to the Senate Standing Committee on Science and the Environment (SFDF 1979). Senate Committee returns for a second visit. States it is powerless to act. Publishes a supplementary report on the woodchip industry (Senate Standing Committee on Science and the Environment 1978).

1981	Second woodchip industry proposed for Albany, based on forests on private land.
1983	New State Labor Government halts logging in the Shannon drainage basin. Federal Senate undertakes its land-use inquiry (Senate Standing Committee on Science, Technology and Environment 1984).
1984	Department of Conservation and Land Management (CALM) formed from the previous Forests Department, National Parks Authority and the Wildlife section of the Department of Fisheries and Wildlife.
1985	Stream reserves logged in the woodchip licence area.
1986	Tasmanian woodchip licence approved by the Federal Government. Shannon drainage basin created a National Park, to be congruent with the South Coast National Park.
1993	CALM releases plans to increase logging in the woodchip licence area. Renewed controversy. Proposals for a pulp mill—originally proposed in 1969 and resurrected from time to time subsequently—re-emerge. Parts of forest listed on Register of the National Estate.
1994	Occupation of logging areas in the woodchip licence area by conservationists.
To 1999	Subsequent forest occupations and arrests related to clear felling and logging in old growth forests and disquiet over the Regional Forest Agreement process. Public rally in Perth draws 10 000 protesters. West Coast Eagles coach Mick Malthouse, prominent medical practitioners and business people sign full-page newspaper advertisements and protest in the karri forests. National Party supportive of conservationists; Liberal and Labor parties apparently unmoved but clear evidence of internal friction. Annual woodchip exports approach 1 million tonnes.

Walpole (figs 2.1 and 2.4), increasingly have realised the income and employment potential of tourism attracted to the area by mature forests. In the 1990s, arguments here (and elsewhere) were tending to focus on the logging of 'old growth' forests (discussed further below and in Chapter 5 in the context of the *National Forests Policy Statement* and *Regional Forest Agreements* between the Commonwealth and the States).

It is important to recognise that this local industry is part of a global problem. Worldwide, forests are being felled at an accelerating rate estimated at around 12 m ha/yr, sometimes for agriculture, but notably to feed pulp and paper mills (Braatz 1997). Plantations (which represent only about 2% of the total forest area) and regrowth are only partly compensating for the loss. In Australia, wood chipping started in the Eden area of New South Wales in the late 1960s, followed soon after by Tasmania (Routley and Routley 1973). Both attracted considerable controversy—and continue to do so—but it was not until the Western Australian industry was proposed (for the second time, 1973) that some hard-to-obtain economic data surfaced.

A significant feature of the Western Australian industry was the intention to replace the forest with indigenous (local) eucalypts, albeit with an increase in the proportion of the preferred karri. In other parts of Australia, indigenous forests clear felled for woodchips have sometimes been replaced either by farmland or by pines. Thus the arguments in Western Australia have tended to be more subtle: hence the questions posed at the beginning of this chapter concerning the preferred use to which the forests should be put, and whether a natural forest ecosystem can in fact be 'managed'.

To complicate matters, not only does the woodchip industry in Western Australia now extend beyond the original licence area to private, forested land, but increasing areas of farmland are being planted to Tasmanian blue gums (*Eucalyptus globulus*) in particular. Farmers have signed various lease agreements, including with Bunnings (the operators of the WA Chip & Pulp Co.), who have proposed the construction of a second woodchip mill at Albany to process an estimated 3 million tonnes by 2005 – 2007 (*The West Australian* 4/3/98:34; see also Commonwealth Government and Western Australian Government 1998b:56). Despite vague assurances that this proposal will reduce the current pressure on the native forests, the suspicion remains that it will be *additional* to the existing industry.

A potential concern here is the prospect of disease or insect attack on these tree monocultures, as was threatening *radiata* pines in the southwest of WA in 1999. Will pressures be re-exerted on native forests? Alternatively, if markets collapse, where does that leave the farmers and others dependent on the tree resource? The Commonwealth Government and Western Australian Government (1998b:54) have commented that if world trade barriers are reduced, it is probable that the price for pulp logs would decline and affect Australia's supplies to Japan. Precisely this situation occurred by early 1999, raising the prospect of the industry applying pressure on governments to reduce royalties. An alternative is to develop a local pulp and paper industry.

Events and conflicts involve people. Thus, the next section considers the attitudes of and arguments used by the 'actors' in the conflict (Table 2.2). Those used by the proponents are outlined briefly first, as this provides a basis for developing the entire controversy. As will be seen, the issue is considerably more complex than was indicated at the beginning of this chapter. Although the discussion is couched in the past tense (cf. Conacher 1983), virtually all the arguments raised are still current and, if anything, sharper in focus.

ATTITUDES AND ARGUMENTS

The first and perhaps main and ostensibly rational argument used by proponents was that the woodchip industry was to be based on 'bush waste'. This fitted in nicely with integrated forest management objectives. It would use material left behind after sawlog quality timber had been taken from the forest. Thus, it was argued, the rate of cutting of the forest was driven by the sawlog industry and not the woodchip industry. Further, it was argued that the industry would be largely based on the marri which, it was claimed, was a 'non-commercial, competitive weed tree'. Its removal would enable the highly valued karri to be regenerated preferentially, to the overall benefit of the timber industry. The terminology used is revealing: the forest was referred to repeatedly as a 'crop' to be harvested (Forests Department n.d.); that is, it was described in purely silvicultural terms.

Table 2.2   The main actors involved in the Western Australian woodchip industry conflict.

PROPONENTS

The 'developer': the WA Chip & Pulp Co., originally 60% Bunnings and 40% Millars, now primarily Bunnings (part of Wesfarmers), including the Directors, it has significant overlap with some major Western Australian companies; plus shareholders.

The Forests Department, now part of CALM: supposedly the forest managers on behalf of the community which owns the State forests, but which has closely allied itself with the timber lobby from which it obtains much of its funding, through royalties.

The Professional Foresters' Association, and the Hoo Hoo Club: through these, Forests Department officers could make public comments, as individuals, which they are not able to do as members of the public service. These associations supported the woodchip industry.

The timber industry in general, and its mouthpiece, the Forest Products Association. PR spokesmen for the Association often wrote to newspapers without revealing their affiliation. More recently (1990s), the Forest Protection (*sic*) Society took over this role.

Consultants for the timber industry, often claiming to be 'independent' experts.

The Timber Workers Union.

The Liberal and Labor political parties; the Manjimup Shire Council; the Manjimup Chamber of Commerce.

OPPONENTS

The Campaign to Save Native Forests (CSNF); the South-West Forests Defence Foundation (SFDF); the Conservation Council (representing the most conservation organisations in the State); the Australian Conservation Foundation; the WA Forest Alliance.

The Beekeepers' Association (a branch of the Farmers' Union, now the Primary Industries Association).

The Trades and Labour Council.

Members of the local community, especially alternative lifestylers.

Key public figures became signatories to prominent advertisements in the local press.

NEUTRAL (but not always)

The Country Party (now National).

The EPA and the Department of Conservation and Environment (now the Department of Environment).

The Senate Standing Committee on Science and the Environment.

The local community, especially townspeople and farmers.

Lawyers, including those in the Crown Law Department and the Legal Aid Commission.

Given the first argument, it was then maintained that it would be more costly to clear fell the forest without wood chipping than with it. This is because the use of the non-sawlog quality timber would yield some financial return, whereas previously that material (including the crowns of trees) was simply left on the forest floor to rot. It also created a fire hazard.

It was further argued that clear felling was necessary in order to regenerate the forest, with successfully regenerated, previously clear-felled areas cited in support (Plate 2.5).

The other arguments in support of the woodchip industry are less fundamental. It was claimed that the introduction of a woodchip industry would bring employment and income to the region. At times, this reached ludicrous extremes, with arguments that as many as 2000 people would be thrown out of

Plate 2.5   a) Nine-year-old karri regenerating from seed from a clear-felled coupe: Quartz road; b) 34-year-old karri regenerating from a clear-felled coupe: Treen Brook State Forest; c) 45-year-old karri regenerating from a clear-felled coupe: the Rainbow Trail, near Pemberton; d) the hundred-year-old forest: karri regenerated from an area previously planted to wheat. Compare with the seed tree in Plate 2.4 (the ranging rod in d), at 2 m, is a little taller than the figure in Plate 2.4). These areas are used to demonstrate the success of regeneration using the seed tree method. But note the comment in the caption to Plate 2.4.

work if wood chipping was stopped. Over-cutting would be reduced, not increased, by replacing the demand for hardwood timber with softwoods from pine plantations (being grown on farmland). And more than 40% of the forest in the woodchip licence area was to be reserved from cutting—in national parks, Management Priority Areas and reserves along streams and roads, and associated with rocky outcrops and swamps.

Responses to the Proponents' Arguments

Opponents of the woodchip industry have disagreed strongly with the above and other arguments, and continue to do so.

The Woodchip Resource

First, it was argued that far from using 'bush waste', including tree crowns and non-sawlog quality timber, wood chipping was a) based on a *log* resource, and b) the logs were often of sawlog quality. The first point is easily demonstrated by inspection of Plate 2.3, which shows logs at the chip mill waiting to be chipped. Indeed the design of the chipper is such that it must have straight logs—curved branches, let alone twigs and leaves, cannot be handled. Only a small proportion of woodchips is derived from sawmill residues (Table 2.8). The second point is less easily proved, and has a number of elements.

Plate 2.6   Logs at the chip mill identified by the late, retired forester John Thompson as being of sawlog quality. Photo by John Sundstrom.

Plate 2.7   a) Log at a Fly Brook 2 log dump, clearly marked as 'pinholes, chip only'; b) the same log illustrated in a), showing the 'pin holes', made by a borer. The pin holes occur only in the central portion of the log. A small mill, or a spot mill, would cut that section out of the log and use the remainder. But the large, mechanised mills will not do so, and the entire log is rejected; c) another log at the same Fly Brook 2 log dump, marked 'chip only'. Again, it is only the central section which makes it unsuitable for sawmilling. The same argument applies as for the log shown in a) and b).

The marri, on which the industry was supposedly based as the 'non-commercial, competitive weed tree', was in fact milled for sawlogs at one sawmill in the region, although it is not a preferred sawlog timber. With regard to karri, plates 2.6 and 2.7 illustrate logs destined for the wood chipper but which were assessed independently as being of sawlog quality. 'Published proof of the chipping of millable logs' was also cited by the Campaign to Save Native Forests (1987:57).

Clear Felling and Forest Regeneration

Conservationists disputed the Forests Department's claims that clear felling is necessary to regenerate the karri forest. Plate 2.8 shows an area of karri regenerated following what is termed a 'group selection cut', where small clumps of trees are felled and the gap in the forest regenerated. This is very different from the up to 200 ha coupes being clear felled under the woodchip program. Even more convincingly, Plate 2.9 shows an area which had been selectively logged twice: 40 and 20 years prior to the date on which the photograph was taken. The argument is further supported by data gathered by members of the Campaign to Save Native Forests (CSNF) (Table 2.3).

Similar measurements made by the CSNF showed that the high-intensity summer regeneration burns are also unnecessary to regenerate karri (Table 2.4). Indeed, research by the Forestry agency revealed that the germination of karri seeds was less effective with an ash bed than without it (Wallace 1970:39).

Since clear felling is not necessary to regenerate karri, and is not needed for the sawlog industry, it follows that cutting rates would be determined by the demands of the woodchip industry (driven by the contracted export tonnages)

Plate 2.8   Karri regenerating after 'group selection cutting', near the Gloucester Tree, Pemberton. The late, retired forester John Thompson (wearing the hat), is demonstrating the site to a group of conservationists.

Plate 2.9   An area of karri in the northeastern part of the woodchip licence area, regenerating successfully after two periods of selective logging: one 40 years before the photograph was taken, and the other 20 years later. Unusually, the site had also been protected from prescribed burning for at least 20 years. Part of a car in the lower left-hand corner gives some indication of scale.

Table 2.3   **Karri regeneration after heavy selection cutting. Counts of karri trees of different diameters (and by implication, different ages, although the relationship is not direct), in an area selectively logged and burnt in 1965. Measurements taken January 1978. Data from Campaign to Save Native Forests.**

Girth Classes (cm)	Stems per Quarter Hectare	
	Plot 1	Plot 2
0–9	320	291
10–19	95	181
20–29	36	64
30–39	11	36
40–49	6	10
50–59	2	1
60–69	1	0
100–199	0	1
200–299	1	2
300–399	0	2
400–499	2	1

and not those of the sawlog industry. This is demonstrated by comparing the rates of forest clear felling before and soon after the introduction of the wood-chip industry (Table 2.5): average annual rates over ten-year periods increased more than fourfold following the commencement of wood chipping. Similarly, different, more recent data, show that most of the logged timber ends up as woodchips for export. Only a tiny proportion is used as 'value-added timber' and a significant percentage of karri is still wasted (Table 2.6).

Table 2.4   **Karri regeneration without fire. Karri stem measurements in Walpole-Nornalup National Park, from an area not burnt since 1918. Measurements taken in 1977. Data from Campaign to Save Native Forests.**

Girth Classes (cm)	Stems per Hectare		
	Plot 1	Plot 2	Plot 3
0–50	0	4	10*
51–100	2	6	1
101–150	2	1	0
151–200	6	2	0
201–250	3	0	0
251–300	2	2	1
301–350	2	1	1
351–400	0	2	2
401–450	1	2	1
451–500	2	5	1
501–550	2	3	6
551–600	0	0	4
601–650	0	1	2
650–700	1	0	1
>700	0	1	0
Total	23	30	30

* Stems were <20 cm and located in gaps left by two fallen mature trees.

Table 2.5   **Rates of clear felling in the karri forest before and soon after the introduction of the woodchip industry. The implication is clear: wood chipping, not the sawlog industry, drives the cutting rate. Source: CALM *Annual Report* 1993–94: Appendix 2. Their reference is to 'areas of even-aged karri regeneration'.**

Year	Area Clear Felled (ha)
1966	160
1967	210
1968	690
1969	1290
1970	80
1971	80
1972	1110
1973	190
1974	140
1975	630
Annual mean 1966–75	458
Start of woodchip industry	
1976	1170
1977	1630
1978	1630
1979	1920
1980	2090
1981	2950
1982	2310
1983	1930
1984	1750
1985	2400
Annual mean 1976–86	1978

Table 2.6   **Use of timber from native forest logging 1996–97. Source: BIS Shrapnel Forestry Group report, cited in *The West Australian* 11/1/99. Note that the annual timber allocation for jarrah (1st and 2nd grade sawlogs) is 1.36 million m³. Under current specifications, harvesting practices and conversion technologies, the sustainable level for jarrah is approximately 300 000 m³ per annum (Commonwealth Government and Western Australian Government 1998b, p. 42).**

	Karri	*Marri*
Total timber cut	351 000 m³	586 000 m³
	use %	use %
Value-added sawn timber	2	0.03
Structural timber	18	0.3
Woodchip exported to Japan	57	99
Waste	23	1

Related to this is the rate of over cutting. Rundle (1996) noted that questions asked in Parliament in 1974 revealed the extent of over cutting that had been allowed for some time (that is, the forest was being logged much more rapidly than it was regenerating). Data on over cutting of both karri and jarrah forest were also presented in Conacher (1983). Much more recently, the Regional Forest Agreement document was also critical of the continued over cutting taking place in the jarrah forests. The annual allowable jarrah sawlog cut was quoted as 490 000 m³, whereas the *sustainable* yield was estimated at only 300 000 m³ (Commonwealth Government and Western Australian Government 1998a:16–17).

Employment and Tourism

Far from creating jobs, conservationists argued that the woodchip industry was part of the capital-intensification of the timber industry (which indeed is now essentially in the hands of one company) and that it would contribute to the loss of jobs. That claim is supported by data, for all the State forests, at the time the argument was at its peak (Fig. 2.2). And according to tourism interests in the Pemberton area, timber industry employment in the region declined from about 3000 people in the 1950s to 128 in 1996 (*The West Australian* 14/4/98).

In 1996, the entire forested region (extending from north of Perth to Denmark on the south coast, near Albany) had a total of 70 000 people employed full-time and part-time. Of these, only 1900 were employed directly in the native hardwood industry—including forestry, logging, forest services, wood chipping, saw milling and timber dressing. This figure was exceeded by those of other sectors: mining, tourism, manufacturing, construction, wholesale and finance (Chapter 11 in Commonwealth Government and Western Australian Government 1998b). Arguments that the cessation of wood chipping would throw 2000 people out of work are manifestly exaggerated.

Conservationists have claimed that the effects of clear felling will be deleterious to the tourist industry in the southwest. That claim was supported by Neville's (1981) research, which found that amongst visitors, 45-year-old regenerating karri forest (Plate 2.8) along the Rainbow Trail near Pemberton

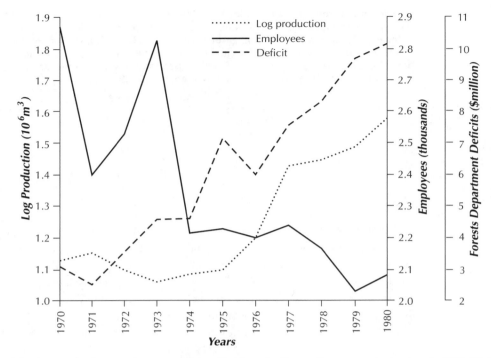

Fig. 2.2   Log production, employment and the subsidy to the Forests Department in the decade 1970–80. The data clearly show the effects of the capital-intensification of the timber industry, with small sawmills closing down and the fewer, larger ones becoming more mechanised and more efficient. As log production increased, so employment declined and the subsidy of the industry by the Forests Department (that is, the taxpayer) increased. From Conacher (1983, Fig. 5.8).

ranked only 4 on an aesthetic rating scale of 1 to 7, with virgin karri forest rating 7: clear-felled coupes rated 0. Visitors want to see mature, diverse forests, not young, even-aged regenerating 'plantation' stands. It has been estimated that more than 2 million people visit Western Australia's State forests each year (Commonwealth Government and Western Australian Government 1998b:68). Given more facilities, those travellers would stay more than one day in the region and spend more money there.

Importantly, tourism provides income to a range of activities (petrol stations, restaurants and other eateries, hotels, motels, bed and breakfast accommodation, hostels, caravan parks, grocers, bus companies). These businesses employ people of both genders and all ages, unlike the predominantly young, itinerant, remnant male labour force in the timber industry. Some 4300 to 5300 people are employed directly in tourism (excluding the self-employed) in the central and southern forest region. The level of private investment in tourism projects in 1997 exceeded $100 million (cf. $76 million for hardwood woodchip exports). Total expenditure by tourists in the southwest forests in 1995/96 was estimated at $250 million; and healthy future growth rates in tourism are predicted. The woodchip industry and its associated clear-felling practices (and associated transportation networks) directly counter such growth. The karri forest region is now criss-crossed with private roads built for the woodchip industry, with the network focused on the Diamond chip mill. That network has had the effect of blocking off extensive areas of publicly-owned forest to members of the public.

Economics

The diseconomies of the woodchip industry do not stop there. Table 2.7 shows the income received by the Forests Department in the early days of the woodchip industry, in relation to the Forests Department's own estimates of the costs to it of managing the industry. Although chip log royalties paid to the forestry agency have increased substantially since that time, data on current management costs are not readily available. One increased management cost has been caused by the predominant need to raise and plant karri seedlings, rather than regenerating them by the proposed seed tree method. The clear-felling regime, driven by annually contracted woodchip tonnages, does not coincide with the approximately seven-year flowering cycle of karri.

Table 2.7   **Income received by the Forests Department from chip-log royalties in 1980, compared with the Department's estimates of the costs to it of managing the woodchip industry. Source: from a compilation by Conacher (1983:Table 5.5) from data in Forests Department (n.d., p. 8 and Attachment 8) and Forests Department Annual Reports, adjusted to 30 June 1980 values using the Consumer Price Index.**

Item	Cost Adjusted to June 1980 Values ($)
Annual running cost to Forests Department for 'planning and supervision of logging, extra regeneration, burning and monitoring, publicity, education and training'	1 209 449
Advance payment for 1974 'to plan for the project adequately'	633 347
Capital costs spread over three years	1 339 773
Chip-log royalties in 1980	324 956

The suspicion is that even in narrow, economic, timber-related terms, the forest is being grossly under-valued by treating it as a source of wood chips for pulping rather than a source of very high grade timber not solely for the building industry but particularly for furniture making (Table 2.6). High quality galleries, such as Boranup in the Margaret River region of the southwest, demonstrate the true value of these hardwood timbers, with dining room tables fashioned from karri and jarrah logs fetching prices of up to $8000 for one setting (the need to value-add is also raised in Commonwealth Government and Western Australian Government 1998b). The same volume of timber chipped would be valued at $40 to $50. Further, an early, independent, narrowly-based cost/benefit analysis showed that not only the Forests Department, but also the railway from the Diamond chip mill to Bunbury, and the new inner harbour at the port, would run at a loss to the taxpayer (Walter 1976). That analysis made no attempt to assess the *environmental* or social costs of the proposed project. It was, of course, the adverse environmental effects of the extensive clear felling that were responsible for the strong opposition to wood chipping—and still are.

Yet, according to the Forests Department's (n.d.) Environmental Impact Statement, a primary objective of the woodchip industry was to help an economically-depressed region. Ironically, the industry supported by the Department would have the opposite effect: 20 years later this had become increasingly apparent to the residents of the region.

Environmental Problems

The main argument raised against the woodchip industry concerned the environmental effects of extensive clear felling. Undoubtedly, the replacement of a forest such as that shown in Plate 2.1 with a clear-felled and burnt coupe (for example, Plate 2.10) engenders an unfavourable emotive reaction. Such emotive responses were derided and rejected by proponents as being irrational. But the effects of clear felling go beyond aesthetics—notwithstanding its crucial importance to the tourist industry of the southwest. Clear felling and burning impinge directly and indirectly on fauna, ecosystem functions, soil and water quality, the value of the replacement forest for future generations (whether for commercial timber or other uses), and contribute greenhouse gases to the atmosphere—although such impacts have been open to dispute (as discussed by Hobbs 1996; Horwitz and Calver 1998).

Ecosystems

For any animal which cannot fly or burrow (particularly small mammals and reptiles as well as the ecologically important litter fauna), clear felling and the intense, summer regeneration burn will result in their diminution or destruc-

Plate 2.10   Clear-felled and burnt coupe, Brockman 7.

tion. Those fauna which can escape to adjacent forest may find that habitat already occupied and could therefore also perish. Eventually, as the forest regenerates (Plate 2.5), fauna will return, albeit with possible differences in abundance and diversity (DEST 1994a:106). But those which rely on hollow logs or holes in old trees for breeding sites and shelter—as do an estimated 15 species of animals and 21 species of birds, as well as reptiles (Department of Environmental Protection 1997, Section 7:33)—will not find such sites even after 100 years of regrowth (Plate 2.5d). This is one of the fundamental ecological arguments currently being raised against the logging of old growth forests, which are many hundreds of years old.

Calver and Dell (1998a, b) conducted an extensive literature review on the impact of forestry practices on the status of mammals and birds in the southwest forests of WA. Poor baseline data hampered any confident assessment of the status of the animals and birds. Thus, while it can be concluded on the basis of circumstantial, anthropological and anecdotal evidence, plus limited experimental data, that forestry practices have had little direct negative impact on the mammals and birds of the region, Calver and Dell (1998a) considered that the question remains open. With reference to indirect effects, Calver and Dell (1998b) found that forestry practices are connected with established causes of loss and decline of mammals and birds, principally through their contribution to changes in fire history. The authors therefore concluded that a precautionary approach is required to the management of the southwest forests.

Driscoll and Roberts (1997) recorded a significant decline in a frog species (*Geocrinia* spp.) following spring fuel reduction burns in the southwest forests of WA, an impact that was still evident two years after the fire. They considered that frequent prescribed burns could pose a serious threat of extinction for very small frog populations. However, it is not known whether there is a long-term effect or whether populations have time to recover between fires.

Fire

Areas of regenerating forest are subjected to a number of threats, including that of being burnt by escapes from one of the summer regeneration burns or from a prescribed spring or autumn 'cool' burn (Plate 2.11), or from other accidental (human-induced), deliberate (arson) or natural (lightning strikes) causes. Karri does not start to flower until some 40 years of age; thus any destruction before that means there is no seed set for subsequent regeneration. Damage to or pressures on recovering habitats could also be critical.

Prescribed burns are designed to prevent major, 'holocaust' fires by removing the forest litter, or in foresters' terminology, 'fuel' (we have even heard of trees being described as 'vertical fuel'!). It is also argued (controversially) that such burns 'simulate nature'. Burns are planned on a mosaic basis over 7–10 year intervals. With about 1.8 million hectares of State forests, between 180 000 and 257 000 ha of forest are burnt every year, mostly ignited by incendiary devices dropped from the air—with variable accuracy. The objective is to keep fuel loads below critical levels. But this removal of the forest litter also results in

Plate 2.11   Area of regenerating karri destroyed by fire: Nairn Road. The karri was too young to have flowered and set seed, consequently no subsequent, natural regeneration took place.

the complete interruption of nutrient cycling and the destruction of important soil and litter fauna, and bacterial and fungal decomposers of the litter.

Nutrient Loss, Erosion

Valentine (1976) used a virgin karri forest site, and stands of varying ages of regeneration after clear felling (including those shown in Plate 2.5), to research the effects of clear felling and subsequent regeneration on soil nutrients. Results for organic matter content, which is highly correlated with the major nutrients calcium, magnesium and, to a lesser extent, potassium, are shown in Fig. 2.3. For surface soils, there is a marked drop in organic matter content from the virgin site to the two-year-old site, and a further drop to the seven-year-old site. As the regenerating forest starts to shed leaves and bark, organic matter contents then start to increase; but do not regain the amounts found under virgin forest—presumably as a consequence of prescribed burning. Deeper soil horizons show a similar but much less pronounced pattern; and it should be recognised that most nutrients are found in the top few centimetres of the soil.

An implication of this work concerns the nutrient content of the soil after repeated cycles of cutting and regeneration, which raises the question: at what age will the regenerating stands be recut? Saw millers have indicated that karri needs to be more than 100 years old, both for size and resilience or strength. But the ideal age of hardwoods for making paper pulp is around 35 years. Since the Minister indicated that the intention was for the woodchip industry to operate 'in perpetuity'—and also given proposals for a pulp mill in the southwest—it is

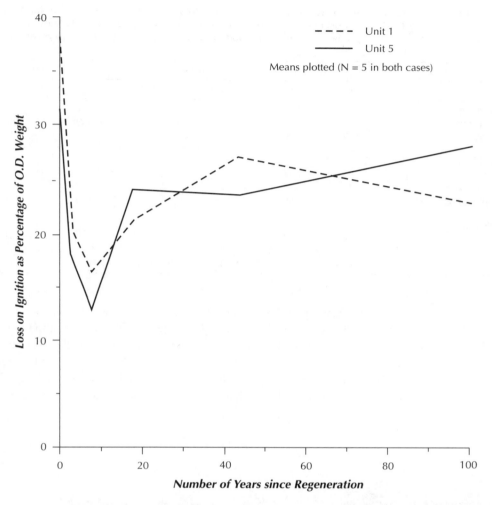

Fig. 2.3   Organic matter content, measured by loss on ignition, in soils under virgin karri and variously aged regenerating stands following clear felling, on interfluves (Unit 1) and midslopes (Unit 5). Source: Valentine (1976, Fig. 7).

a reasonable assumption that the regenerating karri (and other regenerating timber) is likely to be logged after about 35 years. Inspection of Fig. 2.3 suggests that, over time, the soils will become increasingly nutrient-depleted which, in turn, would threaten the health of the regenerating forest.

It was also noted that if regenerating forest were to be recut at an age of some 35 years, then it follows that the sawlog industry (which ironically is the major proponent of wood chipping) will eventually disappear for lack of a resource. However, the age of re-cutting is still uncertain. Increasing sources of material for wood chipping are being obtained from thinnings from regrowth stands (about 16% in 1994), and small proportions from sawmill offcuts and plantations (Table 2.8). This has reduced the pressure on the karri forest, but not removed it.

Table 2.8   **Sources of hardwood materials for the Western Australian woodchip industry 1995–96. Source: CALM *Annual Report* 1995–96, pp. 23 and 83. Note that in 1995–96, only 5.5% of total woodchip volume was produced from plantation forests.**

	Crown Land		Private Land		Total	
	(m³)	*(t)*	*(m³)*	*(t)*	*(m³)*	*(t)*
Marri logs						
– bolewood	492 484	603 430	72 880	89 538	565 365	692 968
– branchwood	0	0	0	0	0	0
Karri logs						
– bolewood	133 385	161 723	15 993	19 002	149 378	180 725
– branchwood	27 821	34 498	229	284	28 050	34 782
– sawmill residues		72 666				72 666
Other logs						
– bolewood	27	34	0	0	27	34
TOTAL	653 717	872 351	89 103	108 824	742 820	981 175

Contributing to the loss of nutrients from the forest is the removal not only of the trunks of the trees for saw milling and wood chipping, but also the bark. It was argued that the logs should be de-barked on site, thereby returning at least some nutrients to the soil. However, de-barking is carried out at the wood-chip mill and was being accumulated in a large bark dump, which subsequently caught fire—accidentally or otherwise (Plate 2.12). Shredded bark (and wood-

Plate 2.12   Burning bark dump near the Diamond woodchip mill.

chips) are also sold in the Perth metropolitan area for mulch and potting mixes, representing a significant loss of nutrients from the forest.

The major part of the nutrient losses measured by Valentine (1976) occurred through accelerated erosion. Research elsewhere has also shown that significant losses of soil materials and nutrients occur within the first two years after forest fires (Conacher and Sala 1998:336–41) or following logging (Rab 1994). Apart from being responsible for the loss of soil nutrients, accelerated erosion is not generally a major problem in the southwest forests. But accelerated erosion by water does occur in particular locations, especially down snig tracks (Plate 2.13), even some time after regeneration has occurred, or associated with poorly designed logging tracks (Plate 2.14a). Research elsewhere has also shown that logging and snig tracks are the main foci of runoff erosion in logged forests (Hauge *et al.* 1979), and the impacts may persist for many years. Such erosion has resulted in the sedimentation of streams, causing the destruction of spawning areas (Plate 2.14b).

Disease

Another threat to the forest concerns the effects of fungal diseases. 'Dieback', caused by the introduced microscopic fungus *Phytophthora cinnamomi*, occurs throughout Western Australia's State forests (and elsewhere) and kills about 1000 species of vegetation in the forests. Conservative estimates in 1975 were that some 15% of jarrah forest was infested, 5% severely. The area affected has continued to increase but the current extent is not known precisely

Plate 2.13   Eroding snig track on a gentle slope of only 4⁰. Such erosion can continue for several years after regeneration.

Plate 2.14 a) Gullying caused by a poorly designed logging track on which the figures are standing, Graphite 8 coupe; b) stream sedimentation, Graphite 8, following clear felling and the erosion as shown in a). Plate 2.14 b) taken by Neil Bartholomaeus.

(Department of Environmental Protection 1997, Section 7:32). Both karri and marri are thought to be resistant to the fungus, but not jarrah. However, there are at least another six fungi as well as cankers which compromise plant health, some of which attack karri and marri.

The Forests Department's (n.d.) Environmental Impact Statement (EIS) made it clear that the increased movement of men and machines in the forest due to the woodchip industry would accelerate the spread of the fungus, and quarantine measures have been applied in different parts of the forests at various times. Yet, even though the Department knew the industry would have that effect on what the Department's successor, the Department of Conservation and Land Management (CALM), recognises as being the major threat to the State's forests, it recommended approval of wood chipping, as does CALM. Since the fungus is spread most rapidly by the physical movement of infected soils, working under wet conditions should be avoided (Plate 2.15). But the annual tonnages contracted for by the woodchip company, in this winter rainfall environment, make it very difficult to avoid logging activities when the ground is wet. It can also be noted that by removing the litter layer, prescribed burns help to create conditions which enhance the spread of *Phytophthora* (Springett 1976). Additionally, plants weakened by a closer sequence of burns are more stressed and vulnerable to disease and insect attacks. There is an inescapable suspicion that the spread of the fungus is linked to forest management practices, despite quarantine measures.

Plate 2.15   Effects of working under wet conditions, Brockman 7 coupe.

Salinity/Hydrology

Whether forests are destroyed or compromised by logging, fire or disease, one of the consequences is the hydrological changes which have been identified elsewhere in Western Australia as being responsible for secondary salinisation of soils and streams (Chapter 3). Permanent clearing of native vegetation in 900–1100 mm mean annual rainfall zones can increase average stream salinities to a marginal quality of 500–1000 mg/L total soluble salts. The risk is low where rainfall exceeds 1100 mm per annum. This was the concern noted by the EPA in its interim report (1973), which observed that much of the woodchip licence area has a mean annual rainfall below 1000 mm and similar soil conditions as those occurring in other areas where salinisation had followed the removal of the native vegetation. The agency's later report (EPA 1975), which in effect cleared the woodchip industry, was based on the assessment that forest regeneration following clear felling would return the hydrology to its previous equilibrium. Nevertheless, the threat of salinity was considered to be sufficiently serious for 40 000 ha in the north eastern, drier (<900 mm annual rainfall) portion of the licence area to be set aside from logging.

Subsequent monitoring showed that groundwater tables rose following clear felling in the woodchip licence area, as did salinity concentrations in streams; but the effects were not sufficient to cause major concern (Steering Committee 1980). It is possible that insufficient time had elapsed to determine the extent of the problem, and logging and other activities are acknowledged as posing ongoing pressures on streams (Department of Environmental Protection 1997, Section 5:3–4). The Warren River, which drains much of the woodchip licence area, is one of the major potential sources of water for future metropolitan use—or for a pulp mill. Its current salinity status is marginal: partial clearing for agriculture of catchments in the drier parts of its headwaters have resulted in significant increases in the salt concentrations of tributaries such as the Tone and the Perup, to as much as 16 000 mg/L total soluble salts. At present, however, tributaries further downstream, flowing from wetter catchments, dilute the salinity of the Warren. Additional uncertainties are posed by the potential consequences of possible climate change.

The Adequacy of Reserves

Given these (and other) real and potential environmental problems associated with extensive clear felling for the woodchip industry, it is not surprising that considerable attention focused on the adequacy of reserves. Existing National Parks at the time the *Amendment Act* was passed in 1973 are shown in Fig. 2.4. All were small, and the Sir James Mitchell National Park, the long, interrupted ribbon along the South West Highway, is wholly unsatisfactory, at least in ecological terms. The Shannon Basin National Park proposed by the Conservation Through Reserves Committee (CTRC) of the Environmental Protection Authority is also shown: it was argued that it was relatively large and compact, extended over a (then) relatively uncut area of State forest, protected the Shannon River which flows into Broke Inlet, one of the very few untouched estuaries in Western Australia, and would serve as a scientific benchmark against which any environmental damage caused by clear felling could be assessed.

Although this National Park proposal was eventually accepted in the late 1980s, the above arguments were rejected at the time and Forest Reserves were proposed instead (Review Committee 1976; EPA 1976). Despite their approval by State Cabinet, even these reserves were in turn further attenuated by the Forests Department in its 1977 *Working Plan* (Table 2.9). But by adding 'reserves' along roads and streams, and around rocky outcrops and swamps, woodchip industry proponents argued that more than 40% of the forest was reserved from logging. This view was still largely maintained in 1996, when it was claimed that 33% of jarrah and 46% of karri forest associations were in existing or planned conservation reserves or in specially protected areas (State of the Environment Advisory Council 1996:6.20).

Management Priority Areas (MPAs) had no status beyond the life of a Forests Department or CALM *Working Plan* and therefore could be changed by the Department at any time, although the security of purpose of these reserves was increased in the late 1980s (see 'Recent Changes' below). Moreover, the Department made it clear at the time that it reserved the right to manage such areas by the use of 'normal' methods including fire and logging (White 1977:21). In other words, there was essentially no difference between the way in which these so-called reserves were to be managed and the rest of the forest.

With regard to the other 'reserves', the SFDF in its non-compliance report of 1979 showed that extensive logging of supposed road and stream reserves, and

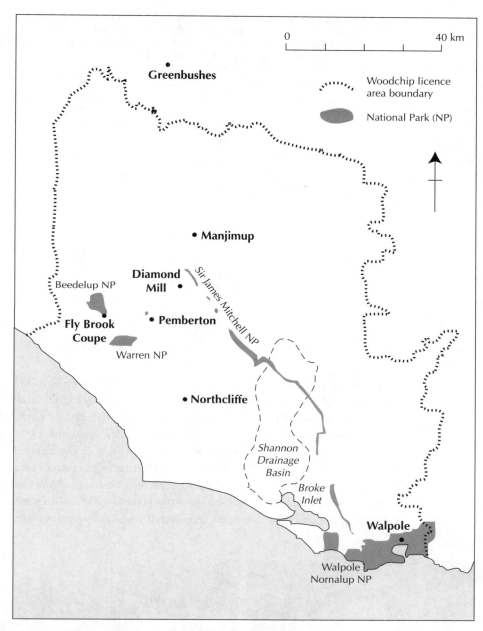

Fig. 2.4   The extent of National Parks in the woodchip licence area in 1973, and the (then) proposed Shannon Basin National Park. From Conacher (1983, Fig. 5.2).

of swampy and rocky areas, was taking place, and demonstrated as much in the forest to the Australian Senate's Standing Committee on Science and Environment. That Committee in turn indicated that it was powerless to act, even though at one site a senior forester admitted to the Committee that an error had been made (Plate 2.16a and b). That was one of nine similar sites documented at that time by the SFDF.

Table 2.9   **Comparison of areas to be set aside as Forest Parks, as recommended by the EPA and accepted by State Cabinet, and areas subsequently listed by the Forests Department as Management Priority Areas (MPAs). Sources: EPA 1976, section 2.4; Forests Department 1977, pp. 84, 96, 97, 103.**

Name	Forest Parks (ha)	MPAs (ha)
Mt Frankland	—	1 212
Dickson	261	261
Iffley	341	385
One Tree Bridge	429	666
Brockman	690	630
Dombakup	144	130
Lindsay	1 086	1 080
Johnston and O'Donnell	9 078	6 202
Soho	5 668	3 236
Beavis	1 755	—
Strickland	1 485	1 276
Giblett	2 849	—
Hawke Treen	1 840	989
Boorara	587	587
Curtin	1 108	1 256
Wattle	2 953	2 898
Lower Shannon	17 054	8 113
Mitchell Crossing	10 990	7 335
Muirillup Rock	353	209
TOTAL	58 671	36 465

Further, the addition of road, stream and rocky outcrop reserves to MPAs to inflate the total area 'reserved' only exaggerates an already highly unsatisfactory, discontinuous reserve system (refer, for example, the map of Jane Forest Block in Department of Environmental Protection 1997, Section 8:59). As the *Australian State of the Environment Report* observed (State of the Environment Advisory Council 1996:4.29), 'high overall percentages in reserves can mask the fact that much internal variation is unprotected'. Fragmentation of reserves places ecosystems under considerable pressure (Hobbs 1996). No matter how well managed, small areas are open to external pressures from fire, weed and

Plate 2.16   a) Deliberate bulldozing of forest debris from the edge of Brockman 7 coupe into a stream (b), presumably to protect the adjacent forest from fire during the regeneration burn. A senior forester admitted that this was a 'mistake'.

feral pest invasion, erosion and other forms of disturbance. Movement of fauna between reserves may be impeded or impossible with consequences for population size, diversity and vigour. Species may be lost or placed under severe threat if the reserve size is too small. For example, one pair of chuditch (the native cat) roams over an area of several hundred hectares (Andre Schmitz, manager, Karakamia Wildlife Sanctuary, pers. comm. 1998), an area larger than many of the small and insecure 'reserves' in the State forests. Thus it has been argued that lands adjacent to reserves must be considered in management strategies to ensure habitat preservation and maintenance.

Adequate reserves have significant implications not only for the ecology of the region but also for the future of the tourist industry.

Recent Changes

There have been many developments since the 1970s, although some would argue that little has changed. In 1987, CALM's regional management plans proposed additional security to Management Priority Areas (MPAs) by upgrading them to National Parks, conservation parks or nature reserves. This saw the transfer of more than 500 000 ha from forest reserves to conservation reserves (Rundle 1996). At the same time the timber industry would be given greater security by issuing long-term contracts (rather than one-year contracts), with harvesting limits applied. Industry was also encouraged to value add to jarrah.

A further assessment of a wide range of values was conducted by the Australian Heritage Commission and CALM in 1990. Following that, the *Forest Management Plan 1994–2003* (approved by the State Government in 1994) continued to propose additions to the reserve system to ensure adequate representation of those values in the reserve system, as well as additional reservation of Crown land as State forests (Department of Environmental Protection 1997, Section 8:55–6). Adopting ecologically sustainable management principles, the plan was to comply with the National Forest Policy (below), with each region to develop a management plan and adopt codes of practice.

Further changes took place in the 1990s (as discussed in State of the Environment Advisory Council 1996:6.42–6.43), although many aspects of the process are still viewed as unsatisfactory by conservationists. One important development was the report of a public inquiry which, although finding insufficient grounds to halt logging in native forests, also found that the logging of 'old growth' forests potentially violates the precautionary principle and destroys an irreplaceable resource (Resource Assessment Commission 1992). This, and the parallel, national, ecologically sustainable development (ESD) process, influenced the 1992 National Forests Policy (NFP) to develop a 'Comprehensive, Adequate and Representative' (CAR) reserve system (Chapter 5).

In summary, conservationists argued that the end result of the woodchip industry would be a depauperate, plantation forest, achieved at considerable cost to the taxpayer, and which would profit only a handful of people. The State forests, it was argued, are there for the benefit of the entire community and not for the private gain of a few. Insufficient attention was paid to economic and social options for the region. The Regional Forest Agreement (RFA) process of the 1990s reinforced the old scenario.

THE REGIONAL FOREST AGREEMENT AND WOOD CHIPPING

Under the terms of the 1992 *National Forests Policy Statement* (Commonwealth of Australia 1992b), the Commonwealth and States were to sign Regional Forest Agreements (RFAs) with the objective of providing greater security to industry and conservation needs over the 10–20 year agreement period. Based on a CAR system of social, economic and environmental forest values, a system of reserves (especially 'old growth' forests) was to be secured from logging. States would be given clear powers over forest use through industry agreements while the Commonwealth would remove itself from intervention in future forest disputes. Various concerns over the RFA process are discussed further in Chapter 5.

LAND-USE CONFLICTS

An important point which emerges from the woodchip case study is that much of the conflict has been over the use to which the land (and/or resource) should be put. In other words, this apparent *environmental* issue is in fact a number of *land-use conflicts*.

In brief, these conflicts are: clear felling for woodchips, versus use of the forest for:

1 the timber industry;
2 recreation and tourism;
3 water resources;
4 honey production (karri produces a prized honey, and is the source of a small but lucrative export industry);
5 scientific study of natural ecosystems, and
6 conservation of flora and fauna—which, it can be noted, is compatible with all the above uses except the timber industry.

LESSONS

From the point of view of this book, there are several lessons to be learnt from the woodchip issue, although some will become more apparent in later chapters.

1 O'Riordan's (1976) decision-making models are all nicely illustrated. The Forests Department was locked into a 'prison of experience', facing an unprecedented situation for which it had no previous experience to call upon. Its environmental impact assessment was the first carried out in Western Australia, with little guidance on how to conduct the assessment. Moreover, it was pressured (by the timber interests and the government) into completing the EIS within a ridiculous six weeks.

2   The complex roles of the conservationists and proponents are well demonstrated. Initially, the conservationists were inexperienced, emotive and
naive. Proponents accused them of inflexibility and questioned their
motives. Conservationists very quickly learnt the need to become skilled,
to do their homework, and to ensure their data were accurate. Proponents
took a leaf from the conservationists' book, forming the Forest Protection
(sic) Society. Amongst other things, its spokesperson has argued that even
without a woodchip industry, the karri forest would have to be clear felled
in order to regenerate it (Sunday Times 17/1/99). One wonders how they
think the forest materialised in the first place.

3   The lack of a database was a fundamental problem for all actors. There
were essentially no data on soils, salinity, hydrology, dieback, forest ecology, wildlife species, tourism, or the economics of the region or the industry, which could be used in 1973 to enlighten the arguments raised. Even
now, deficiencies remain.

4   There was a failure to recognise the *real* problems of land (and resource)
use competition and conflict between groups of people, as distinct from
damage to the biophysical environment *per se*.

5   Similarly, the *real* needs of the region were ignored. Although the EIS
identified correctly the need to boost the economy of the depressed southwest region, no formal attempt was made to identify and assess alternative
means of achieving that objective. The State of the Environment Advisory
Council (1996:6.42) has commented that all States maintain a forest policy which is 'sawlog driven'. That is, logging operations are supposed to be
justified only by the sawlogs they produce. But it is the use of logging
'waste' for woodchips which has made 'the logging of many coupes viable'.

6   The political decision-making process was revealed as being seriously
flawed. First, Parliament passed the *Woodchip Agreement Amendment Act
1973* without considering either the EIS or the EPA's interim report—
which expressed serious reservations. Second, the real (effective) decision-maker was revealed to be the Forests Department, not Parliament
(Conacher 1983). History has repeated itself with the RFA process.
Despite public consultation and a public consultation paper, the final WA
report was not to be made public prior to its signing off, nor were the proposals to be assessed by the EPA, despite previous undertakings.

It is not surprising, therefore, that conservationists recognise the vital
importance of obtaining the support of the electorate for their stance.
There is nothing like an impending election and the threat of losing seats
to engage politicians' attention.

7   The failure of the legal system to assist aggrieved members of the community was clearly demonstrated. The SFDF (which, it will be recalled, was
formed with the primary aim of taking legal action to halt the woodchip
industry) could not become incorporated (until many years later), could
not obtain legal aid despite legal opinion that it had a strong legal case,
could not obtain standing to have its case heard in court, and could not

take a class action. These problems still largely beset the conservation movement in Australia. As pointed out by Fisher (1980), there are many legal remedies available to government agencies and other official bodies to force compliance with their requirements of the community; but few avenues whereby the community can legally force an agency to implement the terms of its enabling legislation (that is, to make it do its job).

CHAPTER 3

SECONDARY SALINISATION IN SOUTHWESTERN AUSTRALIA: A CASE STUDY OF AN ENVIRONMENTAL PROBLEM

Chapter 2 was a case study of an environmental *issue*, characterised by conflict, politicisation and publicity. These are the kinds of situations often associated with 'the environment' in the public mind. But there are other equally important—if not more so—environmental *problems* which are not fought out in the glare of the public spotlight, or at least not to the same degree. These problems need planning and management at least as urgently as do environmental issues.

This chapter examines one aspect of land degradation, an environmental problem which is of fundamental importance to all Australians. The emphasis is on a reasonably detailed examination of secondary salinisation in southwestern Australia, followed by a brief consideration of other forms of land degradation throughout the country. That treatment, of necessity, is much more general, whereas the focus on the salinity problem permits it to be dealt with in sufficient detail to obtain the flavour of the complexities associated with it: this also highlights the nature of difficulties which may be faced in other aspects of land degradation. For planning and management to be effective, these complexities must be understood.

The problem of secondary salinisation of soil and water has been selected for discussion as a case study because it is one of Australia's major land degradation problems, because it has some good, recent data (especially from Western Australia) and because it illustrates well a number of themes in this book. These themes include: the need to define carefully what is being discussed; the problem of data availability and reliability; the need to understand mechanisms; and an appreciation of the ways in which people interact with their environment. Such an appreciation is fundamental to achieving both an understanding of the causes of land degradation and finding solutions.

SOME DEFINITIONS

Salt: in the present context, this refers to salts which are soluble in water. Thus an understanding of *water* movements (or basic hydrology) is fundamental to understanding the ways in which *salts* are translocated through and accumulated in particular parts of the landscape. There are many water-soluble salts. In Australia, close to the coastline, the composition of the salts is generally similar to that of sea water: sodium chloride (common table salt) predominates (about 80%), suggesting an airborne connection from the sea to the land. Salts are carried inland by prevailing winds and deposited in rain and as dustfall at rates of 20–200 kg/ha/yr (the highest rates are close to the coast), accumulating in the landscape over tens of thousands of years. Stored salts in Western Australia's deep, intensely weathered soils are least in high-rainfall and greatest in low-rainfall areas (in the latter areas there is insufficient rain, allied with low gradients, to flush salts from the landscape), with quantities varying from 100 to 10 000 tonnes of salt per hectare in the southwest, depending also on position in the landscape and soil type (Government of WA 1996a:3). Further inland, in the more arid areas and where rock type (lithology) has a greater influence on the composition of salts, as weathering products, than proximity to the sea, carbonates and sulfates become more important than chlorides and the quantities of salts stored in the landscape are greater.

Primary salinity refers to soil or water which had relatively (in relation to surrounding areas) high concentrations of soluble salts prior to European settlement. Areas of internal drainage, characterised by salt lake chains and saline soils associated with them on their leeward sides, are good examples (Plate 3.1). Naturally saline soils may have salt crusts at the surface and often support salt-tolerant vegetation. There are some 29 million ha of naturally salt-affected land in Australia; 14 million ha are salt lakes, marshes and flats, and the remaining

Plate 3.1 Lake Goongarrie, Western Australia: part of a salt-lake chain with leeward dunes: an example of an extensive area of primary salinity.

15 million ha are naturally saline subsoils which are found in arid and semi-arid regions (Ghassemi *et al.* 1995:152).

Secondary salinity, with which this chapter is concerned, refers to soils or water which have become saline since European settlement. The clear implication is that European activities have in some way been responsible for the occurrence of secondary salinity. However, this may need to be demonstrated in some cases; and from time to time people have asked: 'How do you know that area was not going to become salty anyway?' or, more subtly, 'How do you know that human activities have not simply *accelerated* a process (of salinisation) that was already under way, perhaps due to some climatic change?' This chapter does not set out to answer directly these rather fundamental research questions, although some suggestion of an answer will become apparent.

Although the serious nature of the problem is now acknowledged, it is of interest that salinity problems were first noticed shortly after European settlement. In 1853, a farmer in the Western District of Victoria related increasing dryland salting to clearing of native vegetation and overstocking (Office of Commissioner for the Environment 1991:159):

> A rather strange thing is going on now…Springs of salt water are bursting
> out in every hollow or watercourse, and as it trickles down the watercourse
> in summer, the strong tussocky grasses die before it, with all others.

Olsen and Skitmore (1991:47–9) have summarised the history of salinity in Western Australia. They noted that the phenomenon of land and stream salinisation following clearing had been observed as early as 1897, the link between clearing and salinity described in 1907, and the mechanism largely understood by 1920. The first attempt to revegetate a catchment to reverse the increase in salinity caused by clearing was made in 1909. Eight thousand hectares of the forested Mundaring Weir catchment in the Darling Ranges east of Perth were revegetated just seven years after the decision had been made to ringbark the trees in order to promote increased runoff into the dam (refer Fig. 3.2 for location). Nearby, the catchment of Chidlows Wells was also revegetated (reported in 1917) to reverse the increase in salinity following clearing in its catchment. And in 1924, in the first scientific publication on the subject, a railways engineer suggested to the Royal Society that land changes and agriculture were the cause of salinity and presented a detailed and largely accurate account of the mechanisms responsible for the problem (Wood 1924).

Dryland agriculture or more descriptively in the United States, 'rainfed' agriculture, refers to the growing of crops and pastures without irrigation; that is, it relies on seasonal rains, as distinct from **irrigation agriculture**. The problem of secondary salinisation in Australia is associated with both types of farming—the latter primarily and notably in the Murray–Darling Basin of eastern Australia. An estimated 156 000 ha are affected by irrigation-induced secondary salinity around Australia, and a further 650 000 ha of irrigated lands are underlain by groundwater within 2 m of the surface (Ghassemi *et al.* 1995:152). This chapter is concerned with the problem in dryland farming areas, particularly in the

Western Australian wheatbelt (although the problem is also widespread in South Australia and Victoria, and increasing in New South Wales and Queensland); but the mechanisms responsible for secondary salinisation in both dryland and irrigated agricultural areas are very similar.

Secondary soil salinisation refers to soils in which the concentration of soluble salts has increased since European settlement to such an extent that crops or pastures can no longer be grown, or where previously healthy, indigenous vegetation has been killed or severely affected. From an agronomic perspective, salt-sensitive plants die and salt-tolerant plants dominate when the electrical conductivity (EC) of the root zone exceeds 100 mS/m (milliSiemens per metre) and the water table is within 1–2 metres of the surface for most of the year. Soil is considered saline when the EC of a field sample exceeds 400 mS/m or the terrain conductivity measured in the field with geophysical instruments exceeds 50–70 mS/m (Government of WA 1996a:4). Soil salt concentrations vary annually, seasonally and over much shorter periods of time in response to weather conditions.

Salt-affected soils commonly exhibit a zonation (Fig. 3.1) of:

- a *core* area, or what Conacher and Murray (1973) termed a 'salt scald', of bare, often eroded soil, frequently characterised by surface seepages and rills, and salt-encrusted when dry;
- a scald *margin*, where raised remnants of the original surface appear with increasing frequency and size with distance from the core: these remnants carry some relatively salt-tolerant vegetation (such as barley grass), and
- a poorly defined *peripheral outer zone* where crops, pastures or indigenous vegetation may be affected adversely but not necessarily visibly by increased concentrations of soluble salts (Plate 3.2).

The peripheral areas, which may be extensive, are most likely to create problems of data accuracy in the various surveys reporting the extent of secondary soil salinisation, and which may help to explain some of the anomalies in data collection in recent decades: note also the qualification in the caption to Fig. 3.1. Further, estimating accurately the area of irregular shapes of both core and margin areas is difficult for most people (a simple experiment with students, asking them to estimate the size of their class room or lecture theatre will demonstrate this). There has also very probably been confusion in some farmers' minds between areas of primary and secondary salinity. In addition, the desire of some to under- or over-estimate the problem (for different reasons) may have influenced estimates (discussed further below). In any event, the problem underscores the importance of having good data in order to develop sound management strategies (Chapters 8–10).

Secondary water salinisation refers to increased salt concentrations in water bodies—streams, rivers, lakes, dams (both farm dams and larger water-supply impoundments), and also groundwater, since European settlement. The extent to which any such increase presents a problem depends on the use to which the water is to be put (a range of water uses and suggested tolerance levels is presented in Table 3.1).

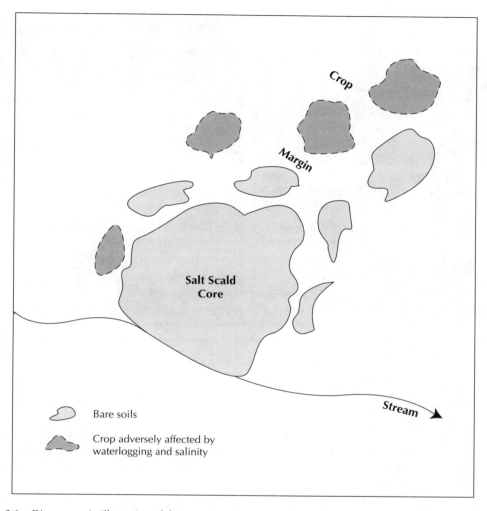

Fig. 3.1   Diagrammatic illustration of the zonation often associated with salt-affected soils; from the severely-affected core, to the transitional margin, to the outer area where some depression of crop yields may be experienced. cf. Plate 3.2 Note, however, the problem of distinguishing between crops whose yields have been depressed by relatively high salt concentrations in the soil and those affected by, for example, waterlogging, water repellent soils causing localised drought, subsoil hardpans causing poor root development, or disease. Indigenous vegetation affected by the fungal disease *Phytophthora cinnamomi*, for example, exhibits identical symptoms to plants affected by excessive salinity.

Public water authorities understandably tend to focus on human consumption, classifying water as: *fresh* (or 'potable') when it has a total soluble salts (TSS) concentration of less than 500 milligrams of salt per litre of water (mg/L) (which is the World Health Organisation's (WHO) 'desirable' limit for human consumption); *marginal*—salt concentrations of 500–1000 mg/L (or to 1500 mg/L, which is WHO's 'maximum permissible limit' for human consumption); *brackish*—from 1000 or 1500 mg/L to 3000 mg/L; and *saline*—where the concentration of total soluble salts exceeds 3000 mg/L. It will be noted that a

Plate 3.2   a) A 'salt scald' (or area of secondary, salt-affected soil) in the Western Australian wheatbelt, showing the progression from the margin (foreground) to the severely affected core (cf. Fig. 3.1); b) and c) show the surface seeps and areas of rilling which characterise many salt scalds on gently sloping sites; although these are rare or absent on the extensive, salt-affected soils in valley floors. d) It is these latter areas in particular for which the term 'saline seep', favoured by CSIRO (for example, Peck 1978) and introduced from the USA, is particularly inappropriate.

stroke of the pen can convert bodies of water from one category to another (and water quality standards do vary from State to State (and in different countries)), something which has not gone unnoticed by harassed (or desperate) politicians.

The above two problems are generally referred to in abbreviated form as **soil salinity** or **water salinity**; although the first reference to either should make clear what is being discussed.

EXTENT OF THE PROBLEM IN AUSTRALIA

Table 3.2 indicates the extent of the secondary soil salinity problem in Australia, noting that the Western Australian data for 1996 were obtained by different methods from those used to obtain the 1955–89 data, and are therefore not directly comparable with the earlier years. As can be seen, for the country as a whole the area of secondary, salt-affected land increased nearly six-fold in only 14 years from 1982 to 1996, with most of the increase occurring in Western Australia and to a lesser extent South Australia. New South Wales has the potential to almost match Western Australia's huge potential area—which, in the latter State, represents some 30% of the total area under crops and improved pastures.

Table 3.1   **Indication of tolerance to various salt concentrations in water for some plants, animals and human activities. Source: Conacher (1982b:Table 9.2).**

Total Soluble Salts Concentration in mg L⁻¹	Use of Water
200	pulp and paper manufacture
250	high-grade steel production, automotive and electrical industries
450	clover[1]
500	WHO standard for human consumption
600	stone fruits, citrus[1]
1050	vines[1]
1500	WHO maximum permissible limit for human consumption
2000	beans[2]
2500	carrots, onions, clovers[2]
3000	lettuce[2], poultry
3800	potato, sweet potato, corn, flax, broad bean[2]
4400	pigs
5000	alfalfa, broccoli, tomato, spinach, rice[2]
5800	soybean[2]
7000	horses
7300	dairy cattle
7500	beets, sorghum[2]
9000	safflower, wheat[2]
10 200	sugar beets, cotton[2], beef cattle, lambs, weaners, ewes in milk
11 200	barley[2]
11 600	bermuda grass, tall wheatgrass, crested wheatgrass[2]
13 200	adult sheep
18 000	adult sheep on green grass
36 000	mean salinity of sea water in Cockburn sound, WA
55 000	camels

Notes

1 The water quality criteria for clover, stone fruits, citrus and vines refer to the quality of irrigation water used and should not be confused with the soil–water criteria for 2. The latter criterion is in many ways more meaningful since the effects of irrigation can vary widely depending on the method of application, the duration of irrigation and the quality of soil drainage.

2 These are approximate values, from saturated soil extracts, during the period of rapid plant growth and maturation, at which 50% yield reductions can be expected.

Table 3.2   **Extent of secondary, dryland soil salinisation in Australia, 1982, 1991, 1996 and potential at equilibrium. Sources: ABS 1992, Table 5.3.2 (where the problem is misleadingly referred to as 'seepage salting'); Government of WA 1996a, Table 1.1; EPA SA 1998, p. xvi. Data in thousands of hectares. 'Equilibrium' is where groundwater recharge is balanced by groundwater discharge (through transpiration and other means), resulting in stable water tables. In the absence of management, the time to achieve equilibrium ranges from a few tens of years in high-rainfall areas to well in excess of 200 years.**

	'000 ha							
	NSW	Vic.	Qld	SA	WA	Tas.	NT	Aust
1982	4	90	8	55	264	5	0	426
1991	20	90	8	225	443	5	0	791
1996	120	120	10*	402	1804	20	minor	2476
Potential area at equilibrium	5000	?	74	892	6109	?	?	>12 075

* Severe salinity only.

The increase (to 1989) 'is as much the result of growing recognition or awareness of salinity effects already existing as it is of a real spread of the physical damage' (ABS 1992:245). However, publicity concerning the problem did not really develop until the late 1970s, and Table 3.3 shows that there were significant increases in the area of salt-affected land reported by farmers in Western Australia in surveys from 1955 to 1989 inclusive. Nevertheless, the large increase in area in the seven years between 1955 and 1962 was not matched by the relatively small increase over the longer, 12-year period between 1962 and 1974; and the huge increase in only five years, between 1974 and 1979, seems to support the above admonition by the ABS. At the time, that increase was widely regarded as an aberration; but the results of the most recent survey, conducted in 1989, do not appear to be consistent with such an interpretation. And, of course, the 1996 data imply that everyone had been seriously underestimating the extent of the problem.

Table 3.3   **The growth of the secondary soil salinity problem in Western Australia. All of the problem occurs in the southwest, and nearly all of those areas are in the wheatbelt. Sources: Burvill 1956; Lightfoot *et al.* 1964; Malcolm and Stoneman 1976; Henschke 1980; George 1990a; Government of WA 1996a. Note that the data for the years 1955 to 1989 inclusive were obtained by farmer surveys conducted by the Australian Bureau of Statistics (except for the first two surveys), whereas the reported 1996 data were obtained by Agriculture Western Australia in 1994 using a range of methods (including field surveys and remote sensing) and not by farmer surveys. Refer to the text for a fuller discussion.**

Survey Year	Hectares
1955	72 219
1962	121 000
1974	167 294
1979	256 314
1989	443 441
1996	1 804 000

The 1955–89 data were obtained by postal questionnaires completed by farmers. The earliest surveys were conducted by the Western Australian Department of Agriculture: they did not achieve 100% coverage of farmers and were focused only on the wheatbelt (Fig. 3.2). In 1962 the return was 86.3%. The three most recent surveys were conducted under the auspices of the Australian Bureau of Statistics; they achieved 100% coverage and also covered all Shires (local government areas) from the entire State. However, over 94% of all reported salt-affected land occurs in the wheatbelt (most of the remaining 6% occurs in Esperance Shire, which is also arguably wheatbelt), and the limitations of coverage by the earlier surveys are therefore probably not particularly significant. Other reservations concerning the data have been mentioned under 'definitions' earlier in this chapter (refer 'secondary soil salinisation'). Unfortunately, the data presented in 1996 were not presented on a Shire basis and therefore cannot be compared directly with the earlier, reported extent of the problem for the smaller areas. These 1996 'best estimates' data were

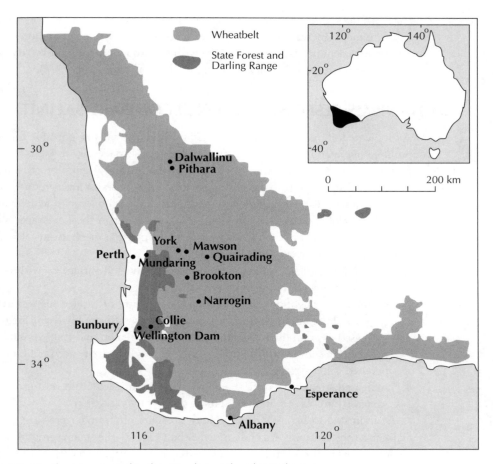

Fig. 3.2   Southwestern Australia, showing places referred to in the text.

obtained in 1994 from air photo interpretation, soil and landform mapping, farm and catchment plans, satellite remote sensing, ground-based geophysics and extensive bore records (Ferdowsian *et al.* 1996). Regardless of possible reservations concerning data accuracy, it is quite clear that significant increases in the extent of secondary soil salinity have occurred.

The annual cost in environmental degradation, degraded water supplies, lost agricultural production and infrastructure damage due to secondary, dryland salinity throughout Australia has been estimated at $270 million (*Australian Landcare*, September 1999:27). In Western Australia, the estimated capital value of land lost to salinity was about $1445 million in 1996 and was predicted to increase by $64 million each year until a new equilibrium is reached (Department of Environmental Protection 1997, Section 4:12). Other costs include: *on-farm*, secondary degradation of saline land; increased susceptibility to erosion; increased salting and silting of on-farm water supplies; increased fertiliser requirements, and loss of aesthetic value. *Off-farm* costs include: damage to buildings and infrastructure (roads, bridges, sewerage pipes, water supply systems); flood damage caused by increased farm runoff; reduced life of electrical equipment; increased water treatment, cooling and steam generation costs, and habitat decline (land and streams) with consequences for biodiversity and loss of aesthetic, recreational and tourism values (Watson *et al.* 1997).

MECHANISMS RESPONSIBLE FOR SECONDARY SALINITY

As indicated above, in order to understand how the salts responsible for the problem move and accumulate in the landscape, it is necessary to focus on the response of *water* movements to clearing of the natural vegetation. Replacement of deep-rooted, natural vegetation, which transpires groundwater throughout the year, with shallow-rooted, crop and pasture species which grow for only part of the year, have resulted in excess water in the landscape—the fundamental cause of the problem. In general terms, four mechanisms of water movements may be recognised (Fig. 3.3):

1   the raising of deep, saline, groundwaters to the soil surface or within its capillary fringe, bringing dissolved salts with it;
2   increased throughflow down valley sides, leading to surface seeps on the slopes and the accumulation of perched water systems in low-lying areas;
3   increased overland flow, leading to increased erosion and the exposure of more saline subsoils at the surface, and also leading to the accumulation of perched water systems in low-lying areas; and
4   increased streamflow from the above three inputs, which in turn provides increased inputs of water and salts to down-valley locations.

As indicated in Fig. 3.3, deep groundwaters are often under pressure, especially in the Western Australian wheatbelt. Here, the main groundwater aquifer occurs in the 'saprolite grit' of weathering granite, below the pallid zone of the intensely and deeply weathered 'lateritic' soils. The kaolinised soil materials of the pallid zone were previously regarded as the main groundwater aquifer; but they are fine-textured and poorly structured. Thus their porosity and permeability are low. Hydraulic gradients induced by the topography cause the groundwaters in low-lying positions in the landscape to have a tendency to rise to the potentiometric head if the overlying soil materials are penetrated with a bore. It is a type of artesian situation, although here it has nothing to do with sedimentary layers. With reference to secondary salinisation, it has long been argued that groundwaters have risen towards the soil surface following clearing as a result of both increased recharge (deep infiltration) and reduced water loss due to tree removal. Yet the widely quoted but probably little-read paper by Wood (1924) in fact presented a rather more complex story.

The throughflow mechanism is widespread in the Western Australian wheatbelt (Conacher 1975b) and was recognised as early as 1924 by Catton Grasby, in a review of Wood's (1924) paper. There does not appear to be any literature relating an increase in overland flow to the development of secondary salinity, with the partial exception of the study by Bettenay *et al.* (1964) in Western Australia's wheatbelt. In contrast, as reviewed by George (1991a),

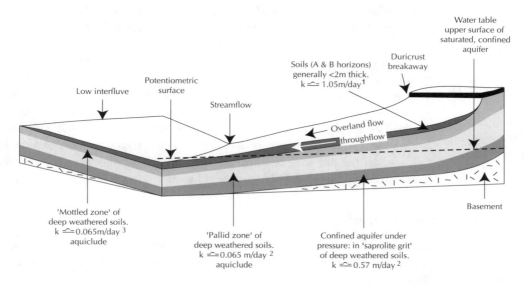

Fig. 3.3   Diagrammatic illustration of the mechanisms involved in the process of secondary salinisation, based on real landsurfaces in the Western Australian wheatbelt. Modified from Conacher (1982b, Fig. 9.4).

[1] Hydraulic conductivity (k) data from Table 3.4.
[2] From Table 3.4, air blast drilling (k from rotary-drilled holes is an order of magnitude less).
[3] k for mottled zone estimated as being similar to the pallid zone.

there is now experimental data showing that streamflow has increased in catchments in which secondary salinity has developed following the replacement of native vegetation with agricultural crops and pastures; and, as indicated above, streamflow is itself derived from the other water movement mechanisms. However, most of these data come from high rainfall areas and not the drier, agricultural region.

Since there is little question that the problem of secondary soil salinity results from the interaction of people with the environment (primarily through clearing and agricultural practices)—as is of course true for the overall problem of land degradation—the remainder of this case study focuses on that particular aspect.

THE INTERACTION OF PEOPLE WITH THE ENVIRONMENT IN RELATION TO SECONDARY SALINITY

People are involved with the salinity problem, and there are disagreements amongst them. But unlike the previous case study, there are no *proponents* and *opponents* as there are with an environmental issue. There is no 'industry' which people are objecting to. Controversies over salinity were not a question of calling for an end to farming on account of environmental damage, but concerned differences of opinion with regard to the technicalities of managing the salt problem. In a limited sense this sometimes amounted to 'conflict' involving farmers, agencies, scientists and politicians. But it was not an 'environmental issue' as defined in Chapter 1 and illustrated with the woodchip case study in Chapter 2.

There are two broad aspects to the interaction of people with the environment in relation to the salinity problem: a) people's actions in causing the problem, and b) people's responses to the problem. The former aspect is reasonably evident—clearing, and the replacement of native vegetation with exotic crop and pasture species. It is the various ways in which people have responded to the problem which is of particular interest. These responses are considered below under the following headings: is there a problem? what are the physical causes of the problem? and what are the solutions to the problem?

Is there a Problem?

Although this question today may seem unnecessary, it was not always so. There are two aspects of human responses which are of interest: agency response and responses by farmers.

When Conacher and Murray (1973) and their students started researching the secondary salinity problem in the late 1960s and early 1970s, Department of Agriculture officers did not consider salinity to be a problem. They stated, correctly (according to the farmer survey data—Table 3.3), that the area of Western Australian wheatbelt land rendered unproductive by secondary salinity comprised less than 1% of all cleared land in the wheatbelt (by 1994 the figure had risen to nearly 12%). Therefore it was considered to be insignificant; and one officer thought that gully erosion was a more serious problem.

However, secondary soil salinisation does not occur uniformly over the landscape. Some farmers may have no salt land on their properties at all, while others may be severely affected. Thus, at the time the 1970s' work was carried out the majority of wheatbelt farmers did not have saltland on their properties, but 30% or more of the land of an unfortunate few had become unproductive due to encroaching salt.

Some of the early farmers built their homesteads adjacent to creeks. The stream water was fresh and used for household purposes and on vegetable gardens and orchards around the house. But these low-lying areas were amongst the first to become saline; and some farmers had to abandon their homes as salt both destroyed their home gardens and weakened the mortar which cemented the walls of their houses (Plate 3.3). Spennemann and Marcar (1999) have considered the broader impacts of salinity on historic buildings, gardens and other plants in Australian country areas and the urgent need for planners and heritage managers to develop mitigation plans. In 1999, the Federal Minister for Agriculture noted that some 80 regional towns around Australia were affected by secondary salinity (*Australian Landcare*, September 1999:27).

Secondary salinity also affected streams, farm dams, wells and bores. The York–Mawson area east of Perth, for example, is not on the comprehensive water scheme which pipes potable water from the forested ranges to parts of the

Plate 3.3   Western Australian wheatbelt house near Mawson abandoned due to the encroachment of secondary salt.

wheatbelt and the eastern goldfields. Thus farmers in this and other areas have to rely entirely on their own water supplies both for domestic purposes and to water their sheep. Conacher and Murray (1973) found that 7% of farm dams (n = 306) and over 20% of bores and wells (n = 88) had been abandoned in the area due to increased salinity. Some considerable expenditure was required of the affected farmers to replace their water supplies. Coincidentally, the former percentage is similar to the figure quoted for all wheatbelt farmers by the Australian Bureau of Statistics in 1989. And in the northern wheatbelt, a survey of the salt content of groundwater in 76 usable bores found that the salt concentration increased at a rate of 80 mg/L per year over an eight-year period from 1968/69 to 1976/77 (Government of WA 1996a:8).

Not surprisingly, farmers affected in one or more of these ways—a high proportion of their property rendered unproductive, loss of homestead, or the need to replace dams and wells/bores—were unimpressed by the attitude of the government agency. This problem extended further, to control measures, as will be discussed in the following section.

Nevertheless, there were also some interesting responses by farmers to interviewers' questions as to whether they considered salt to be a problem. Some answered 'no' to the question, even though an extensive salt scald was clearly visible from the homestead. (It is interesting that farmers also react defensively to the same question with regard to soil erosion, although they are quick to point out that a neighbour has a 'problem'.) The difficulty here was that the farmers anticipated correctly that a follow-up question was designed to determine what remedial actions they were taking. When they had taken no or very minor and ineffective action, for whatever reason, the farmers were understandably reluctant to acknowledge that there was a problem. One of the farmers who had denied having a problem was forced to sell his property a few years later, as it had become economically unviable following the loss of one-third of his productive land to secondary salinity.

Yet others, who had only a few, incipient patches covering less than two hectares, responded very positively to the question, stating that they had a major problem. These farmers were reacting with strong concern to seeing salinity develop on what was often their most productive land. Thus farmers' responses to the question, 'is salt a problem on your property?', varied not according to the extent of the salt-affected land so much as to how recently it had appeared, how rapidly it was spreading, and whether they were taking any remedial action. (Of course, farmer decisions are shaped by many factors, such as the age of the farmer, level of formal education, sources of advice, attitudes, degree of indebtedness and social constraints: Vanclay 1997.)

Although the damage to rural infrastructure was not seen as an agency concern in the late 1960s and early 1970s, it is encouraging to find that the 1996 WA *Salinity Action Plan* and its 1998 update (State Salinity Council 1998) paid full attention to the problem, noting that it has become a heavy financial burden on some rural communities. Almost $500 000 was allocated to country towns under a Rural Towns Program to help manage their salinity problems (*Australian Landcare*, September 1999:28). Another aspect raised in the 1997 draft *State of the Environment* report (Department of Environmental Protection 1997, Section 4:4) concerns the social context of the salt problem and its causes and solutions. Management issues are still not seen as a broad social responsibility, and it is difficult for farmers to appreciate that they must contribute to managing a problem beyond their farm gate.

What are the Physical Causes of the Problem?

A fundamental issue in managing the salinity problem was disagreement over its causes, which in turn was a major stumbling block to finding solutions. The main dispute was over the roles and importance of the four water movement mechanisms outlined above (Fig. 3.3: rising groundwater tables; increased throughflow leading to perched soil water tables; increased overland flow, and increased stream runoff, with all mechanisms having similar consequences of increasing the extent and duration of waterlogging and salinisation of particular areas in the landscape). There have also been other disagreements, with some farmers denying a causal connection between clearing and salinisation, and others considering that the problem is primarily one of waterlogging, not salinisation.

The Western Australian Department of Agriculture has long considered the cause of the problem to be a relatively simple hydrological one of rising groundwater tables. Whilst this view goes back probably to the time of Wood (1924), S.T. Smith's 1962 Masters thesis was particularly influential. It contained the statement that:

.... it is quite certain that land development has resulted in a very much accelerated rise in watertables which, in some cases, have risen from 80 feet to within 6 feet of the surface within 10–15 years of initial clearing (p. 28).

A detailed search through the thesis reveals no evidence to support the assertion; but Smith subsequently became the State's Commissioner of Soil Conservation (within the State Department of Agriculture), and versions of the above quote appeared in Departmental Technical Bulletins as well as in the wider literature (for example, Holmes 1971). It was not until 1979—55 years after Wood's paper—that detailed evidence supporting the rising groundwater table hypothesis of secondary soil salinisation was published (Williamson and Bettenay 1979): and that evidence was in relation to *water* and not soil salinity—although the findings are undoubtedly applicable to both problems.

Subsequent work by Agriculture Western Australia (AgWA) (the current title of the State's agricultural agency) and CSIRO researchers provided substantial support for the rising water table mechanism of soil as well as water salinisation. Groundwater tables in experimental catchments rose dramatically after clearing, by as much as 20 m in 20 years, reaching the ground surface 11 years after clearing (Government of WA 1996a:3). In addition, data from more than 2000 bores established throughout the agricultural areas by AgWA showed that in areas away from discharge areas, groundwater rose at annual rates of between:

0.02 and 0.3 m/yr in <350 mm/yr rainfall areas,

0.05 and 0.5 m/yr in 350–500 mm/yr rainfall areas, and

0.15 and 1.5 m/yr in >500 mm/yr rainfall areas (Government of WA 1996a:7).

However, Conacher and Murray (1973) and later Conacher (1975b) were puzzled by the complete lack of evidence of rising water tables according to farmer interviews in the York–Mawson area. In the Dalwallinu–Pithara area (Fig. 3.4), where valley-floor salinisation extended rapidly in the early 1960s following several wet seasons, farmer interviews yielded only partial supporting evidence. Later data from drilling by the Department of Agriculture, reported in Conacher (1982b), also made it difficult to fully support the global explanation of rising groundwater tables.

Direct field observations and dye tracing experiments demonstrated the ubiquitous presence of throughflow perched above the higher clay-content subsoils of the duplex soils; and Conacher (1975b) outlined a model of throughflow accumulation in valley floors, leading to waterlogging and secondary soil salinisation. In essence the model states that ephemeral throughflow occurs during and shortly after winter rainfall events. The water is largely fresh, but carries some soluble salts. Throughflow water and salts accumulate in lower lying positions in the landscape, or emerge as seepages on valley sides. Over time, the water is evaporated from these areas and, as throughflow continues to transport salts to these locations, salt concentrations increase. Eventually, the salt concentrations become high enough to kill the plants.

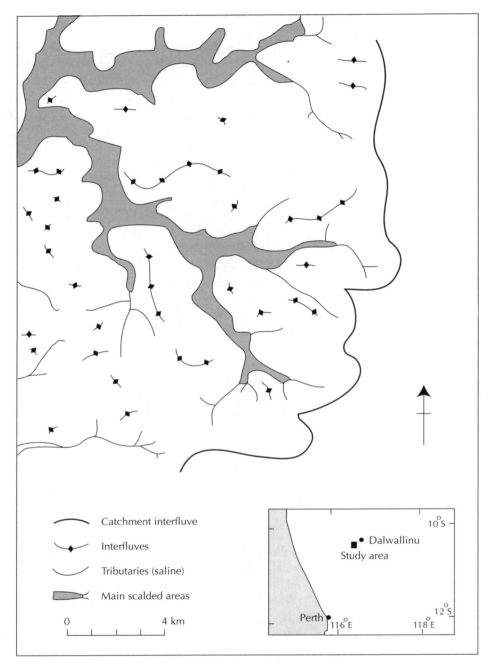

Fig. 3.4   Salt scalds in the Dalwallinu–Pithara area. Source: Conacher (1975b, Fig. 2).

A wheatbelt farmer, H.S. Whittington, located near Brookton, had previously come to similar conclusions, based on many years of observations and drilling on his property. His work eventually led to the formation of an influential group of farmers promoting a particular method of salt control, discussed further in the following section. The point is that the considerable friction which ensued between that group of farmers and the Department of Agriculture was based on these differing explanations of the causes of the problem.

Later work by Conacher (1982a) and particularly by Conacher, Combes *et al.* (1983) and Conacher, Neville and King (1983) found that throughflow alone could not account for most salt scalds; and that a combination of all four mechanisms needs to be considered and researched in any particular location. Deeper groundwater under pressure is often, but not always, involved. While these findings were made known to the farmer group, the results did not appear to influence their views on the causes of the salt problem.

More recently, George (1991a; refer also George and Conacher 1993) undertook a detailed hydrological research program to measure and quantify the relative roles of the throughflow and deeper groundwater aquifers in causing secondary salinisation in a range of situations in the Western Australian wheatbelt. Shallow aquifers occur in surficial soil materials (0.3–0.8 m thick) which are sand-textured and have a high mean hydraulic conductivity of 1.15 m/day. Throughflow also occurs in thin (<0.3–1.0 m), sand-textured, colluvial, alluvial or aeolian sediments which have been partly altered by pedogenesis. In contrast, the underlying mottled and pallid zones of the deeply (up to 15 m) and intensively weathered soil profile have a much higher clay-sized content (21–35%) and a correspondingly lower mean hydraulic conductivity of 0.009 m/day. This difference in hydraulic conductivities provides the potential for infiltrating water to be impeded above the mottled zone and to generate throughflow.

The main, regional groundwater aquifer (cf. Fig. 3.3) is usually located at depths of 20–40 m below the surface and as deep as 60 m. It is characteristically about 10 m thick, ranging from 1–2 m when developed in relatively fine-grained bedrock, to more than 18 m in coarse-grained granites and gneisses. Groundwater salinities in this saprolite grit aquifer are generally high, but they vary with a range of factors—especially position in the landscape, geology, salt storage, rainfall and recharge characteristics. On the upper valley-side slopes, for example, the proximity of deep sand and bedrock outcrops usually results in relatively low salinity (<10 000 mg/L) groundwaters beneath them. Relatively high salinity (10 000–30 000 mg/L) groundwaters occur below areas of exposed pallid zone and beneath lateritic residuals or 'breakaways'. Groundwaters are commonly very saline (50 000–300 000 mg/L) in the lower valley floors, especially when located down-gradient from saline playas.

A third aquifer is present in the central and eastern wheatbelt. This is the 'perennial sandplain aquifer', which occurs in 1.0–8.0 m of yellow, clayey sands overlying the deeply weathered basement materials. These relatively thick slope sediments are of aeolian and colluvial origin.

A wide range of hydraulic conductivities occurs within specific aquifer and aquitard materials and catchments (Table 3.4). Both the saprolite grits and the thick, valley sediments are relatively permeable, with mean conductivities of 0.57 m/day and 0.55 m/day respectively. Pallid zone materials are the least permeable, with a mean hydraulic conductivity of 0.065 m/day. Note the variation in hydraulic conductivities shown in Table 3.4 arising from different drilling methods. Standard deviations (not shown) also indicate that there are large spatial variations in the hydraulic conductivities of pallid zone, saprolite grit and sedimentary materials. The high permeability of the near-surface soils in which ephemeral or seasonal winter throughflow develops is characteristic of wheatbelt soils. Finally, the perennial, sandplain aquifers also have relatively high conductivities (0.15 m/day).

Table 3.4   **Hydraulic conductivities (slug test) in various materials in the central and eastern wheatbelt of Western Australia. Drilling method affects the results by approximately an order of magnitude. From George (1992, Table 1).**

Aquifer Domain	No. of Catchments	No. of Observations	Mean Conductivity (m/day)
Perched	2	43	1.05
Sandplain			
– rotary auger	1	11	0.05
– air-blast drill	1	7	0.15
Sediments	5	30	0.55
Pallid zone			
– rotary auger	4	24	0.01
– air-blast drill	4	28	0.07
Saprolite grits			
– rotary auger	3	21	0.06
– air-blast drill	8	39	0.57

Recharge of the deep, regional aquifer system was estimated to be between 6 and 15 mm/yr—that is, water moving beneath agricultural crops and pastures and through macropore channels in the mottled and pallid zones. Calculations from one 4200 ha catchment showed that the amount of water leaving it through the deep aquifer is equivalent to a recharge rate of between 0.05 and 0.3 mm/yr across the entire catchment. Comparing these figures with the above recharge rates, it can be seen that recharge under agricultural crops and pastures is as much as two orders of magnitude (100 times) greater than the maximum possible discharge from that catchment's deep aquifer system. The effect of increased recharge since agricultural development, and the inability of the deep aquifer to cope with the increased recharge, are reflected in rising water levels in the deep bores. At current rates of water table rise of 0.05–0.25 mm/yr, groundwater discharge is likely to develop at the mouth of the catchment in a few decades and could affect most of the valley floor by the middle of the twenty-first century unless appropriate management systems are introduced.

In summary, the comprehensive work by George, which has extended over a number of years (and is still continuing), showed that secondary salinisation is caused by both deep and perched aquifers. Their relative importance varies considerably from place to place. Few locations do not have some form of perched aquifer; in contrast, sandplain seeps are not caused by a deep aquifer (yet, to complicate matters still further, they actually contribute water to it).

What are the Solutions to the Problem?

The ways in which this question is answered by the various 'actors' involved are determined to a large extent by their perceptions of the causes of the problem.

The 'Learn to Live with It' Response

As discussed earlier, the rising groundwater table mechanism was possibly the first and certainly the most widely accepted explanation of secondary, dryland soil salinity in the Western Australian wheatbelt. Since the underlying reason for rising groundwater tables is the replacement of native vegetation with agricultural crops and pastures, it was considered that the cause could not be treated. Clearly, a wholesale revegetation of much of the wheatbelt with native woodlands would put the farmers out of business. Thus the early response to the salinity problem was that farmers should learn to live with it. In particular, the State's Department of Agriculture recommended for many years that farmers fence off their salt-affected land (so that grazing can be controlled) and establish salt-tolerant plants.

Considerable research was carried out to obtain plant species growing in naturally saline soils in other countries—North America, Israel, the Middle East—to determine their suitability for the Western Australian salt-affected soils. Trial plots were set up throughout the wheatbelt, on different soils, in different climatic environments, and in areas with various degrees of severity of salinisation and waterlogging. Propagation methods were also evaluated (Malcolm 1969, 1986). Farmers responded with varying degrees of success, and by 1988–90 they were establishing 9000 ha/yr of indigenous saltbush (Government of WA 1992:93).

This approach had several consequences. Some farmers were under the impression that the establishment of salt-tolerant plants was intended to *rehabilitate* the salt land; although the Department has indicated that this was not the intention. It was stated by farmers that transpiration by the plants would lower the groundwater table, thus removing the cause of the problem. Some even considered that the salt-tolerant plants would in some way 'use up' the salt. Of course, establishing plants on the salt land alone would be unlikely to affect the hydrological imbalance to any significant extent, especially when it is appreciated that it was the extensive clearing of the much more extensive *catchments* of the salt-affected land that was the fundamental cause of the problem. The fact that naturally vegetated valley floors became saline following catchment clearing should have made this evident. Inevitably, disillusionment set in amongst many farmers once it became apparent that saltland revegetation was not having the anticipated result (Plate 3.4).

Closely related to the above was a feeling of 'let down' by farmers. They would see what was often their most productive land gradually becoming salt-affected (low-lying, often wetter parts of the property). The Department of Agriculture would be asked for advice, and an officer would visit the property and suggest that the affected area be fenced off and salt-tolerant plants established:

Plate 3.4 This salt scald was fenced off to control grazing, and salt-tolerant vegetation (saltwater couch and tamarisk trees) established. As can be seen, the salt-affected area has extended upslope well beyond the fence. This demonstrates that the establishment of salt-tolerant vegetation does not treat the *cause* of the problem, merely the effect. One farmer was heard to comment that 'they must have used the wrong kind of fencing wire'.

in fairness, it needs to be recognised that the agency was also operating on limited knowledge, limits to the training of their personnel, and a relatively low priority given to the problem; the situation was vastly different by the 1990s. Most farmers knew that at best (assuming successful plant establishment which, as can be seen in Plate 3.4 may not be the case), this would: a) improve the aesthetics of the scald; b) reduce erosion by water and wind, and c) provide some grazing for sheep. However, the latter has to be controlled very carefully, and rarely can grazing continue beyond six weeks in any one year. Moreover, grazing on salt-tolerant vegetation must be compensated for by providing fresh water, which is often not possible, or the sheep will scour. Above all, the land would never be returned to its previously productive state. It is a 'band-aid' approach.

The Change: the Hydrological Balance Response

Still assuming a rising groundwater table process, this response sets out to tackle the mechanisms causing the problem. It can be subdivided into: a) actions designed to increase discharge from the deep aquifer; and/or b) actions designed to reduce recharge to the deep aquifer.

The first alternative has been more widely implemented than the second, although the latter would seem to be preferable in that it emphasises prevention. Option a) has mostly involved the excavation of deep drains on salt-affected land in order to drain groundwater away from the site and lower the water table. There are fundamental problems with this approach. Early experi-

Plate 3.5   A ditch at Crocodile Creek, Yalanbee, a CSIRO experimental farm at Bakers Hill on the western edge of the WA wheatbelt. The ditch was designed to dewater the salt scald by draining off the groundwater into an adjacent stream. Twenty years after this photograph was taken in 1977, there was even less vegetation on the scald than can be seen here.

ments did not work (Plate 3.5), because the hydraulic conductivity of the kaolinised, pallid zone materials of the deeply weathered soils is too low (cf Table 3.4 above). Later work in some other areas, presumably characterised by more permeable materials, achieved better results; but a relatively closely spaced network of ditches is still required as the effects of groundwater draw-down are restricted to an area close to the drain.

This leads to the second problem: the method is not practicable in areas used to graze stock or for movement of machinery. A response was to install slotted pipes at the base of the ditch and then to backfill with gravel; but this led to the third difficulty—cost. At $1600 per salt-affected hectare in the 1970s, this was far more than the land is worth (P.R. George and Lenane 1982). (In irrigated areas, where the land is much more valuable and productivity much greater, deep drainage is a fundamental requirement for the prevention of waterlogging and salinisation.)

The fourth problem with this approach concerns the disposal of the saline groundwater. It is usually drained into the nearest stream, which then affects the downstream neighbour. If the stream is fresh and likely to be used for public or private water supply, then drainage is not a practicable option. It also raises potential issues regarding common law. In some areas, salt lakes provide suitable outlets; but even these 'sinks' overflow and possibly breach if over-filled. The low relative relief of the WA wheatbelt landscape compounds this difficulty.

Most of these problems—low hydraulic conductivities, cost and disposal of the saline water—also apply to the alternative to deep drainage, namely mechanical pumping. However, drainage or pumping may have specific application in areas with high-value crops, or valuable conservation reserves containing rare and endangered species, or around towns where buildings are at risk. An alternative solution is to find a commercial use for the saline groundwater. Purifying the water and selling the salt is one such option, but is currently associated with numerous, interrelated disadvantages: the extensive spatial distribution of saline groundwaters; developing markets for the salt in competition with the very large industrial salt works in the State's north west, and economics.

A third means of increasing groundwater discharge from salt-affected land is to plant vegetation on the saline land to 'pump' water by transpiration. This alternative was rejected in the previous section; but some researchers consider that there are species of high-transpiring and, of necessity, salt-tolerant trees available which would be successful. However, even if water tables are lowered sufficiently by this method, the fact of the trees' presence in large numbers on the (previous) salt land means that the land will not have been returned to its previous state of agricultural productivity, such as growing a wheat crop.

Option b) concerns actions designed to reduce recharge to the groundwater aquifer. Essentially, this option emphasises using rainwater where it falls on the landsurface, before it can infiltrate to the deep aquifer. The obvious approach is to revegetate the land with the original vegetation (that is, using local species) which, if done extensively, would put the wheatbelt farmers out of business. A variation on this theme is to use trees with commercial value which will perform the same hydrological task as that previously performed by the indigenous vegetation. Early work along these lines was carried out in water supply catchments, notably in catchments of tributaries of the Collie River which supplies Wellington Dam. Research focused on two aspects: finding trees which will transpire most water (Plate 3.6), and determining the optimum spatial distribution of replanting (Plate 3.7) (Schofield 1990). Schofield and Bari (1991) reported that more than one-third of cleared areas in the Wellington Dam catchment were reforested on the slopes and valley floors between 1979 and 1989. The groundwater table dropped by 1.5 m and groundwater salinity decreased by 30%. One of the most successful reforestation projects in the Wellington dam catchment, using dense planting, achieved a groundwater lowering of 7.3 m between 1979 and 1993 (Mould and Bari 1995). The CEO of the Water and Rivers Commission was reported as stating that salt concentrations in Wellington Dam have been reduced from 1400 mg/L to 1100 mg/L, although he attributed this to clearing restrictions, not replanting (The West Australian 18/2/99).

In general, only groundwater table levels immediately beneath treated areas near discharge areas, or in high-rainfall catchments, have shown a falling trend. Some writers have suggested that as much as 80% of catchments need to be replanted, whilst McFarlane and George (1992) considered that strategic planting in the landscape may be effective. One site near Bridgetown (high rainfall), where a saline seep no longer flows in summer, has eucalypt plantings

Plate 3.6   Research into transpiration rates of trees, a) upslope from a salt scald, b) using the ventilated chamber technique by Greenwood and his colleagues (Greenwood and Beresford 1979; Greenwood *et al.* 1985).

Plate 3.7   Revegetation in the Collie catchment. Research here has examined the spacing of tree planting, whilst the research illustrated in Plate 3.7 is designed to find trees which will both grow successfully in the Western Australian environment and transpire large quantities of water.

Plate 3.8   Dewatered seep near Bridgetown, and the eucalypts planted immediately upslope in the recharge area.

covering as much as 35% of the seep's catchment in a close belt configuration (Plate 3.8). This has eliminated the possibility of grazing by sheep in the second half of the production cycle.

The State Government's *Salinity Action Plan* (Government of WA 1996b:7–8) focuses on various vegetation management options and set a 30-year target for the State of an additional 3 million hectares of deep-rooted, woody perennials and revegetation schemes as follows:

> 0.75 m ha commercial tree crops;
> 1.25 m ha land conservation and biodiversity;
> 0.50 m ha forage crops (such as tagasaste);
> 0.50 m ha new tree crops (such as oil mallee).

Protection of native vegetation was also encouraged. The *Action Plan* also covers other water management (hydrological) practices, including collection, reuse, pumping and disposal options. Public and private investment of about $100 million per year for 30 years were required (that is, a total of $3 billion), which was said to represent only 2% of the annual gross production of the agricultural region (State Salinity Council 1998:49).

Because of the obvious agricultural disadvantages of replanting catchments with trees, the alternative of finding commercially useful plants which perform the same hydrological role as the indigenous vegetation but which can be farmed as part of the grains farmers' rotational pattern, or as fodder in mixed farming operations, is an attractive proposition. This has reportedly been done in parts of the Great Plains of the United States, which have a similar, secondary dryland salinity problem. Here, it was found that planting deep-rooted and high-transpiring alfalfa (lucerne) on only part of saline seep catchments, not necessarily every year, was sufficient to dewater the seeps and return them to full productivity (Black *et al.* 1981). But alfalfa will not grow in the Western Australian wheatbelt without being irrigated (due to the long, hot, dry summers)—which rather defeats the purpose. Sedgley *et al.* (1981) researched the transpiration rates of various crop and pasture species, finding that wheat and lupins are to be preferred (from this point of view) to clovers: pastures use only about 40% of rainfall. But wheat alone does not transpire sufficient moisture, using about 70% of rainfall (nor is it deep-rooted)—and only does so for part of the year. A much wider search for hydrologically appropriate, commercially useful plants (including trees and shrubs as well as crops) amongst those currently being farmed in Mediterranean climate environments overseas would be amply warranted. Nevertheless, it is unlikely that the solution will be found amongst plants alone.

There is considerable irony in such a search—to look for plants which will use *more* water in a semi-arid region where the lack of rainfall is the biggest single limiting factor to crop yields.

Valley-side Interceptors

This solution to the secondary salt problem assumes that the cause is not rising groundwater tables, but throughflow (as discussed in a previous section of this chapter). If throughflow is responsible, then it follows that intercepting it and diverting or using the water *before* it reaches the low-lying areas, where it 'ponds' on low-permeability subsoils, leading to waterlogging and eventually salinisation, will solve the problem (Plate 3.9).

The interceptor method was suggested in the early 1950s to H.S. Whittington of Brookton by Hugh Bennett, Director of the United States Department of Agriculture. Whittington had previously tried salt-tolerant plants, and had been advised by the Western Australian agricultural agency that there was nothing else he could do to solve the severe salt problem on his property. To raise funds, Whittington sold part of his farm and experimented with interceptors. Over a period of many years, through to the 1980s, he gradually extended his interceptor system until it covered his entire property (Plate 3.10).

Plate 3.9   a) This shallow ditch across a slope near Mount Barker demonstrates the effectiveness of throughflow interception. The area upslope from the ditch is waterlogged (b) and characterised by hygrophilous vegetation. Downslope, the soil is dry and has good quality pasture.

Plate 3.10   a) Shows the method of construction of an interceptor on Whittington's farm. A bulldozer is used to cut into the clayey B horizon and to construct a clay barrier on the downslope side of the interceptor. The sides are gentle and can be traversed easily by stock (b). These early interceptors were constructed on the contour; thus spaces could be left for vehicular access; Plate b) shows part of the system holding water after a rainfall event. Note the extensive interceptors on the side of the hill in the background.

Early interceptors were constructed along the contour. Thus it was necessary for each interceptor to have sufficient capacity to hold the rainwater which runs off from the upslope side. Consequently, the first interceptor is constructed as close as possible to the top of the hill, with subsequent ones constructed progressively downslope. If this is not done carefully, an interceptor may be overtopped. The water which is intercepted evaporates or infiltrates vertically and laterally into the soil. Some of that water is used by crops or pastures; but some of it must infiltrate to greater depth.

For this latter reason, several observers—and the State Department of Agriculture—have stated that interceptors may *exacerbate* the salt problem by *increasing* deep infiltration to the deep aquifer, thereby raising groundwater tables and pressures in the valley floor. In response, Whittington pointed to the success of the method on his and other properties. Additionally, farmers increasingly have constructed interceptors on gentle gradients, feeding the water to streams or, often, to farm dams (recalling that the throughflow is essentially fresh), thereby both minimising the problem of deep infiltration and drought-proofing their properties. However, Whittington argued that the wheatbelt is a semi-arid environment and farmers should be conserving and using water on the slopes, not getting rid of it. He also claimed that many of the agronomic problems were often due more to waterlogging than salt (although the two are closely related) (Whittington 1975).

Neighbouring farmers observed Whittington's work and some were sufficiently impressed to construct their own systems. In 1978 an organisation known as WISALTS (Whittington Interceptor Salt Affected Land Treatment Society) was formed by a group of farmers at Quairading (Fig. 3.2), and by 1980 the organisation had grown to about 1000 members. At the time, this was almost a third of all wheatbelt farmers reporting a salt problem, demonstrating the extent of disillusionment with the State agricultural agency's advice. Since then, many farmers have constructed interceptors throughout the wheatbelt (the actual number is not known). WISALTS has also worked on Kangaroo Island off the coast of

South Australia, near Swan Hill in Victoria, and even in Canada (where the prairie provinces have a similar salt problem). (In the 1990s the words 'sustainable agriculture' replaced 'salt affected' in the WISALTS acronym.)

Research in the early 1980s showed that farmers constructing WISALTS interceptor systems reported mixed success (Table 3.5). The degree of success did not seem to be related to expenditure or the severity of the salt problem (it should be noted that some claimed successes were verified by interviewing non-WISALTS neighbours).

Table 3.5   **Responses by WISALTS farmers to interviews concerning the effectiveness of their interceptor systems. From Conacher and Combes et al. (1983, Tables 1 and 2).**

Farm	Farm Size (ha)	Area Cleared (ha)	Area Salt Affected (ha)	Lag from Catchment Clearing to Salt Appearance (yr)	Cost of Interceptor Construction to 1981 ($)	Success Score*	Improvement Salt to Land	Future Interceptors
Brookton	1227	1093	55	?	?	4	yes	no
Brookton	1215	1215	40	30	8000	2	yes	possibly
Brookton	967	809	60–80	30	1200	2	yes	possibly
Brookton	1457	1149	121	40	2000	1	no	yes
Brookton	576	568	40	40	800	3	yes	possibly
Brookton	2077	1672	49	?	9800	1	no	yes
Brookton	1327	1000	150	5	14 735	4	yes	possibly
Quairading	526	526	16	30	3300	4	yes	yes
Quairading	955	931	121	10	545	1	no	yes
Quairading	2024	1740	485	19	9800	2	no	yes
Quairading	785	783	137	31	2750	2	no	yes
Quairading	1734	1349	161	20	11 000	3	no	yes
Quairading	561	544	105	15	10 000	2	yes	yes
Wickepin	2834	2672	80	30	3000	1	no	yes
Gingin	1102	726	283	7	5600	3	yes	yes
Tammin	3036	2834	486	20	28 000	4	yes	yes
MEAN	1376	1250	150	22	7369	2.4		

* 'Success score' refers simply to the number of the following four criteria identified by the farmer as having improved. A maximum score of 4 does not indicate complete rehabilitation (the adjacent column asked specifically for any improvements to the salt-affected land itself and not just adjacent areas). The criteria were: 1. Decreased area of waterlogging; 2. Improvements to soil properties; 3. Improvements to pasture, and 4. Improved crop production.

Two factors seemed to affect success. The first concerns the care with which interceptors were constructed. There is ample field evidence that unless the interceptors cut into the clay subsoil and the banks are lined carefully with clay, then throughflow will continue downslope relatively unimpeded (and not visually evident unless drilling is undertaken during appropriate weather conditions). There are also situations where clay content is not sufficient to provide an adequate seal, either at the base of the interceptor or in the bank materials.

The second factor concerns the spatial comprehensiveness of the interceptor system. The systems shown in Fig. 3.5 illustrate some of the problems—which may include lack of commitment or cash, or lack of control over the catchment of the salt-affected land.

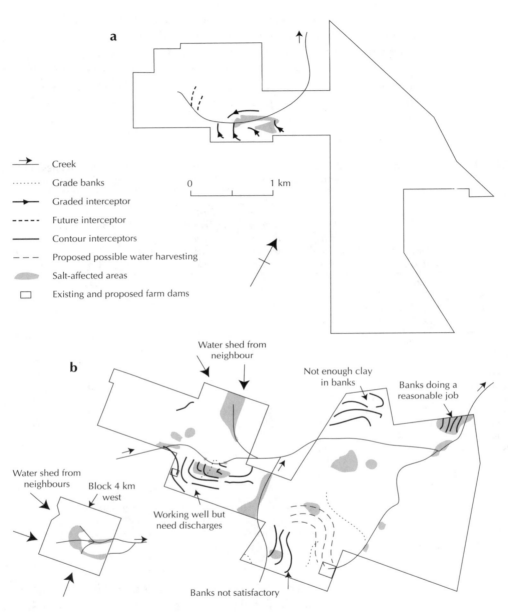

Fig. 3.5   Two WISALTS interceptor systems: a) is clearly inadequate to control water movements over the catchment; b) is much more comprehensive but, as can be seen from the farmer's own annotations, it still has some inadequacies. Not least of these is the fact that some water is shed from a neighbour's property: in other words, the farmer does not have control over the catchment which supplies water to his salt-affected land. From Conacher, Combes *et al.* (1983, figs 2 and 4).

Nevertheless, the researchers also concluded that, in general, control over throughflow alone was not sufficient to solve the secondary salinity problem. This was further confirmed by detailed hydrological research on one of the WISALTS properties (Conacher, Neville and King 1983), which showed that both throughflow and the deep aquifer were contributing water and salts to the salt land; and, subsequently, by the research reported above carried out by R.J. George.

By and large, WISALTS either has not acknowledged the e research findings or has rejected them. By the same token, however, the government agency can be seen to have been somewhat inadequate in its early approaches to the problem and in its responses to the people most directly involved.

Nevertheless, the agency's own research showed that interceptors have had some success in reducing waterlogging. Negus (1981, 1987), from surveys of a large number of interceptors, reported that on average they dewatered a strip 40 m wide downslope of the interceptor, resulting in improved wheat yields. However, there was no reduction of salt content in the soil. In relation to sand-plain seeps (which are fed by throughflow and not the deep aquifer), R.J. George (1990b, 1991b) reported success both with tree planting and gravity-fed pipe drainage. The eucalyptus trees were planted in strips upslope from the saline seeps, intercepting the throughflow and assisting in dewatering the seep. The gravity-fed pipes also intercepted the throughflow upslope of the seep; in this case the fresh water was supplied to farm dams. The quantity and quality of water so provided was more than sufficient to supply the stock being carried on the farm.

DISCUSSION

In relation to the various responses to the problem of secondary soil salinity, and as might be expected, the truth of the matter lies somewhere in the middle of the extreme responses. The problem is complex and needs a variety of solutions. All the water movement mechanisms summarised in Fig. 3.3 operate and are important. But each site is different: the relative importance of the various mechanisms varies from one site to another; and in some locations, one or more of the mechanisms may be absent. Significantly, similar comments are applicable to many other forms of land degradation as well.

Only when the combinations of mechanisms at a particular site have been identified and quantified, is it appropriate to design remedial measures *for that site*. This provides another theme for a 'precision agriculture' which recognises the need for strategic planning and management at appropriate scales. It underlines the importance of site-specific and holistic farm planning, but it also emphasises the necessity of evaluating a site in a catchment and indeed regional planning context. Nevertheless, having made that point, and while all the remedial measures discussed above are appropriate *in their relevant situation*, the most promising approach overall is to establish hydrologically beneficial, commercial plants in the *catchments* of salt-affected land. Such plants should have the potential to tackle the cause of the problem (regardless of the specific water movements involved) and maintain farmers' economic viability by providing a wide range of fruits, nuts, flowers, grains, stock feeds, oils and resins, as well as tannins and other timber-related products. Indeed, should Australia's difficulties with international wheat (or meat) sales continue or worsen, then on purely economic grounds it is time to search for alternatives, including more appropriate crops.

Regional implications of such shifts include the sociology and economics of an area's people and the nature of the towns supported by rural hinterlands. For example, the socioeconomic characteristics of areas primarily reliant on timber (or pulpwood) production are very different from the (rapidly changing) characteristics of a region based on a grains/sheep economy (Plate 3.11). There is an urgent need for careful, long-term observations and research in relation to the types of extensive tree-planting schemes illustrated in Plate 3.11 and under way throughout the southwest of Western Australia (see also Black 1999). The problem lies not only with the socioeconomic consequences, which do not appear to have been considered at all; but more obviously with the trees themselves. The Tasmanian blue gums (*Eucalyptus globulus*) preferred by the lead agency involved, CALM, are non-endemic and are not matched with local ecological or hydrological conditions. They provide minimal habitat for local insects, birds and animals. The monocultural aspects of the program contradict the message the agricultural agency, Agriculture WA, has been driving home to farmers for decades: the risk of 'crop' failure through pests, diseases, drought or fire is high. Furthermore, these species are relatively shallow rooted and prodigious users of water; ironically, this results in local soil drying and the death or poor performance of the trees, exacerbated by their overly close spacing in the

Plate 3.11   Tasmanian blue gum plantings on (previous) farmland to support a future woodchip export industry (or a pulp mill), on the northern slopes of the Porongorups, facing the Stirling Ranges National Park in the southwest of Western Australia. The probable social upheaval this and other extensive plantings will cause in previously agricultural districts has not been investigated.

plantations, but leaving the deeper groundwaters relatively unaffected, dropping perhaps up to 2 metres. Consequently, there is likely to be a minimal effect on the salinity problem the tree planting is purportedly resolving. Instead, it appears that economic imperatives—providing a source material for existing and planned woodchip and pulp industries—are driving revegetation strategies, not environmental priorities.

Extensive research is continuing into a number of important aspects related to the salinity problem. These include the efficacy and economics of deep drainage, tree planting on and off salt-affected land, pumping (George 1990c), methods for rapidly assessing recharge and discharge areas (Engel *et al.* 1989), and the role of geological structures in facilitating or obstructing the flow of subsurface water (Busselli *et al.* 1986). Remote sensing techniques are extending the results from laborious, time-consuming and expensive deep-drilling and monitoring programs. In contrast, there is comparatively little research into the economics of salt land and its rehabilitation, farmer responses and regional integration.

Chapter 15 discusses the important Land/Soil Conservation Districts, Landcare and ICM developments which have swept Australia. But in Western Australia at least, secondary salinisation, and the varied responses to it, were important forerunners of these movements.

OTHER ASPECTS OF RURAL LAND DEGRADATION IN AUSTRALIA

Secondary salinisation is but one example of an environmental *problem*: a much studied and highly visible expression of an imbalanced ecosystem. There are numerous other manifestations in Australia's agricultural areas, including accelerated erosion by wind and water, soil and water degradation and invasions of exotic plants, animals and diseases. Table 3.6, for example, lists some of the more important problems of rural land degradation in Western Australia's agricultural areas, and provides some indication of their severity and/or extent. Data are available for the other States, for both agricultural and pastoral areas, in the various *State of the Environment* (SoE) reports (Table 8.3); but the data are often dated or not comparable and it is difficult to obtain a clear, Australia-wide perspective. This is attempted to a degree in the 1996 Australian SoE report (State of the Environment Advisory Council 1996), but with shortcomings of unevenness in presentation, assessment and data (refer to ABS 1996a and SCARM 1998 for some additional summaries and data). Concerted national efforts are under way to address this data deficiency (Chapter 8).

It is therefore very important that the emphasis on problem-solving should not be solely on a particular aspect of land degradation, such as salinity; and this is perhaps the major shortcoming of WA's salinity strategy, or Action Plan. Indeed, in a sense it is partly misguided, in that it should be looking at the total land degradation problem, using comprehensive natural resource management in an integrated or total catchment or regional approach (Chapter 15).

Table 3.6   **Some land degradation problems in Western Australia. Sources: Government of WA (1992), Department of Environmental Protection (1997).**

Condition	Area Affected 1997 (ha)	Potential Area	1997 Cost (A$ million)	Management Comments
Dryland salinity	1 804 000 ha	6 109 000 if no action taken	$1445m capital value of land lost; plus continuing loss of $64m per yr; plus $46m paid in compensation to farmers not permitted to clear land	>$20m per yr being spent on State Salinity Plan; 3 million ha revegetated
Waterlogging	Approx 2 million ha in average years; 2.3 to 2.5 million ha in wet years	Increasing, especially with trend from wool to crops	$952m in 1988; in 1997, $23m in 4 shires crop loss in wet year. In southern areas, estimated >$100m crop loss and similar amount pasture	Water harvesting; aquaculture; drainage; surface water control measures; revegetation
Erosion—wind	50 000 ha (variable) in 1988; in 1997, 10–20% of susceptible west coastal plain areas affected every 5–10 years; south coast highly susceptible	Stable, but figures are probably significantly under-estimated	$21m in 1988 (loss of cereal and pasture productivity); no recent estimates	Improved management; revegetation; windbreaks; minimum tillage; stock control; Landcare
Erosion—water	750 000 ha in 1988; no recent estimate		$21m in 1988; no recent estimate	
Soil structure decline in 1988	3.5 million ha in 1988; no recent estimate	Reducing	$70m in 1988	Minimum tillage benefits
Compacted subsoils	8.5 million ha in 1988; no recent estimate	Stable: all susceptible areas are now affected	$153m in 1988	
Water repellency	3–5 million ha	Increasing	$150m in 1988. Cost of clay amelioration $100–200 per ha	
Acidification	4.7 million ha highly acidic; 4.7 million ha moderately acidic	Increasing. Up to 11 million ha susceptible. 90% of soils inherently acidic in some areas	Loss of production estimated at $150/ha with high acidity. Total cost estimated at $70m per yr	An estimated 750 000 t of lime needed each year
Land contamination (pesticides)	1200 rural properties subject to notices for organochlorine (OC) contamination, especially old potato and orchard sites	Generally declining OCs in rivers but increasing heavy metals above EPA guidelines		Chemical stocks Recalled and stored
Loss of fringing vegetation	Over half the rivers surveyed have less than 50% of their fringing vegetation retained			

Table 3.6 (cont.)	Some land degradation problems in Western Australia. Sources: Government of WA (1992), Department of Environmental Protection (1997).			
Condition	Area Affected 1997 (ha)	Potential Area	1997 Cost (A$ million)	Management Comments
Sedimentation	More than half the former pools in the Swan–Avon River sedimented; also some in Kimberley region			
Rangelands vegetation decline	19 600 000 ha in 1991. Estimate still current. 7 300 000 ha wind and water erosion included in the 19.6 million ha affected by vegetation decline	More than 50% in fair to poor condition		Control stocking; rehabilitation; control feral animals

Although multiple processes and therefore remedial actions should be involved in relation to the salinity problem, the situation discussed above in relation to the extensive tree-planting program which is under way is an excellent illustration of the probable adverse consequences of taking a too narrowly focused approach to a problem.

As with secondary salinity, problems of land degradation have not attracted a great deal of attention from the conservation movement—perhaps reflecting urban/rural dichotomies. An early exception was the publication by the Australian Conservation Foundation of the book: *What future for Australia's arid lands?*, which brought together a number of scientists from a range of disciplines, posing some thoughtful questions (Messer and Mosley 1983). Later, the Australian Conservation Foundation's role in working with the National Farmers' Federation to frame a national Landcare policy in the late 1980s was particularly influential (Chapter 15). Other writers have also attempted to raise community awareness of and popularise the land degradation problem (examples: Smith 1990; Beale and Fray 1990; Barr and Cary 1992; Flannery 1995; White 1997).

For Australia's 'Outback', it is important to distinguish between impacts on perennials and annuals, and to differentiate the effects of drought from overgrazing. These distinctions are not always easy to draw, providing scope for proponents of alternative viewpoints to present arguments unhindered by facts. Only in some parts of the country has detailed relevant research been undertaken; for example, by Agriculture Western Australia in some of that State's 'rangelands' (for example, Pringle 1991). Some of the issues have been addressed in a draft *National Rangelands Strategy* (ANZECC/ARMCANZ 1996), discussed in Chapter 5.

The term 'rangelands' is a particularly unfortunate borrowing from North America. It implies that the region's primary purpose is pastoralism—to provide grazing for introduced stock—whereas in fact it has a number of economically and socially important functions. Other land uses in the Australian Outback include Aboriginal homelands, tourism, mining and wilderness preservation (some of which are economically much more important than pastoralism, and they are recognised by the *Strategy*). Not only are these land uses not implied by the term 'rangeland', they are often incompatible with pastoralism.

Despite data inadequacies, it is beyond question that serious, increasing land degradation is occurring throughout the country, not only in the Outback (Table 3.6; Conacher and Conacher 1995; State of the Environment Advisory Council 1996). It was and is being caused directly by inappropriate management practices, and indirectly by a range of social and economic considerations, market forces, the nature of the Australian environment and even government policies. There is an important distinction to be made between on-farm problems and those which may occur at (sometimes considerable) distances from the main cause. Stream, river and dam salinisation are good examples of the latter, as are eutrophication and sedimentation of waterways and estuaries. Yet cause and effect are not always clear, direct or obvious. In the case of the eutrophication of the Murray–Darling system, for example, initial explanations shared the blame between agriculture and inadequately sewered towns along the rivers. More recent evidence suggests that the main cause is not so much the addition of phosphorous to the system by farmers (P was already present in the subsoils in sufficient quantities—unlike other parts of Australia) as the considerably reduced flow regime of the rivers following the construction of numerous impoundments to store water for irrigation (Olley 1995).

The above (and the example of secondary salinity) stresses the need for a thorough, scientific understanding of the problems. This is essential before solutions can be identified and promoted.

Nevertheless, sufficient is known about many of the problems and their causes for solutions to be recommended in general terms, and in detail at some specific sites. The US Department of Agriculture, in particular, has long developed many specific means of tackling a wide range of soil degradation and erosion processes; these are incorporated in a number of excellent texts (for example, Morgan 1995). More recently, there has been a move towards agriforestry, whole farm planning and integrated (or total) catchment management, and the incorporation of some of these approaches into the regional planning process (Part 5).

But at this point, given that certain solutions are available, there is nevertheless a range of difficulties which raise questions concerning the adequacy of the legislative and administrative framework. How does one persuade a farmer who is barely breaking even or seriously in debt, to undertake conservation measures, however inexpensive they may be? How does one persuade a farmer to modify practices which are having an adverse effect on adjacent or downstream properties but not on the farmer's land? How are offsite costs of treatment to be met? Are these individual or societal responsibilities? How can local governments be persuaded to improve the treatment of their towns' sewage, or to maintain reserves under their jurisdiction more adequately in order to prevent eutrophication, weed infestation or the depredations of feral animals? How can State and federal governments be persuaded to modify policies which are having adverse effects on the environment? And who has the responsibility of doing the persuading, providing education and information, encouraging and enforcing compliance, and resolving conflicts amongst competing or conflicting land users? How adequate are existing enactments and administrative structures for these purposes? These and other questions are considered in various ways in the remainder of this book.

Part 2

The Development of Environmental and Resource Management Legislation and Agencies

To an unusual extent, the establishment of white communities in Australia has been characterised by the efforts of the several tiers of government to initiate, maintain and manage the occupation and development of rural and urban areas. In the process, governments have had to respond to the problems and demands of the populace, and complex interactions between official and popular appraisals of social and material aspirations and environmental values have been critical ingredients throughout (Powell 1988:14).

Environmental problems and issues are not new: in the case of secondary salinisation, for example, there were reports of developing problems in Western Australia and Victoria in the nineteenth century, soon after clearing; while issues associated with the allocation and management of land and/or resources go back to the time of first settlement.

Part 2 looks at the development and modification of Australian legislation and agencies designed to resolve environmental and resource management problems. **Chapter 4** considers public interest legislation and agencies such as those relating to public health and the biophysical environment. Resource and land management legislation is discussed in **Chapter 5**, referring to urban and rural land use, water, forests, minerals (mining and industry) and energy. **Chapter 6** considers the evolution of environmental law in Australia, and the importance of international conventions, treaties and agreements for the development of environmental protection measures in the country. The limitations of the above Acts and other measures with regard to the maintenance of environmental quality are discussed, leading to consideration of environmental protection legislation in **Chapter 7**. These Acts were introduced specifically to deal with the deficiencies of the existing fragmented legislation, but they in turn also have an extensive array of weaknesses and are increasingly subject to revision and reform. Discussion of those shortcomings overlaps with consideration of environmental and land assessment methods, which are dealt with in **Part 3**.

To provide a basis for the discussion, it is helpful first to consider the nature of the environmental problems the legislation was intended to address. They may be classified as follows, noting that some overlap occurs:

1   sectoral, intersectoral or regional problems (such as land degradation, greenhouse gases, the Murray–Darling Basin);

2   project-specific problems (such as those relating to the woodchip industry, or a proposal to construct a new dam, nuclear power station, bauxite or uranium mine, or a polluting industry): these types of proposals often come under the category of special State projects, with particular consequences in law and management;

3   cumulative problems: numerous, dispersed, discrete actions which individually have relatively minor environmental effects, or are only of local concern, but which collectively may have significant adverse environmental consequences. Examples include: the use of household pesticides or fluorocarbons; noise and air pollution caused by motor vehicles; water pollution resulting from domestic sewage or waste from farms or small-scale industries, or the incremental loss of wetlands, remnant vegetation and biodiversity due to land-use changes, and

4   environmental issues: characterised by conflict, publicity and politicisation. Many are related to project-specific problems (category 2 above); but not all such problems become fully blown issues, and some issues are unrelated to specific projects (such as whaling, the culling of kangaroos, the live sheep export trade, or the interstate movement of hazardous wastes).

Australia's legislative and administrative responses to the environmental problems and issues classified above can be summarised in a simple, overlapping chronology:

1   'public interest' legislation (Chapter 4);

2   amendments to, or developments of, resource development legislation (Chapter 5);

3   environmental protection and impact assessment (Chapters 6, 7 and 11), and

4   integrated, regional, environmental and land-use planning (Part 5).

Parallel to this is another sequence identified by Fisher (1980; refer also Buckley's (1988) progression discussed in Chapter 16). Fisher considered that the origins of environmental protection can be traced to the incorporation of 'nuisance' (common law torts or wrongs) as a statutory concept. This has now been much extended, as follows:

nuisance —> public health legislation —> land use control —> protection of the environment for its own sake.

These various themes are can be developed a little more fully, as attempted in Table 4.0. The table indicates the progression of environmental planning and management in Australia since the 1700s; it also provides a framework for the remainder of this book.

Table 4.0   Progression of Environmental Planning and Management in Australia: 1700s to 1990s

Public Interest	Resource	Resource and Environmental Protection	Environment Protection Acts and EIA	Integrated Natural Resource Management	Integrated Land Use, Environment and Natural Resource Planning and Management
		1960s – 1980s	1970s – 1980s	1980s – 1990s	1990s
Late 1770s Health, common law	Resource development (utilitarian)	Reducing environmental impacts of resource development	Specific environment protection legislation and agencies	Consolidated natural resource legislation and agencies	Integration of national, State, regional and local levels of plans/policies/roles
Pollution	Private rights		Pollution control	ICM and Landcare increased regional focus	Comprehensive and strategic planning/policies incorporating natural resource management, land-use planning and environmental management
Parks and reserves, conservation			EIA	Increased powers of EP legislation (penalties)	Integrated regional planning
			Environment protection policies	Incorporation with planning	National and State planning strategies
		Regional economic development planning land use and resources integrated but low environment priority	Environment clearly defined	Broader responses to EP legislation	Increased local government responsibilities in planning and environmental management
			Some reference to planning (still narrow/specific focus) ESD EMS BMP environmental values	Environmental priorities raised	Bioregional planning? ->2000+
		Resource laws amended or new laws (utility still paramount)			
Settlement	'Development'	Responses to environmental and public pressures: integration			
		Increased public awareness/participation, international agreements national legislation, policies, measures, strategies ESD; national, State, local roles clarified environmental values recognised social, cultural, heritage and equity values recognised			

89

CHAPTER 4

PUBLIC INTEREST LEGISLATION AND AGENCIES

TRADITIONAL PUBLIC INTEREST LEGISLATION

Surprisingly, one of Australia's earliest attempts to regulate environmental mismanagement in the name of the public interest took place soon after colonial settlement at Sydney Cove in 1788. For a number of years, a succession of governors and public servants had tried to preserve the water quality of Tank Stream (used for drinking water) from pollution by domestic and industrial wastes and from the effects of erosion arising from excessive clearing. So desperate was Governor King in 1805 that he promised offenders that anyone throwing filth into the stream would have their houses torn down. In addition, a fine of £5 would be paid into the Orphans' Fund (Powell 1976:18). Australia had to wait almost 200 years before serious penalties for environmental pollution would be applied again.

Echoing English public health and welfare legislation in the mid and late nineteenth century, some of the earliest Australian legislation to protect the public interest in a general sense was the *Public Health Act 1902* (NSW) and the *Local Government Act 1919* (NSW)—and their equivalents in other jurisdictions. These Acts dealt with a range of activities in the public interest, usually by conferring discretionary powers on central or local public authorities. It is significant that such public interest Acts contained provisions which absolved public authorities from situations where the 'nuisance' is perpetrated by them and not by private individuals (Fisher 1980:161–2). Conversely, agencies today are increasingly required to operate under a 'duty of care' and avoid environmental harm. The public's 'right to know' about environmental hazards and risks is also a developing issue (Pearse 1996).

What is apparent about this type of public interest legislation is that human beings are central and not the environment. Most Acts concern: people's direct or indirect use of land, water and air; food quality; nuisance (such as noise, odour and dust); control of noxious industries; use and spillage of toxic substances; abattoirs; burial grounds; municipal waste disposal and sewerage; public parks; vermin; flooding and drainage; noxious plants; clean water delivery, and disease control. These Acts (and local regulations and codes) often have a public health component. But they are also controlled in terms of human occupation and use of land and therefore subject to land-use planning and development legislation. Many of these matters are administered at the local government level but generally operate within laws, regulations, guidelines or policies set by State or federal statutory authorities. Most State agencies/laws (and in some instances, local governments) now have substantial powers to prosecute or control pollution and other nuisances under a system of licences, penalties, incentives, EIAs or best practice codes. But laws have been fragmented and administration diffuse and complex, so that developers and planners complained about the myriad requirements needed to satisfy all the different approving agencies. In recent years, consolidation of the relevant legislation has dealt with some of these complaints, such as the 'one stop shop' and fast track development approval systems, although not always with environmentally sound consequences.

New Public Health Legislation and Agencies

In a more crowded world, with its extensive array of environmental problems, new approaches to protecting human welfare have been needed. These developments have also reflected changing knowledge, technology and lifestyles. Significantly, the World Health Organisation (WHO) has described health as 'a state of complete physical, mental and social well-being and not just the absence of disease or injury' (Evans and Stoddart 1990, cited in EPA NSW 1995a:167).

From an environmental management standpoint, various environmental hazards arise from human activity and, in turn, may impinge on human health. Since the 1970s, an increasingly sophisticated range of mechanisms has evolved to respond to such hazards. The mechanisms have been expressed as new policies and standards, legislation, and administering agencies. The development of standards, limits, quality criteria and codes of practice is supported by scientific studies relating to human and environmental exposures to the various hazards. Those affecting public interest, health and quality of life include: sun exposure (ozone thinning increasing the rate of skin cancers); occupational exposures (such as asbestos causing fatal mesothelioma; the effects of long-term exposures to noxious or damaging substances such as lead, pesticides and radiation, possibly linked to cancers and other disorders), and industrial and urban noise linked to stress and deafness (EPA NSW 1995a, Chs 1.10 and 11.14). A number of these issues now have legislation attached to them—such as the Commonwealth *Ozone Protection Act* of 1989. Thus, WHO (under the Ottawa Charter) encourages changes to *physical*, *social* and *economic* environments to achieve health goals. Although this is still human-centred, importantly it reflects the more comprehensive and desirable definition of 'environment' which was discussed in Chapter 1. Australian national health goals strongly reflect this change. For example, the National Health and Medical Research

Council (NHMRC) has recommended that environmental impact assessments now include health impact assessments and has endorsed a national framework for such assessment (NHMRC 1992, 1993).

Guided by international and national standards and codes, occupational health, safety and welfare laws and agencies now operate in most States and consider environmental impacts on health. For example, South Australia's *Occupational Health and Safety Welfare Act 1986* established exposure guidelines and standards to protect worker safety. The South Australian Health Commission's primary responsibility is to ensure that people live and work in a healthy environment. The Commission carries out this function through a number of subsidiary organisations which include the Division of Public and Environmental Health, and through liaison with other supporting agencies (Environmental Protection Council of South Australia 1988). There is similar legislation in the other States.

However, and of particular interest to the environmental management of non-urban Australia, the self-employed, such as farmers, often are not as closely bound to occupational health and safety laws. Farmers, farm workers and families are particularly vulnerable; for example, in areas relating to the handling of farm chemicals (and unknown long-term effects including genetic defects and cancers) as well as their improper use and disposal. The latter are increasingly addressed by public education programs on safe practice, including the wearing of protective clothing and avoidance of off-site impacts from aerial drift or water contamination (refer also the following section on 'Pollution Control').

Since the mid-1970s, and adopting WHO guidelines, the NHMRC, with other organisations, has extensively revised codes of practice and standards relating to air and water quality. In turn these have impacted on related legislation and policies in most States. For example, national water guidelines have been influential in setting and harmonising health and environmental standards around Australia for water consumption and use (NHMRC/ARMCANZ 1995; ANZECC 1995; ANZECC/ARMCANZ 1999): Table 4.1.

Food is also an important source of human exposure to environmental contaminants arising from agricultural practices. A range of federal and State laws and regulations deal with human exposure. While State agricultural and health agencies are still responsible for surveillance, licensing, and ensuring compliance with chemical use guidelines and requirements, national organisations are responsible for registration, establishing residue limits and some major monitoring programs. The registration of agricultural chemicals, previously a State agricultural agency function, now rests with the *National Registration Agency*, which was established in 1993.

Food monitoring for pesticide, veterinary chemical and heavy metal residues is carried out in three ways. The *National Residue Survey* was administered by the Department of Primary Industries and Energy (DPIE)—and is now the responsibility of AFFA (Agriculture, Fisheries and Forestry Australia)—and monitors raw foods, primarily those for export, in order to comply with overseas requirements and standards. Second, the *National Market Basket Survey* (ANZFA 1996)

Table 4.1   **Some Australian indicators and guidelines for quality of drinking water. Source: NHMRC/ARMCANZ 1996.**

Indicator	Guideline Level	Reason for Concern
Microbiological indicators		
Total coliforms	0 CFU/100 mL	Indicator of faecal pollution
E. coli or thermo-tolerant coliforms	0 CFU/100 mL	
Chemical indicators		
Aluminium	0.2 mg/L	Aluminium flocs. Affects appearance and should preferably be reduced to 0.1 mg/L (potential health problems for dialysis patients)
Chlorination by-products (trihalomethanes)	0.25 mg/L	By-products induce tumours in rats and mice
Nitrate (as NO_3)	100 mg/L(50 mg/L in infants)	Can cause methaemaglobinaemia (blue-baby syndrome) especially in infants
Aesthetic indicators		
Turbidity	5 NTU	High levels affect palatability and appearance
Colour	15 HU	As above
Taste (as salt)	500 mg/L	Affects palatability
Supply amenity		
Calcium and magnesium (as carbonate)	200 mg/L	Requires increased use of soaps; causes pipe corrosion or deposits
Iron	0.3 mg/L	Can cause staining of clothing and affects taste
Manganese	0.05 mg/L	As above

Notes
CFU—colony forming units.
NTU—nephelometric turbidity units.
HU—hazen units.

sporadically screens a range of domestic foodstuffs for a range of commonly used chemicals and other substances. Third, *State health authorities* conduct regular surveys of metropolitan market products. In general, compliance violations do not exceed 2%, with occasional exceptions for individual substances (an organophosphate and cadmium are two of concern). Surveillance for the import of exotic pests and diseases is also the responsibility of AFFA under its Quarantine Service—which, incidentally, is under increasing pressure with reduced resources and the freeing up of world trade. Finally, where inappropriate use of pesticides and fertilisers has caused land or food contamination, quarantine notices may be issued to producers under agricultural chemicals residue laws.

POLLUTION CONTROL

Environmental protection laws are discussed in some detail in Chapter 7. Legislation related to public interest is touched on here.

Pollution control, with reference to air, water, noise and waste, is another important aspect of public interest environmental legislation insofar as it impinges on human health and welfare. The purpose of pollution legislation is to ensure that natural qualities of the environment do not fall below certain standards as a result of human activities, and that human and other life forms are not harmed. Increasingly, pollution control is becoming standardised and subject to national and international harmonisation, and it is generally the responsibility of one agency, such as the Environmental Protection Authorities. Other government agencies and local authorities may also have responsibilities under a range of environmental laws, but the responsibilities are generally exercised within the provisions of State environmental protection Acts and the policies of the environmental protection agencies. Some of the latter often include the development of statutory environmental protection policies (EPPs). In Western Australia, for example, EPPs have been formulated to control eutrophication in a large estuary, to protect wetlands and a major groundwater mound, and to control industrial pollutant emissions into the atmosphere. Infringements of these policies can incur major penalties.

New South Wales provides a recent illustration of reforms to this type of legislation—changes which have their parallels in the other jurisdictions. The State consolidated its related though fragmented pieces of pollution legislation under one regulatory, integrated framework. The Act of 1997, the Second Stage of the *1991 Protection of the Environment Administration Act*, removed a number of duplications and inconsistencies. It repealed the Clean Air and Waters, Pollution, Noise and Environmental Offences and Penalties Acts. It also aimed to meet the objectives of the *1995 Waste Minimisation and Management Act*, and included human health (impacts of pollutants) in its scope (Lipman 1997). The latter is similar in approach to the US EPA and the OECD Directorate in that human health is considered with the biophysical environment.

In non-urban areas, land users are also subject to a number of environmental, agricultural, food and health laws and regulations, as well as environmental protection legislation, with regard to causing pollution or public nuisance. Areas covered include pesticide drift and contamination, nutrient enrichment of water bodies, feedlot operations (noise and odour as well as effluent), and off-site runoff and sedimentation. Increasingly, land users are expected to exercise a 'duty of care' and avoid deliberate environmental harm. Failure to do so may invoke punitive responses (Chapter 6 in Clement and Bennett 1998). The aggrieved may also take action under common law.

Amongst other reforms which influence public welfare and environmental management in non-urban areas are farm chemical recall programs and the encouragement of 'best management practices' (BMPs) such as integrated pest management (IPM). Considerable revision of national and State agricultural chemicals Acts and regulations was a major outcome of the Commonwealth Government's Agricultural and Veterinary Chemicals Inquiry (Senate Select Committee 1990). An earlier inquiry into toxic and hazardous substances (House of Representatives Standing Committee 1982) was also responsible for a range of reforms and initiatives relating to the handling, transport and disposal of these and other toxic materials. The substances impact both on the environment (contaminating soil, air, water and natural habitats through spillages, dumping, leachates and inadequate storage and disposal) and on human health. Increasingly, the responsibility for dealing with highly polluting substances is being placed in the hands of environment protection authorities, which have powers to penalise and enforce compliance as well as to audit and monitor. Specific legislation, registers of contaminated sites, stricter codes of practice, and the establishment of dedicated disposal areas and methods, have been some of the outcomes of the national inquiries. A number of disputes have arisen in country areas over proposed sitings of intractable waste incinerators and toxic dumps away from major cities (Pearse 1996). These instances of 'out of urban sight, out of mind' have been opposed vigorously by rural communities, who are no less NIMBY-minded than their urban cousins.

The 1992 Intergovernmental Agreement on the Environment (IGAE) (COAG 1992) provided the basis for a co-operative national/State/local government approach to environmental management issues in Australia, and 'public interest' matters are included. Amongst its tasks, the (former) Commonwealth Environment Protection Agency (CEPA) was required to develop a national pollutant inventory (NPI) and a national waste database, and ambient standards and guidelines for environmental protection. Most States and Territories have now adopted the relevant national guidelines on standards and developed their own policies or legislation on these matters.

BIOPHYSICAL ASPECTS OF THE ENVIRONMENT

The biophysical environment is a further 'public interest' concern. The idea of natural resources (air, water, land, minerals, sea, wilderness) as common property has its origins in Roman Empire law (Bates 1995). Such resources are seen as being held by governments 'in trust' for common, or community benefit. The concept recognises public rights over seashores and, to some extent, in the establishment and protection of public parks, reserves and wilderness areas. This view is again human-centred; however, conservation values have always had an important role in the management of these areas, although they are dependent on the particular status of the land set aside.

The protection of flora and fauna in Australia is achieved through the creation of a system of national and conservation parks and other reserves and sanctuaries, and through protected and threatened species laws. Larger areas (mainly Crown lands) are administered by federal and State agencies under parks and wildlife, nature conservation, flora and fauna, biodiversity or heritage legislation; or for 'public purposes' by local governments under planning legislation. Some protection of flora and fauna is also provided under the umbrella of forestry (multiple purpose management), fisheries (but usually for commercial purposes), water (catchment protection) or customs and quarantine legislation, as a requirement of development projects or as an adjunct to other legislation. A number of important Acts date back to the nineteenth century, reflecting an early concern with the protection of natural resources and demands for public recreation areas.

Of interest in this context was a claim in the *Sydney Morning Herald* on 16/9/96, page 3, that Australia's Royal National Park to the south of Sydney, gazetted in 1879, was the world's first. Although US authorities claim the proclamation date of Yellowstone National Park as 1872, the legislation which established Yellowstone did not use the term 'national park', referring instead to it as a 'public park' and 'pleasuring ground'. National park status for Yellowstone did not come until 1883, four years after Australia's Royal National Park (information from Frank Vanclay in an email discussion group for the International Association for Impact Assessment, 5/11/96).

The main objectives of the Acts are the protection of flora and fauna in their natural state for recreation, scenery, wilderness, habitat protection, scientific study and educational purposes. Commonwealth and State co-operation occurs in areas of national or international significance, particularly where the Commonwealth is a signatory to international conventions such as those concerned with biodiversity, illegal trade, wetlands, endangered species or special ecosystems (Chapter 6). Some reserves are totally protected (except for minerals exploration) whereas others are subject to different levels of conservation and access according to their qualities and uses. Privately owned lands can also be dedicated to conservation through voluntary or compulsory measures.

Management plans are central tools and are generally a statutory requirement for larger parks, and relate to access arrangements, education programs, fire control, protection of habitat, issue of permits, erosion and pollution control, and the control of exotic pests (plants and animals) and diseases. Regulations are enforced by rangers who have wide-ranging powers. Various measures are used to control visitor pressures, especially in high-use areas. They include entrance fees, structures and barriers, track marking, education and zoning. Queensland's *Nature Conservation Act 1992* (which replaced the *National Parks and Wildlife Act 1975*, the *Fauna Conservation Act 1924* and the *Native Plants Protection Act 1972*) is a good example of the integration of nature conservation with other land uses, including voluntary agreements with landholders. It stresses the need for habitat protection and the conservation of nature for all the State and not just in national parks or for particular plant and animal species.

National and World Heritage legislation also affects unique or significant natural and cultural areas, including historic and Aboriginal sites. The setting aside of such areas can also be considered as being in the public interest, and it may be accomplished (and sometimes controversially secured) through special legislation including World Heritage listing. Australia has eleven World Heritage listed sites, including the Great Barrier Reef, Tasmanian wilderness, Kakadu National Park and Shark Bay (Chapter 6 and Fig. 6.1).

Under the Commonwealth Heritage Act, 'national estate' registration may include natural areas of significant value (as wilderness, special habitats and scenic areas—again, with a public interest focus), as well as historic and social elements (AHC 1998). The object of the register is to inform people of the sites. There is no legal power to enforce protection although every effort is made to ensure that any planning activity takes a listing into account. Proposals to develop a registered site may be subject to an EIA process, and Ministers have some discretion over actions taken. Heritage and conservation agreements may also be made with private land owners, sometimes with incentives or penalties attached.

Controls on native vegetation clearance, habitat disturbance and loss of biodiversity are arguably another 'public interest' matter, insofar as maintenance of environmental quality for present and future generations is related to human welfare in some way. The protection of remnant vegetation and biodiversity, and clearing controls in rural areas, have been major concerns in rural land management in recent decades. These concerns have been reflected in specific legislation and are discussed further in Chapter 5. Biodiversity agreements, strategy and legislation at national and international levels (chapters 6 and 7) are aimed at the regulation and protection of Australia's biological diversity in a comprehensive manner. Consideration is also given to the more utilitarian aspects of potential uses of native vegetation for food, fibre, medicine and industry.

COMMENT

This chapter has shown how laws evolved to protect the 'public interest' in Australia in relation to the biophysical environment. There were basic laws of direct benefit to people which ensured the protection of air and water quality primarily for health reasons and to maintain living standards, the passage of specific environmental protection legislation in the 1970s, and the development of broader environmental policies (and attitudes). Environmental protection laws were to gain additional strength with a greater emphasis placed on the environment, but not ignoring human needs.

Less direct actions in the public interest related to the protection of the biophysical environment for human enjoyment and other purposes. These, too, had an early history. While the value of conserving natural environments appears not to have been forgotten, their role in human well-being has always been recognised in the formulation of policies and laws. This recognition continues, but often in conflict with over-riding economic demands, as Chapter 2 well demonstrated.

The next chapter examines the ways in which traditional natural resource management with its utilitarian focus has undergone some important transformations since the 1970s. Balancing human and environmental needs has been an increasingly important objective.

CHAPTER 5

NATURAL RESOURCE AND LAND MANAGEMENT POLICIES, LEGISLATION AND AGENCIES

> The allocation of resources and the preservation of environments are always political decisions swayed by economics, powerful interests and electoral timetables (Dargavel 1998a:30).

INTRODUCTION

Land-use planning and management are concerned with existing and future land uses and operate at national, State and local levels through various agencies and supporting legislation. Ideally, the 'best use' for land is sought, incorporating considerations of social and economic equity amongst affected communities as well as the technical and economic suitability of land for alternative uses, and avoiding conflict where possible. In reality, political and economic priorities often skew outcomes away from best-use options.

Increasingly, environmental aspects of planning and management have been built into more comprehensive planning legislation and practices. This shift has in part been driven by public demands to be involved in decision-making processes which affect the quality of life and the communities in which people live: public participation and the integration of planning and environment are discussed further in Parts 4 and 5.

As discussed by Grinlinton (1992), however, in traditional natural resource management, any search for integration has been hampered by factors such as: unco-ordinated policies amongst various levels of government; diverse and conceptually isolated decision-making arrangements for allocation and use of the various resources; rigidity and resistance to change or shared responsibilities by administering bodies; lack of full public participation, and difficulties in enforcing compliance where rights and duties exist. There is no over-arching natural resource management legislation in Australia directly equivalent to New Zealand's 1991 *Resource Management Act*, although Tasmania has a close approximation in its Resource Management Planning System (RMPS) (Chapter 20). Moves for integrated natural resource management legislation at national and State levels appear to have been thwarted generally, although there are several interesting new developments (Part 5). In addition, some structural reform has taken place in several States with amalgamation of natural resource agencies into one management department (Table 6.1). It would appear, however, that agencies often still operate independently within these mega departments, thus perpetuating largely traditional arrangements.

The following sections look at legislation which deals with the allocation and use of natural resources (land, water, forests, minerals and energy), and some of the moves to improve the laws and management arrangements. Historically, laws affecting the use of these resources reflected the colonial, pioneering nature of Australia's settlement by Europeans primarily of British origin, and the orientation of the new colonies to supplying the needs of the mother country. Railway networks focused on the main port of each State, drawing raw materials and food from the hinterland to the point of export. Planning took the form of *ad hoc* responses to situations as they arose. Interest now lies in how the earlier development-orientated legislation has been modified in recent decades to take into account problems of environmental stress and non-utilitarian values and to incorporate elements of forward planning.

Surprisingly, despite the strong development ethic of the nineteenth century, some strong elements of environmental protection or conservation were taken into law late in that century. This period also saw the first parks and reserves set aside. Public debate had arisen over growing pressures on the colony's natural resource base. The increasing prevalence of soil erosion, sedimentation, flooding, feral animal damage and failure of water supplies (some of which was exacerbated by severe droughts) indicated that something was amiss. A general political response to these and other environmental problems was to use a series of Royal Commissions, well into the twentieth century, to investigate causes and propose solutions—sometimes with little success. Nevertheless, it was in this context that a suite of legislation was enacted relating to forestry, water use, soil conservation and land use.

LAND USE—URBAN

Early urban settlement was subject to the powers first of colonial administrators, town trusts and district boards, and later, municipalities, which controlled the release of land, development of essential infrastructure and public interest needs. Contemporary local governments perform essentially the same role, but with expanded responsibilities incorporating environment and planning (Part 5).

In discussing early land-use planning in Australia, Heathcote (1994) noted that little account was taken of physical, environmental or social needs, the focus largely resting on land survey and disposal (see also Powell 1976 and Berry 1992). Urban land was allocated for various administrative, military and public purposes and took place along geometric lines. Based on a rigid British system of subdivision into counties, hundreds and parishes, this method proved inappropriate for Australia's harsh environment with its vast, unpopulated spaces. Survey units became, in effect, useful only as reference points for land survey, or provided the basis for data collection. A basic problem was that local governments had little control over detailed land use once land transfers had taken place.

One inheritance of this period is that, unlike the more flexible urban land-use planning techniques in the United Kingdom, those in Australia tend to be formal and relatively rigid, principally because of their legal structure and greater enforceability. There is an emphasis on plot ratios, setbacks, height restrictions and zoning; in some jurisdictions this extends to permitted building materials and other elements specified in restrictive covenants. However, the system is capable of contributing to environmental control if the administrators so decide.

Town Planning

Town planning, with specific planning schemes and regulations, did not come into existence as a discipline or government entity until the early twentieth century, when the first town planning principles were applied in planning for Australia's new capital, Canberra. The notion of rational planning was particularly promoted by two individuals: one, C.C. Reade, who was to become Adelaide's first town planner in 1916, introduced the idea of Garden Cities to Australia; the other was J.J. Sulman, whose 1921 book, *Town Planning in Australia*, dealt with issues such as minimum lot sizes and healthy living conditions (serious plagues in crowded, unsanitary conditions had afflicted Perth and Melbourne early in the century). By 1913, the country's first town planning association had been established in Sydney, followed by its first town planning conference in 1917.

Between the wars, several States enacted local government (NSW 1919, WA 1934, Qld 1936) and planning legislation (1920 in SA, 1928 in WA). Curiously, the Western Australian *Town Planning and Development Act 1928* (still in force, though amended and undergoing consolidation in 1998) includes the conservation of natural features and preservation of trees; but, as in the later *Metropolitan Region Planning Scheme Act 1959* which also asks for consideration of environmental factors, no particular legislative directions are given. However, the latter scheme, with considerable foresight, set up a fund for the purchase of land for public open space.

With the onset of the Great Depression and World War II, interest in planning waned, to be revived during the 1940s in Sydney and Melbourne to deal with new city expansion. However, rapid population growth and urban and industrial development showed existing laws to be inadequate. To this point, planning functions had generally been separated from environmental protection functions (with a secondary role where they existed at all) and were mainly concerned with orderly development.

Planning laws of subsequent decades were increasingly forced to become broader in scope as well as answerable to other environmental protection laws—and in fact developed into quite sophisticated regulatory regimes. This was reflected in some early moves to incorporate environmental considerations into the planning process. For example, three paragraphs added in 1972 to the Third Schedule of the Victorian *Town and Country Planning Act 1961* introduced an environmental dimension to the concept of a planning scheme. The crucial phrase in each paragraph is 'conservation and enhancement' (including both built and natural environment). These terms, however, were open to wide interpretation: later legislation has been more specific. For example, the 1987 *Planning and Environment Act* incorporates into planning the protection of natural resources, ecological processes and various social and cultural values at State, regional and local levels.

The *Environment Planning and Assessment Act 1979* in New South Wales was also progressive with its incorporation of environmental, social and economic objectives, achieved through State, regional and local planning policies and plans. The State's *Local Government Act 1993* (replacing the 1919 Act) gives local Councils quite specific environmental responsibilities. Each Council is given a charter to 'properly manage, protect, restore, enhance and conserve the area for which it is responsible' (Section 8(1)); however, if Councils fail to fulfil this charter, they are not liable to civil action. Each year, Councils are required to prepare reports detailing achievements in relation to management plan targets. The reports are required to provide details on the state of the environment in relation to areas of environmental sensitivity, important wildlife and habitat corridors, unique landscapes or vegetation, and any development likely to affect the community or environmentally sensitive land.

Queensland's *State and Regional Development Public Works Organisation and Environmental Control Act 1971* and its later *Local Government (Environment and Planning) Act 1990* (now replaced—see Chapter 19) also emphasised local (as well as State) authority responsibility for attending to the environmental effects of proposed developments. Later legislative reforms emphasised local government responsibilities in regional frameworks (Chapter 19).

But for all the apparent progress, planning still appears beset by Lindblom's (1959) 'science of muddling through'—a reactive, incremental decision-making framework moulded by institutional and individual forces, despite efforts to pursue forward planning. Even where plans exist they are not always adhered to. Comprehensive regional and strategic planning is a more recent attempt to be integrated and proactive, to allocate resources amongst competing uses, and to acknowledge environmental and social needs and constraints. But some planners fear that the prevailing economic rationalism of the 1990s threatens the core values of planning which had motivated practice in the past (Part 5).

LAND USE—RURAL

During the nineteenth century, land beyond the town was progressively claimed for the Crown and subject to various Lands and Pastoral Acts (see below—'Crown Lands'). Heathcote (1988) noted that as in urban areas, rural lands were generally surveyed along grid lines with little regard for land form.

This system posed administrative problems with respect to differing standards and regulations across State boundaries.

Private landholders in rural areas are subject to various local authority and State regulations relating to services, rates, development and nuisance (common law). A range of legal environmental management constraints apply in the States through vegetation, water, wildlife and agricultural chemicals legislation in respect of clearing, biodiversity and pollution (Chapter 4). Concerns over soil erosion control have a longer history (Bradsen 1988). Enforcement provisions have been strengthened in some States in recent years, although they are seldom invoked. In Western Australia, for example, under the *Soil and Land Conservation Act 1992*, the Commissioner for Soil Conservation has the power to issue 'notices', levy fines and even to carry out works or take over severely degraded land from recalcitrant land owners if necessary. Action can also be taken over off-site movement of soil (drift, sand, soil or dust) or water. More comprehensive land management procedures, requirements and incentives now assist landholders to combat degradation and implement sustainable agriculture, relying on voluntary rather than mandatory measures (Chapter 15).

On leased Crown lands, users (including governments) can be legally required to control weeds, vermin, disease and fire, or take measures to prevent further degradation through destocking or revegetation. Examples of Acts relating to the above are listed in Table 5.1.

Table 5.1   **Some examples of types of environmental protection Acts relating to the use of rural land.**

Quarantine Act 1908 (Commonwealth)
Rabbit Act 1919 (ACT)
Soil Conservation Act 1938 (NSW)
Noxious Weeds Act 1963 (NT)
Fruit and Plant Protection Act 1968 (SA)
Agricultural and Veterinary Chemicals Act 1988 (Commonwealth)
Soil and Land Conservation Act 1988 (WA)
Pastoral Land Management and Conservation Act 1989 (SA)
Native Vegetation Act 1991 (SA)
Catchment and Land Protection Act 1994 (Vic.)

Vegetation clearance controls or remnant vegetation protection Acts are other forms of land-use control and habitat protection—such as South Australia's *Natural Vegetation Act 1992* or the NSW *Native Vegetation Conservation Act 1997*—though sometimes subject to criticism. In New South Wales there have been reported concerns amongst farmers about the effects of its native vegetation controls on rural property values. The NSW lobby group, NSW Farmers, stated that since the introduction of *State Environmental Planning Policy (SEPP) No. 46* in 1995 (which the above 1997 Act replaces), there had been 'an effective ban' on farm development and many farmers had been prevented from bringing new land into production (*The Australian* 19/6/98:38). It would appear that the legislation is having its intended effect. There are also overlapping laws (bushfire, roads and lands legislation) which cover the protection of flora and fauna in road reserves and from the activities of public utilities.

Problems of widespread land degradation in Australia and the need for better land management have been recognised increasingly at government levels. Historically, concerns focused on specific problems as they arose—loss of natural vegetation, soil erosion, water allocation, rabbits or marginal lands (Heathcote 1965; Powell 1976). The first major report since the Second World War, despite its inadequacies, adopted a broader view of land degradation and provided a basis for a national soil conservation policy (Department of Environment, Housing and Community Development 1978). Later investigations included government inquiries into land use, land degradation and agricultural chemicals, national and State conservation strategies, soil conservation strategies, the Decade of Landcare and the Ecologically Sustainable Development (ESD) strategy—discussed in Conacher and Conacher (1995) (refer also to Table 15.1 and Appendix 3). These have had varying degrees of influence on land use and natural resource management policies.

Queensland's traditional soil erosion legislation is narrowly focused; but new and more comprehensive natural resources legislation was in preparation in 1997, and a Landcare and Catchment Management Council has been set up (Chapter 15). Nevertheless, the State continues to engage in large-scale vegetation clearance, notably in the Brigalow Belt (between 1991 and 1995, 138 830 ha were cleared each year: Peter Bek, Qld SoE Reporting Unit, pers. comm. 1999). South Australia is more progressive with its network of Soil Conservation Boards, native vegetation and pastoral lands legislation (some were under review in 1998). This has been particularly so in lower rainfall areas which traditionally are severely affected by drought, imposing constraints on vegetation clearance and pastoral land use. New South Wales and Victoria are noted for their pioneering work in total and integrated catchment management (Chapter 15), and have provided some useful national models. In most States, such issues increasingly are being drawn into State and regional planning strategies (Part 5).

The approach to land use planning in predominantly rural areas differs from that in urban areas. 'Town' planning was upgraded in the 1940s and 1950s following some controversy between those supporting the need for more regulated planning and those who preferred a more *laissez faire* form of governmental guidance. To some extent that disagreement still exists, encapsulated in part in the philosophies underlying Australia's two major political parties. But in general, the need for town planning, to deal with some of the more obvious problems associated with living and working in large, crowded cities, is now widely accepted. In contrast, there was no apparent, equivalent need for this type of planning in rural areas—although, as is argued in this book, this is no longer the case.

Crown and Unalienated Lands (see also Rangelands, below)

A large proportion of Australia—about 40%—is owned by the (anachronistic) 'Crown'. Over a third of the areas of five States is under Crown lease but, unlike the USA, is under State not Federal control (the Commonwealth is only a relatively minor land owner in the States but is the primary owner of land in the Territories).

By virtue of their extent and location, Crown lands are often important regions of biodiversity and other natural resources. Each State has its own legislation dealing with the management of Crown lands: for example, the NSW *Crown Lands Consolidation Act 1913*, the SA *Crown Lands Act 1929* and Western Australia's *Land Act 1933* (now the *Land Administration Act 1997*). As indicated, the Commonwealth has no direct powers over Crown land except in the Territories, or through some of its constitutional powers, or where it owns land directly. Management is generally at the discretion of State government authorities, administered through their Crown or Lands Departments, which permit or reserve uses for development or conservation, or leave the land idle. Where land is not otherwise allocated, it is an offence to graze, remove material, drill or remove water from Crown land without permission. Interference to reserves can be an offence under law, and clearing bans may also apply.

In general, Crown land is managed in the interests of 'development' as conceived by the public administrators. This is done through the allocation of leases and conditions attached to those leases, through legislation originating in the nineteenth century. Wider environmental implications were usually ignored by the legislation, except in some specific instances. This resulted in considerable controversy in the case of pastoral leases, where in certain areas (such as the Ord and Gascoyne catchments in WA, and parts of South Australia), overstocking by leaseholders resulted in considerable and extensive damage to native vegetation, and consequent accelerated erosion of the exposed soil. Increasingly, the sense and ethics of permitting a handful of people to degrade thousands of square kilometres of country for relatively insignificant economic returns to the country have been questioned. But rural mythology associated with the 'kings in grass castles' (Durack 1959) is still supported by powerful pastoral industry groups and the political, rural-based National Parties. There is little political will to take effective action in many instances, although native title (see below) may be forcing the issue in some cases—and leading to pressure from some 'grass kings' to convert leasehold to more secure tenure.

Ownership and use of Crown lands have also been the subject of conflict in other areas, such as: competing claims on the use or allocation of resources (particularly minerals); economic development; development of public utilities; tourism, or Aboriginal land claims. Development projects may take place through State agencies (such as public gas or water pipelines), or by means of government concessions, licences, grants or industry agreements which may apply to private developments.

Economic development of Crown lands is increasingly subject to scrutiny. Issues such as environmental impacts, indigenous rights and public participation are now expected to be part of decisions affecting the use of these lands. In some instances such decisions may override State and local planning schemes, such as reservations for public purposes or coastline protection. In other cases, the Crown may be required to carry out beneficial practices to keep land in good condition, such as through the control of feral animals and weeds, or bushfire management. Stricter controls on mineral exploration and mining now also apply (discussed later in this chapter).

In recent decades, revisions of various Crown and public lands Acts have required more detailed assessments of land allocation and use. The NSW *Crown Lands Act 1989*, for example, expects that certain principles of land management such as environmental protection and land capability evaluation, will be incorporated in any assessment of future land development, subject to the provisions of NSW's *Environmental Planning and Assessment Act 1979*. The Victorian *Crown Land (Reserves) Act 1978* also provides powers to reserve land for public purposes including any of ecological, scenic and historic significance. Bates (1995) provides details concerning other States' requirements in public land management.

'Rangelands' or Outback Management

The draft *National Rangelands Strategy* of 1996 (ANZECC/ARMCANZ 1996) was a serious attempt to address a range of complex environmental, social, economic and cultural problems in Australia's Outback (Fig. 5.1). Although its contribution to Australia's economy is relatively insignificant (mining in the Outback accounts for about $12 billion or 3% of GDP, tourism $1.7 billion or 0.4% and livestock $1 billion or 0.2%), the physical extent of the area (some 70% of Australia's land mass) and its various non-economic values are of considerable importance to the nation. The Outback, of which about two-thirds is pastoral country, is rich in thousands of species of flora and fauna. Conversely, there are huge 'costs' associated with: isolation from services such as education, health, transportation and banking; greatly modified ecosystems; natural resource degradation exacerbated by economic pressures, and the dispossession of the original Australians.

In 1975 it was estimated that more than half the Outback area was moderately to severely affected by land degradation (Department of Environment, Housing and Community Development 1978; see also Payne *et al.* 1998). Since then, the condition of some areas has worsened, notably parts of central Queensland (Queensland Department of Lands 1993), whereas others, such as parts of the Gascoyne drainage basin in Western Australia, and southwestern Queensland, have improved in response to better management. Problems include loss of ecosystem integrity (species and habitat loss, woody shrub invasion), loss of water quality and resource, increased wind and water erosion, and loss of productivity and aesthetic value. These problems have been exacerbated by pressures such as overgrazing and/or predation by livestock and feral animals, the increased incidence and extent of fire, vegetation clearance, impacts of drought, demands on water resources, and economic factors leading to declining terms of trade for pastoralists.

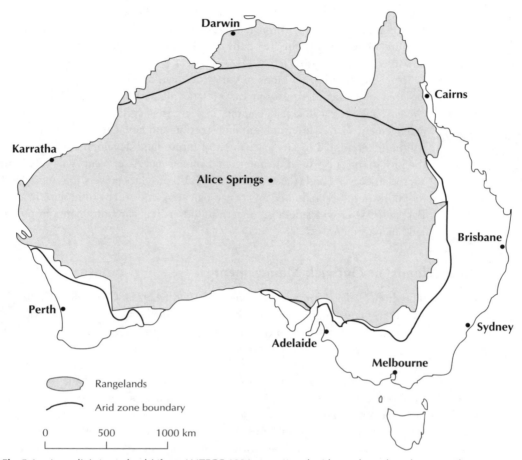

Fig. 5.1   Australia's 'rangelands' (from ANZECC 1996, map 1) and arid zone boundary (from Woods 1983).

Some of the direct and indirect consequences of land degradation have included depopulation, social disruption (withdrawal of services), loss of productivity and vitality (which in turn may increase pressures on land through lack of management or increased stocking rates) and land-use conflict. Aborigines are significant landholders and managers of about 20% of the Outback. Conflicts over the need to recognise their traditional rights have arisen with mining and pastoral interests (discussed below). Native title legislation has attempted to resolve some of the issues by clarifying land tenure and land access rights by offering more equitable and flexible land tenure and use arrangements.

The *Rangelands Strategy's* 25-year 'vision' is to recognise the value and significance of the rangelands for diverse ecological, economic, aesthetic, cultural and social values and a commitment to ESD (ecologically sustainable development) in management. Its goals are to achieve: integrated policies; administrative practices which will ensure ESD; community participation; clear property rights and responsibilities; improved management, diversification and 'best practices'; rehabilitation and protection of degraded resources; improved information and monitoring; preservation of biodiversity and the development of reserves and 'bioregions'; acknowledgement and protection of the rights of indigenous people; accommodation of non-pastoral uses, and identification of social, cultural and heritage values.

The Strategy sets out various proposals to recognise the complexity of the environmental, economic, social and cultural problems facing the rangelands, noting that the problems are not easily resolved under existing legislation and institutions, or with present attitudes. Reform and the means to alter current attitudes and behaviour towards and management of the natural resource base of the rangelands by miners, pastoralists, Aborigines and tourists, are needed urgently. States and Territories are to be asked to review existing regulations and controls. The Strategy favours a regional approach to management, with consultation amongst users, managers and interest groups. Various policy instruments and reforms can be used, including: incentives (pricing structures) and restrictions (legislation and regulations); research, development and education (through grants and land care); improved information (computers, business advice); government intervention (infrastructure, regional development); links to other strategies (ESD, biological diversity, water and drought management, land care, greenhouse), and co-operation amongst and within States, agencies and groups of individuals (refer also Hoey 1994 and Holmes 1994).

In Western Australia, a proposed Regional Development Policy also acknowledges the range of issues confronting remote areas—economic and social disadvantage, reliance on natural resources, low skills, isolation, loss and high cost of services, global influences and land degradation. Goals are suggested to: revitalise and enrich lifestyles and livelihoods; recognise cultural heritage; develop and diversify new opportunities for creating employment and wealth; protect the environment and natural resources, and manage land degradation in an integrated and sustainable manner (Department of Commerce and Trade 1999).

Western Australia's *Land Administration Act 1997* replaced the *Land Act 1933* which had facilitated the development of unused and vacant land and had been directed particularly to pastoral leases. The new Act differs from its predecessor in several respects. Lease renewal periods are shorter and management policies and penalties more clearly spelt out. There are also provisions for permission to pursue non-pastoral activities. The terms of the *Soil and Land Conservation Act* with regard to powers to serve soil conservation notices remain (although they were rarely used in the past).

A Pastoral Lands Board has greater powers and representation than the former equivalent. In the spirit of the National Strategy for Rangelands Management, the Board is to: ensure that leases are managed on an ecologically sustainable basis; develop policies to prevent rangeland degradation; rehabilitate degraded land; establish monitoring sites, and conduct relevant research. A lease may not be granted if the Board considers a unit to be economically or ecologically unsustainable. The onus is on the Board to see that lease conditions are upheld.

The new body has been given a wider range of discretionary powers. But to an extent, the ultimate power lies with the Minister who can appoint and terminate Board positions. In this regard, political pressure can still affect the Board's operations (Baston 1999). As far as public involvement in the decision-making process is concerned, there is no formalised process for third party appeal. Neither are the Board's proceedings subject to public scrutiny. Lessee appeals are also limited.

Baston sees many of the changes in the new Act as positive, particularly in terms of the greater environmental focus. However, there are several weaknesses which concern him, not least the presence of discretionary powers for the Board and Minister and the absence of a third party appeals process. Considerable uncertainty still surrounds the new leasehold arrangements, complicated by native title issues. Problems exist for lessees of marginal properties, who may be forced to neglect good land management practice: the application of punitive measures in times of hardship might be inappropriate under some circumstances when other measures could be used. The failure to provide for reviewable lease conditions also appears to diminish the flexibility of tenure arrangements. Baston concludes that unless the legislation can help align pastoral leasehold goals with those of the broader community, and with the concept of sustainable development, then legislators will be fighting a losing battle.

Aborigines and Land Management

As is now widely recognised, Aborigines have a close, spiritual and practical relationship with the land (Chapter 8 in Aplin 1998). Although having a deep understanding of their environment, this relationship has not been entirely without impact, particularly with regard to the use of fire (Hallam 1975). On the other hand, strong elements of 'sustainability' in their culture ensured more than 40 000 years of land use, as compared with the damaging effects of European settlement in a mere 200 years. Indigenous Australians are important land users, but only in recent years have their rights of access and use been acknowledged formally and, unfortunately, controversially. The Northern Territory and mining issues provide good examples of land-use conflict and the importance of acknowledging broader needs and values in managing natural resources and land.

The 1992 Mabo and 1996 Wik High Court judgments and passage of the *Native Title Act 1993* were attempts to redress the enormous harm wrought by Europeans, by protecting and rebuilding that which had been lost to Aboriginal culture. Although the *Native Title Act* removed large areas of freehold land and some areas appropriated for public uses from Aboriginal land claims, native title may exist over leasehold land (particularly pastoral areas) and some other areas such as parks and reserves, marine environments and rivers. But the Act, which is based on the key principles of non-discrimination and equity, does not stop land management or economic activities.

The Mabo ruling essentially stated that negotiations should be held concerning future land use with Aboriginal groups who could demonstrate an unbroken line of occupation of or spiritual involvement with the land in question. The High Court found that the previous doctrine of *terra nullius*—that Australia was unoccupied at the time of European settlement—was clearly wrong in fact and was legally null and void. The High Court's unfortunately split majority decision in the Wik case handed down the ruling that pastoral leases should co-exist with Aboriginal land rights (as has indeed long been the practical situation on many stations). Pastoralists and mining interests were unhappy with the judgment, leading to extraordinary objections in some quarters, including calls for the States to have a greater say in the appointment of High Court judges. On the other hand, fears amongst many Australians that their land might become subject to Aboriginal land claims have not been assuaged by some of the more extreme land rights claims.

The central objective of the *Native Title Act* 1993 was to emphasise indigenous participation in innovative land management regimes in traditional lands of unalienated, pastoral, riverine and marine environments, through planning and decision-making processes of mutual recognition and negotiation (Lane *et al.* 1997). Native title claims over land are subject to specified processes under the Act, and land and resource planning is seen as essential to the native title process. Aborigines may initiate claims and develop management plans which are then subject to mediation and negotiation processes. Claims may include the protection of the environment as well as cultural values. Planners may be required to work within the native title process, which might not always be particularly rational where 'rights' are indistinct. Planners now need to account for native rights in areas subject to claim, which poses a significant challenge as rights may vary widely. Flexible planning strategies in an atmosphere of mutual co-operation are sought in order to meet complex demands and objectives for managing these lands.

Lane *et al.* (1997) raised some of the difficulties in new land management strategies for native title lands where stakes are often high, processes lengthy and frequently in conflict. They identified a range of barriers relating to language, lack of education, sense of disenfranchisement, lack of resources and problems of reconciling scientific rationality (and top-down planning approaches) with Aboriginal cultural perspectives. A community-based planning process was recommended, conducted at local levels with locally-derived goals, allied with formal recognition of Aborigines' rights to manage their own affairs and communities (see also Thackway and Brunckhorst 1998).

In 1997, further efforts were made by the Federal Government to reconcile some of the differences, through development of what was known as Prime Minister Howard's '10-point plan'. This plan in effect came close to extinguishing native title and giving pastoral leaseholders rights similar to those of land owners (it should be noted that some pastoral leaseholders are Aborigines— and some others are foreign-owned corporations). While the plan received grudging support from grazing interests, Aborigines were largely opposed to it. A key point at issue is the right of Aborigines who can demonstrate unbroken occupation or spiritual involvement with land to be involved in negotiations concerning the future uses of that land. The 10-point plan proposed to remove this right and was therefore in direct opposition to the High Court's Mabo ruling. A modified version of the plan was eventually approved by the Senate in 1999 with the support of independent Senator Brian Harradine.

Native Title and Mining

As discussed by McAuliffe (1998), registered native title claimants have a right to negotiate over the granting of licences or leases to mining companies. Agreements may include a share in future profits, access to traditional lands and provision of services and employment (conditions may also be imposed on mining companies requiring adequate environmental protection and rehabilitation of mine sites—see 'Minerals and Energy Resources' below). Negotiating parties develop a management plan to deal with the resource and environmental management of an area. However, the secret nature of negotiations prevents scrutiny of the effectiveness of such agreements. Nevertheless, mining Acts and codes of conduct should ensure that environmental concerns are considered and acted upon.

Since passage of the *Native Title Act 1993*, a large amount of confusion, manipulation and mistrust between parties has developed and considerable overlap has occurred. In mid-1995, there were three native title claims in an area of the Western Australian goldfields. By 1996 the number of claims had increased to more than 80, with layers of claims eight deep in some areas around Kalgoorlie (Department of Minerals and Energy 1996). While the Department of Minerals and Energy was of the opinion that some of these claims were vexatious, it is believed that Aboriginal groups of northwestern Western Australia are genuinely concerned about the land itself and not monetary reward. Native title has given Aborigines considerable leverage to ensure their rights to traditional lands are recognised, but there are concerns over the outcomes of the Act—including mining companies' use of funds to manipulate negotiations, hostility among Aboriginal groups over competing claims, and long delays in reaching decisions.

Nevertheless it should also be noted that despite reported claims by the Western Australian Premier that native title threatens to stifle $14 billion of minerals and energy investment (*The West Australian* 19/6/98), the Registrar of the Native Title Tribunal responded that since May 1995, when the Western Australian Government began working within the *Native Title Act*, more than 8000 'or 81.87% of mining industry applications for exploration, prospecting, miscellaneous and retention licences have been cleared for grant without objection' (*The West Australian* 22/6/98).

WATER MANAGEMENT

Useful overviews of water management in Australia have been presented in Sewell *et al.* 1984; Pigram 1986; MDBC 1990; Powell 1991, 1993a, 1998; and Smith 1998.

Often described (correctly) as the world's driest inhabited continent, Australia paradoxically experiences drenching rains and severe flooding as well as frequent droughts. The central 70% of the continent generates negligible sur-face runoff while the northern tropics and temperate Tasmania are well endowed with water. Only a small proportion of the divertible national water resource has been developed. Significantly, much of the remainder is either poor in quality or relatively inaccessible. It could be said that one of Australia's main water management problems relates to its quality rather than quantity.

Water managers face an array of difficult and complex problems, exacer-bated by growing human demands and environmental pressures. Water quality and flow have been compromised by clearing of native vegetation in catch-ments and the loss of riparian flora, pollution from point and non-point sources, and interference to natural hydrological regimes (over-extraction, diversion, drainage and construction activities). Some of the consequences have included rising water tables, increased soil and water salinity, waterlog-ging, loss of or damage to aquatic habitats, erosion and sedimentation, and physical, chemical and biological pollution of surface- and ground-waters (examples in Office of the Commissioner for the Environment 1989; State of the Environment Advisory Council 1996).

Until recent decades, the focus of water management (and legislation) was on the development and use of the resource. This was achieved through provi-sion of infrastructure (pipes, dams, sewage works and drainage schemes) which supported the extraction, storage and delivery of water to the consumer. Water quality issues, in a limited sense, were dealt with by health agencies, which was sometimes the source of inter-agency conflict. Along with policy development, such activities remained largely in the public sphere, although there was a brief and unsuccessful flirtation with the private development of irrigation schemes in Victoria and private water supplies for Perth; for example, both during the late nineteenth century (Powell 1976; 1998:16). The eventually failed Humpty Doo scheme in the Northern Territory and the Camballin irrigation scheme on the Fitzroy River in the Kimberley region of Western Australia were other pri-vate projects in the 1950s and 1970s; and in the 1990s, the Western Australian Government was encouraging further private irrigation schemes in the north, again on the Fitzroy and its tributaries (abandoned—for how long?—in 1998 following strenuous opposition).

Elsewhere, also in the 1990s, there were concerted moves towards European models of partial privatisation and corporatisation of sectors of public water agencies (Chapter 6 in Smith 1998). It would seem that the European lessons have not been grasped. The New Scientist (No. 1936, 30/7/94:43) provided some salutary information. Following privatisation of water companies in Britain:

- disconnections because customers could not pay their bills were up 50%;
- charges were up 65%;
- profits were up 125%, and
- chairmen's pay was up 130%.
- Also up were river pollution and the water companies' contributions to Tory party funds.

The Federal Government's role in water management has tended to be minimal, with the notable exceptions of interstate matters such as the Snowy River Scheme in the 1950s, the various Murray–Darling Basin Agreements (below) and, most recently, the Great Artesian Basin and Lake Eyre Basin Agreements. Federal powers—drawing on international agreements—were also used to halt the controversial Gordon-below-Franklin hydro-electric power scheme in Tasmania in the 1970s.

Disputes over water use were generally dealt with under common law, water being considered a public resource (or part of the 'commons'). Property owners, for example, had rights of domestic and agricultural use, provided these did not adversely affect downstream users. Over the years, such rights have been modified gradually by statute in the public interest. Now, all significant abstractions are controlled under licence from the Crown. The Crown also exercises statutory powers with regard to public water supplies, land drainage, sewage disposal, flood prevention and water diversion (Bates 1995; refer also Chapter 4 in Clement and Bennett 1998). It might be argued that such prerogatives occasionally generated conflict and public concern, such as the controversies initiated by the generous federal support given to WA's flawed major irrigation scheme on the Ord river, or the 'fuzzy' economics of the Snowy scheme (Davidson 1969). This question of government largesse was raised more recently in connection with the Queensland Government's support of the large Burdekin project (Powell 1991).

In common with other natural resources, water has also been the victim of an *ad hoc* planning and legislative process, one outcome being the proliferation of laws and agencies dealing with administration and management. For example, in 1986 Pigram estimated there were over 800 local and regional agencies dealing with water management around Australia, and 375 in Victoria alone (Pigram 1986:58).

Until the 1980s, little attention had been given to the protection and conservation of water in its own right, or to environmental considerations such as the care of aquatic habitats (Fisher 1980). Interestingly, however, early forests legislation often incorporated water catchment protection within their multiple use objectives, even if in a subservient role. Unfortunately, closed catchment policies are now rather more difficult to maintain due to pressures for increased logging and public access.

Thus, up to the 1980s, water management in Australia was essentially utilitarian in its approach. Stewart (1997) engaged in a spirited discussion on the comparisons between the old and new styles of water management. He noted that, traditionally, water authorities were ill-fitted to meet the new environmental challenges facing them, particularly since decisions increasingly needed to be made under conditions of high variability, complexity and uncertainty. The earlier agencies were too centralised, technical and rule-bound to implement solutions effectively. For example, Sewell *et al.* (1984) had felt it necessary to dispel the myths that water was free, deserts could bloom and management was purely a technical issue. There was a growing awareness that natural resource management could no longer be compartmentalised. New approaches were needed to bring together different and more comprehensive management strategies.

Table 5.2 tracks some of the key changes to water management in recent decades. It shows a marked shift away from purpose-specific legislation and agencies towards broader-based organisations and strategies which incorporate other natural resources as well as environmental and social needs. These changes have been supported by restructured central agencies and legislation, with more clearly defined roles (separating environmental from service functions) and objectives particularly emphasising regional and local management through strategic and integrated planning. Four important developments over this period can be summarised as: recognition of the environment as a legitimate user of water; new market arrangements which reflect the full cost of water delivery; increased interaction and co-operative arrangements between agencies and the public, and the development of integrated catchment management (ICM).

The Murray–Darling Basin: a Model in Changing Management

The vast Murray–Darling Basin (Fig. 5.2), extending over one-seventh (> 1 million km²) the area of Australia's mainland, provides a good example of the change in approaches to water management; namely, the switch from a utilitarian focus on the resource to broader-based, integrated resource management. The sequence of events which led up to the late 1980s' agreements is unique in Australia (Powell 1993a; Wilcher 1995; Doyle and Kellow 1995; State of the Environment Advisory Council 1996).

In 1884 a Royal Commission on the Conservation of Water in NSW proposed a treaty between New South Wales and Victoria to govern water control to the South Australian border. Understandably, South Australia was unhappy about being excluded, particularly since the Murray River was that State's lifeline for river trade and water needs (and still is for the latter), and objected to the upstream extraction of water. A compromise was eventually reached amongst the three States in the 1915 Murray River Waters Agreement, which provided for various river entitlements, navigation, infrastructure and irrigation needs. A Murray River Commission was also set up. No fewer than nine Agreements were made subsequently, culminating in the establishment of a new Murray–Darling Basin Commission in the late 1980s, as follows.

The Snowy River hydro-electric power scheme commenced in 1949, adversely affecting water flow to the Murray; and again South Australia expressed displeasure over increased irrigation entitlements to New South Wales and Victoria. These disagreements extended to the late 1960s. But as Doyle and Kellow (1995) pointed out, up to this time the issues were always over water *quantity* and not its quality—until Adelaide's water supply became increasingly affected by problems of salinity and turbidity. Another important tributary, the Darling River (and thus Queensland) had been generally ignored in the river's management.

Table 5.2   Changing water resource management in Australia 1970s – 1990s.

Some Key Features Pre-1970s	Post-1980s
Closed systems—rigid	Open systems—flexible
Centralised, top-down management	Structural reforms—corporatised/privatised, devolution of power
Formal, skills- and task-based	Increasing federal/State/local/inter-departmental co-operation
Utilitarian/development focus	Clearer definition of roles/policies
Ad hoc, reactive legislation, fragmented (water, sewerage, drainage Acts etc.)	More comprehensive, pro-active legislation (integrated resource management/planning Acts)
Little attention to 'conservation'	User pays—true costs of water delivery recovered
Inequitable allocation of rights	ESD in planning process Environment recognised in its own right

Some Key Developments 1970s – 1990s

1960/70s	Whole farm planning (Potter scheme) Eppalock catchment management (Vic.) National/State Conservation Strategies (Chapter 6)
1980s	Various States' water, pollution Acts revised (chapters 4, 7) International conventions (wetlands, pollution, biodiversity—Chapter 6) Catchment Acts (NSW, WA, Vic.—Chapter 15) Water 2000 Report (Australian Department of Resources and Energy 1983) Australian Water Resources Council reports on quantity, quality (AWRC 1977, 1982) Increased federal funding to States and Territories: Federal Water Resources Assistance Program Decade of Landcare announced (ICM, revegetation—riparian management—monitoring—Ribbons of Blue) (Commonwealth of Australia 1991) (Chapter 15)
1990s	Agenda 21, ESD reports (Chapter 6) Increased interstate, inter-agency co-operation (Chapter 7 and Part 5) Integrated planning and management (Part 5) Separation of agency functions to service, cost recovery and environmental/resource management (Chapter 5) Inter-governmental agreement on the environment (COAG 1992) PM's Statement on the Environment 1992 (Keating 1992) (includes water) Transboundary agreements—Murray–Darling Basin strategies (MDBC 1990) Great Artesian Basin rehabilitation program (Industry Commission 1998) National water quality management (ANZECC 1995) National guidelines on drinking water (NHMRC/ARMCANZ 1995) Council on Australian Government water reform agreement (COAG 1995) Allocation and use of groundwater (ARMCANZ/SCARM 1996) Research Land and Water Resources Research and Development Corporation (LWRRDC); CRCs—freshwater ecology, catchment management; CSIRO Strategies wetlands; salinity; algal blooms; marine/estuarine environments (Appendix 3) Monitoring National and State rivers health program (CEPA *et al.* 1994) Waterwatch, Frogwatch, Streamwatch, Algalwatch, water indicators (schools, community involvement, NHT, Landcare, ICM) (Chapter 15) State of Environment reports (most States plus Commonwealth and Territories) (Chapter 8) National Land and Water Audit (Chapter 9) Conservation Recycling, waste reduction, sludge treatment, 'best practice'

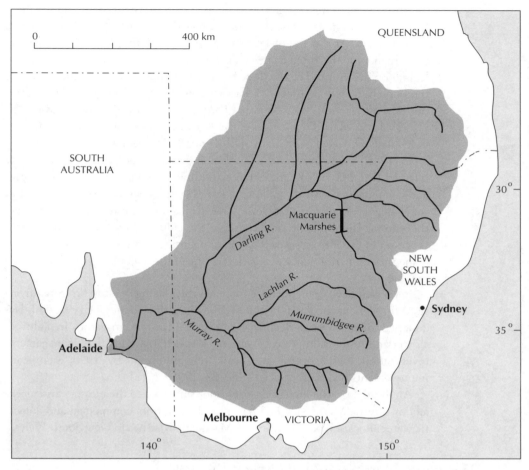

Fig. 5.2   Murray–Darling Basin and location of the Macquarie Marshes (from Australian Academy of Science 1994, Fig. 5.51 and EPA NSW 1995a:81).

In 1973, South Australia requested a three-States meeting, and after three years' hard bargaining it was agreed that water quality be considered ahead of quantity; although final agreement was not secured until 1982. Also, for the first time, the Commonwealth recognised the importance of other values—recreational and environmental.

In 1985 the Federal Government supported an interstate agreement and set up the Murray–Darling Basin Ministerial Council (Queensland became a signatory in 1992), which ushered in a period of substantial change and broke decades of gridlock. A new administrative body, the Murray–Darling Basin Commission, was also established, replacing the original Commission. A Community Advisory Council provided for public participation. The Commission receives directions from the Ministerial Council in relation to water, land and environmental resources in the Basin and co-ordinates efforts between communities and governments. The States' entitlements to water abstraction are set out in Schedules to the Agreement.

An Agreement was reached on a common plan to address water salinity problems, and to formulate research and management solutions and cost-sharing arrangements. At the end of 1987, a new *Murray–Darling Basin Agreement* was signed amongst the three States and the Prime Minister, and complementary legislation was enacted by the three States. An important factor at this time was the unprecedented alignment of Labor governments federally and in all three States, with no rural-based National Party to contend with. Doyle and Kellow (1995) also attributed the success of the Agreement in no small part to some key individuals who were particularly skilled and strong negotiators.

In 1988, a key report, the *Natural Resources Management Strategy*, was released (MDBC 1990). The Strategy outlined a framework for governments and communities with the following key objectives: to maintain and improve water quality; control and reverse land degradation; protect and rehabilitate the natural environment, and conserve the cultural heritage. The two main programs in the Strategy covered research/education and integrated catchment management. Performance evaluation of one of the major programs is discussed in Chapter 15.

Important developments have included the placing of a cap on water diversions throughout the Basin by the MDB Ministerial Council in 1995. This was designed to reduce water decline and provide security to irrigators. In addition, 17 wetland restoration projects were established, involving flow management, revegetation, pest control and bank rehabilitation. By 1998, significant improvements had been noticed (EPA SA 1998:97, 121).

An example of the increasing attention being paid to the idea of *environmental* flows in the Murray–Darling Basin (as distinct from commercial and domestic water allocation) comes from the Macquarie Marshes in New South Wales.

Macquarie Marshes: a Management Problem

The Marshes (located in Fig. 5.2) cover an area of more than 200 000 ha, including an 18 000 ha reserve which is a Ramsar site (Chapter 6; Fig. 6.1). The Marshes are heavily degraded from the effects of clearing for agriculture, adjacent cotton farming (using irrigation and pesticides), modified water regimes and river regulation (control of flooding and river flows), resulting in increasing salinity, nutrient enrichment, erosion and sedimentation, loss of vegetation, weed and pest invasions, and an altered fire regime. The Marshes are estimated to have contracted by 40–50% since the 1940s as a result of water abstraction (Comino 1997; Stewart 1997; EPA NSW 1995a).

There is a plethora of inadequate legislation to protect or manage the Marshes on a regional basis. Water quality matters come under various Acts—such as pollution, water, environmental protection and regional plans—but they are little used. Current legislation has little impact on the Marshes. However, a nature reserve management plan was developed in 1993 to preserve the habitat values of the area, and a water management plan was released in 1996. Most importantly, under the plan the State Government proposed to improve water flows through wetland areas by increasing the environmental allocation of water from an average of 50 000 to 450 000 megalitres per year. Average diversions for irrigation were to be reduced by 55 000 megalitres per year (NPWS 1993; NPWS/DLWC 1996).

NATIVE FORESTS

Much of the following discussion could have been relocated to Chapter 2 (which includes discussion of land-use conflicts and the adequacy of reserves; Chapter 4 also briefly covers parks and reserves as matters of 'public interest'). On the other hand, forestry legislation provides a good example of how governments and laws have responded to management issues.

Australian forestry was strongly influenced by American and European experience (Powell 1976), with the introduction of the concept of 'multiple use' at Australia's first forestry conference in 1911. This concept was adopted subsequently into State and federal forest policies. However, forestry legislation in all States, such as the NSW *forestry Act* 1916, WA 1916, Tas 1920, set economic exploitation of State forest timber as a management priority. Although specified in the legislation, other forest values generally are assigned secondary roles despite the enunciation of multiple use management principles in the various State Acts. The other uses usually include catchment and soil protection, recreation, tourism, and flora and fauna management. This position appears to be largely maintained in the 1992 National Forest Policy and subsequent Regional Forest Agreements (RFAs), albeit with more clearly defined management goals incorporating resource allocation, sustainability and environmental needs (see RFAs below and Lane 1999). There is also the additional and important constraint that forest legislation does not apply to privately owned land, which accounts for a significant proportion of the native forest estate. Indeed at about 11.5 million hectares, privately owned forests in Australia cover the same area as State forests, with an additional area of about 20 million hectares of forests in conservation reserves or classified as 'other Crown land' (Fig. 4.2 in State of the Environment Advisory Council 1996).

Explicit *preservation* and *conservation* of forests as near as possible to their natural condition (for public enjoyment, recreation, wilderness experience, scientific study or ecological purposes) are the prime responsibility of federal and State national parks and wildlife agencies, operating under appropriate legislation (for example, the federal *National Parks and Wildlife Conservation Act 1976*—replaced by the new *Environment Protection and Biodiversity Conservation Act 1999*—and the New South Wales *National Parks and Wildlife Act 1986*). While the roles of the two agencies (forests, and parks and wildlife) often overlap, or co-exist, they are often in conflict. Under Tasmanian forestry law, for example, in some situations forestry management plans may take precedence over those of parks and wildlife. In the case of Western Australia's Department of Conservation and Land Management (CALM)—formerly existing as separate forests, wildlife and national parks agencies—the agency inevitably finds itself in conflict amongst its differing intra-agency goals in natural resource management. There are political moves in WA to seek a re-separation of these roles.

While native forest management continues to be driven by economic imperatives and bound to the political process, utilitarian goals in forest management will dominate the management process. Unless there is a political will to recognise that other forest values can generate jobs and income or exist in their own right, then major changes will not occur. The RFA process of the late 1990s illustrates the pull between economic and environmental values, while the quote at the beginning of this chapter highlights the realities of natural resource management.

Regional Forests Agreements (RFAs): New Strengths, Old Flaws?

The Regional Forests Agreements (RFAs), which were under negotiation for several of twelve major forest regions around Australia in 1998/99, were an inter-governmental attempt to negotiate regional forest management plans. These were intended to fulfill a range of economic, conservation, social and cultural goals. But, as the discussion will show, the process has sometimes been complex, controversial and confrontational.

RFAs grew out of the *National Forest Policy Statement* (Commonwealth of Australia 1992b), which was agreed to by State and federal governments and supported by the *Intergovernmental Agreement on the Environment (IGAE)* (Chapter 7). As the largest environmental management exercise ever undertaken in Australia, the RFA process attempted to resolve some of the ongoing and contentious forest debates of the previous 25 years. RFAs between federal and State governments were intended to provide a sustainable and secure resource base for industry. They were also intended to ensure the protection (from logging and other pressures) of biodiversity, old growth forests, wilderness and heritage through a Comprehensive, Adequate and Representative (CAR) reserve system, and complementary off-reserve management.

Twelve RFA agreements are to be negotiated by 2001 in Victoria, Western Australia, New South Wales, Queensland and Tasmania. The Agreements are to run for 10–20 years, with reviews at five-year intervals.

The Process

In developing an RFA, the full range of forest values and scientific data is gathered and assessed in *Comprehensive Regional Assessments* (CRAs) (for example, Commonwealth Government and Western Australian Government 1998b). CRAs cover timber production and industry, tourism, mineral resources, social and heritage values, biodiversity, old growth forests, ecosystems and wilderness. The information is then summarised for the RFA and open to a public consultation period prior to signing of an Agreement (for example, Commonwealth Government and WA Government 1998a).

The legislative framework for RFAs operates at two levels. The Commonwealth Government has several pieces of legislation which impact on forest policy. They cover matters such as export approval, wildlife, environmental protection, national/world heritage, native title, international strategies and ESD (some of this legislation is under review). In addition, the *National Forest Policy Statement* sets out criteria for forest management along ecologically sustainable lines. State legislation provides the second tier of laws which operate through various forest, mining, flora and fauna, Crown land and environmental protection Acts.

Legislative Implications

RFAs are legally binding to a degree. However, in order to increase the statutory backing of the agreements, it is expected that governments will introduce complementary legislation or issue a new RFA Act. Where existing legislation needs to be changed, this will be pursued by both governments. In effect, the Commonwealth government accredits State governments with forest management and discharges its own responsibilities. Thus in 1998, in a politically expedient move, the Commonwealth proposed environmental law reform (Commonwealth of Australia 1998—enacted in mid-1999) in order to remove legal impediments or triggers which draw the Commonwealth into local issues, such as the conflict over the woodchip industry. Considerable environmental powers will reside with the States, although the Commonwealth will retain control over matters of national significance (undefined). There are concerns that industry and economic interests will be best served by such reforms and that environmental *protection* will be weakened considerably. The RFA process is no exception (see further, Tribe 1998; Forsyth 1998).

Some consolation may be found in the RFA commitment by the Federal Government that no woodchip licences would be issued after 2000 to any industry not deriving its resources from defined RFA areas. But it should be noted that the Commonwealth has amended its export requirements so that no controls apply to the export of hardwoods or woodchips sourced from the East Gippsland, Tasmanian and Western Australian RFA regions. On the other hand, the Victorian timber harvesting Act exercises control over annual yields to forestall over-cutting (Tribe 1998). Others have argued that these controls have been inadequate, with a clear escalation in woodchipping rates (Redwood, in Forsyth 1998:340). According to the Environment Institute of Australia's *Newsletter* No. 11 of December 1997, Tasmania's biggest woodchip exporter will be able to almost double its exports, by another 1 million t/yr.

In Western Australia, other changes to laws affecting forestry and the environment are anticipated, possibly (or predictably?) in favour of industry and economic development. Although the RFA is subject to a five-yearly performance review of commitments made with respect to forest industries, forest use and conservation, there is no indication of the extent to which the findings of such reviews will be legally binding or how compliance with commitments will be enforced. Given the past history of over-cutting, the level of confidence in the process is not high. An EPA report in 1998 on over-cutting saw the Environment Minister agree to the appointment of two review panels; one to determine sustainable logging levels, and the second to ensure that the ensuing guidelines are adhered to (*The West Australian* 25/1/99:4). It would appear that she does not trust either CALM or the EPA.

Stakeholder Interests

The matter of stakeholder interests is an important one as Western Australia's public consultation period demonstrated. As in other States, this was marked by some bitter public confrontations, opening up sharp divisions between industry, government, community (including local governments) and conservation groups, unfortunately quite contrary to the consultative and co-operative intent of the RFA process. Dissatisfaction with the process was made clear at a national forest-planning symposium held in Canberra in July 1998 (Dargavel 1998b:133). The difficulties of trying to incorporate public consultation and participation in the political process were raised frequently, and it was believed that the RFA process did not reflect the public's emphasis on conservation. Despite 'consultation', the formal agreements were brokered between Commonwealth and State officials only. Given that the RFA process was founded in part on the need to reduce public conflict over forestry issues and to increase opportunities for public participation in future planning, then various participants and stakeholders in the process understandably felt a strong sense of betrayal.

The fact that specific public/stakeholder concerns were relegated to Appendix I of WA's RFA public discussion document (Commonwealth Government and Western Australian Government 1998a), with its clear industry/government bias, was offensive to many; even more so since signing of the RFA was to proceed without further public exhibition after the public consultation phase. That such an outcome can occur suggests that something was fundamentally wrong with the process. It may reflect the complexity of that process, its ambitious and unprecedented nature, and the inexperience of the facilitators; or perhaps community expectations are higher than they were 25 years ago. As Dargavel (1998a) commented, the fact that the consultation process occurred at all and included important environmental parameters was a major advance on the forest debates of the 1970s. But is this too kind a response to a process which may only be paying lip service to environmental values and merely upholding the *status quo*? More recently, Lane (1999) observed that the RFA process has not so much succeeded in reconciling conflicting ideologies in the forests' debate as helping the Commonwealth to manage the politics of forest conflict.

In fairness to the Western Australian RFA Steering Committee, it clearly (and realistically) recognised the difficulties it would face when it stated in the objectives of its document that:

> difficulties will arise in seeking to meet a number of objectives concurrently...the objectives themselves do not identify the appropriate balance between environment, heritage, social and economic values (Commonwealth Government and Western Australian Government 1998a, Section 2:12).

In response to strong public opposition, early in 1999 the WA Environment Minister delayed signing the RFA (to mid-1999) pending further independent consultation. However, complaints over the lack of openness in this process remained (*The West Australian* 22/1/99:4; 25/1/99:4).

The following lists some of the issues that were raised by the Council of the Royal Society of WA (*Proceedings* July 1998:4), conservation groups and individuals during the public RFA consultation period in Western Australia in 1998. Many are applicable to the RFA process in the other States.

- The WA Steering Committee of eight people was composed entirely of government representatives.
- Initial criteria and assumptions were flawed. They did not always have local or regional applicability, or were too inflexible or inappropriate. This distorted outcomes, potentially impacting on future management.
- The process was too brief (driven by politics, funding and time constraints). Some projects were dealt with hastily, others not at all.
- There was too much top-down, bureaucratic involvement. The WA forestry agency, CALM, was said to have 'captured' the RFA process.
- There were inter- and intra-government frictions over policy and jurisdictions.
- Public/stakeholder views were not properly acknowledged.
- Industry/government/economic imperatives (forestry, mining, employment) outranked other values (environmental, cultural, social). Economic 'costs' were used to close off debate.
- Aboriginal values were poorly represented and the implications of native title treated cursorily.
- There was a lack of openness: some documents or data were not available; others were incomplete, misleading, biased or ignored (see Horwitz and Calver 1998). No public scrutiny or parliamentary debate was to take place before final signing of the agreement.
- Questions on notice in the WA Parliament failed to produce satisfactory information on the scientific reports.
- There was concern over the transfer of Commonwealth environmental powers to the States and the perceived weakening of those powers. Some devolution could prove misplaced or unpopular with future governments.
- Doubts were expressed over the legal enforceability of RFAs.
- The three options set out for Western Australia were narrow, biased, unclear or inappropriate and in particular did not reflect the views of the community.
- Alternative forest uses were not actively evaluated.
- Some important old growth forest areas were still not secure (despite this being a principal objective of the RFA).
- About one-third of the RFA-defined reserves were not forest but other vegetation types.
- Insufficient attention was paid to the ability of plantations to provide employment and meet timber needs.

- Privately owned forests are not covered under the RFA.
- Insufficient attention was paid to the precautionary principle and the ability to respond to changed circumstances.
- The RFA (WA) Committee was unable to appreciate or acknowledge the depth of anger and frustration in the community over forest management issues.

During the public consultation period, comments by the Forest Industries Federation Executive Director were revealing of the industry/economic view of forest management, and the indications are that these views will prevail. In defending the continued use of old growth forest, he was reported as stating that: 'If the industry was denied access it would lose one-third of Western Australia's sawn timber production and cost about 4500 jobs (direct and indirect)'. (In fact the numbers are closer to 200.) He believed tourists were also attracted to regrowth forest areas. The whole aim of the management of timber production in native forests is to 'sustain the ecosystem'. He did not want 30-year-old plantation pine to replace old growth karri because old growth timber provided a unique product which was not really available in plantations (*The West Australian* 15/6/98). In February 1999, a State government minister holding a marginal seat in the heart of the RFA region, threatened to resign over the issue—eliciting one heartfelt but unfulfilled one-word response in a Letter to the Editor: 'goodbye'. It would appear that the 'forests debate' is far from over.

MINERALS AND ENERGY RESOURCES

Mining and energy industries are important income earners and employers in Australia, with the minerals sector accounting for more than a third of the value of all goods and services exported. In contrast, their activities occupy a small proportion of land compared with other major, primary resource uses (pastoralism, agriculture and forestry). But this comparison belies the sometimes considerable on- and off-site impacts that the exploration, extraction and development of mineral and energy resources have on the environment. Problems include: vegetation disturbance or clearance and loss of biodiversity; pollution of air, soil and water by particulates, gases and toxic wastes; other disturbances such as increased flooding, noise, subsidence and erosion; costs of dealing with abandoned sites (there are thousands around Australia), and damage to aesthetic, cultural and social values.

Exploration and Mining

It is important to recognise that exploration, as well as mining, may cause controversy and environmental damage. Exploration methods involving the grading of wide tracks for seismic techniques, are particularly invasive in fragile environments. Individual prospectors, too, may cause considerable damage by removing the top 50 cm of soil over extensive areas and in dry river beds to expose subsoils to their metal detectors. Even national parks and world heritage areas are not sacrosanct. In Western Australia, exploration has been approved in some 50% of land set aside as national parks, and oil and gas exploration was approved in 1997 in the Shark Bay world heritage area, one of the few such areas which meets all the criteria for inclusion on the world heritage register (Chapter 6). There is little point in allowing exploration if mining approvals cannot follow, and parts of the Rudall and South Coast National Parks in Western Australia have been excised to permit uranium and mineral sands mining, respectively.

Historically, resource developers have enjoyed a relatively free rein in their use and development of Australia's natural resources, as shown by the nineteenth-century gold rushes. However, there is now a recognition in the legislation of the impacts of such development. Increasingly, companies are being required to meet environmental obligations in response to community concerns and legal requirements, or through voluntary commitments and incentive schemes.

The Crown exercises its proprietal rights over minerals on and below the ground in all public and private lands in Australia. It grants rights and licences to explore and develop, but with conditions attached. Generally, the powers are exercised through various mining or resource development Acts in the various States and are administered by Departments of Minerals and Energy or their equivalents. Under the terms of these mostly revised Acts, companies are required to minimise the environmental impacts of exploration and development, and to undertake restoration work. Various measures to achieve this can apply, including preparation of environmental management plans, issuing bonds, use of the EIA process, environmental audits and performance reviews, the development of contingency, insurance or safety plans, and the application of industry codes of practice. As with other natural resource sectors, voluntary approaches to environmental management are preferred over punitive methods. For example, Environment Australia's Supervising Scientist Group (now the Science Group) and the Australian Minerals Council developed 'best environmental practice in mining' codes and information to aid the industry (CEPA 1995; Chamber of Minerals and Energy 1995). However, concerns have been expressed over the vagueness of some codes, that they are market-driven and unenforceable, and that smaller industries tend to resist adoption (see the further discussion on EMS in Chapter 8).

According to the OECD's *Environmental Performance Review* of Australia (OECD 1998), many ESD objectives in mining have not been met and there is a significant problem relating to indigenous people's rights (Munchenberg 1998). Approval processes are often rushed and there is a lack of uniformity in assessment procedures. The public still has negative perceptions of the industry despite its efforts to improve its environmental performance: in 1993/94, for example, mining spent a total of $186 million on environmental protection, with the largest expenditures being $20.5 million on minesite rehabilitation, $16 million on air protection and $14 million protecting water quality. Even so, environmental protection expenditure still accounts for less than 1% of mining's total income—although the proportion is rising (ABS 1997b).

Mining legislation in the various States can be illustrated with reference to Western Australia whose *Mining Act 1978* places conditions on tenements to minimise environmental impacts. *Mining Regulations* (1981) incorporate environmental inspection powers which may stop work or impose fines. Other government Acts also require 'consideration' of environmental matters, notably the *Environmental Protection Act 1986, CALM Act 1984, Wildlife and Conservation Act 1950, Bushfires Act 1950,* the *Soil and Land Conservation Act 1945–92* and the *Aboriginal Heritage Act 1972–80,* although the Acts are not binding in all situations. Other agencies may also be involved in the permit issue process, such as soil and water authorities. New projects must issue a *Notice of Intent* prior to commencement, detailing proposed activities and environmental rehabilitation measures. Projects over a certain size are referred to the EPA, although in theory the Authority can be called on in relation to any project proposal (Chapter 11). Mining and exploration in national parks and other sensitive areas are open to assessment by the EPA and CALM and in some instances are subject to Commonwealth laws (Department of Environmental Protection 1997).

There is increasing pressure on governments by industry over security of access to land for exploration and development, and the sector feels constrained by environmental requirements and native title claims (discussed earlier in this chapter). In response, some governments are moving to allow unrestricted access. For example, in 1993 the Western Australian Government announced that all of the State, including environmentally sensitive areas and national parks, were potentially open to exploration and development, although special conditions apply: national parks and A class reserves require parliamentary approval.

Smaller mining activities, such as quarrying, generally operate under local planning procedures, although these can be overridden by appeal processes (and often are, in the context of local power politics). *Derelict sites* are an ongoing problem. There are 245 in New South Wales alone, with a gradual rehabilitation process costing an average of $120 000 each year (EPA New South Wales 1995).

Energy Resources

Mining for coal, petroleum and natural gas, and the construction of dams for hydro-electric power (HEP), fall under 'production' and have specific legislation which regulates the development and delivery of the resources (for example, Petroleum Acts). Some are subject to environmental considerations similar to mining Acts, such as the NSW *Coal Mining Act 1973* (Fisher 1980).

The environmental dangers of *petroleum exploration* (oil and gas) beneath the sea bed beyond territorial waters is recognised in provisions in international continental shelf conventions (Appendix 2) passed into Commonwealth and State legislation in 1967. State jurisdiction over offshore resources was overridden by the Commonwealth's Declaration of Sovereignty in 1973. Following discussions, in 1979 there was agreement that States had sovereignty within the three nautical miles limit under various States' Coastal Waters Acts. The effect was to grant the States jurisdiction to control petroleum and other operations within coastal waters, with the States and the Commonwealth sharing the administration of areas outside but adjacent to coastal zones under the provisions of the *Petroleum (Submerged Lands) Act 1980*. Common rules apply with regard to petroleum operations in these areas. Additionally, in parallel seabed minerals Acts, there are provisions which prevent any 'unjustifiable interference' with the living resources of the sea, control waste, and require the repair of any seabed damage. Similarly, there are onshore oil exploration provisions to minimise damage to the natural environment. Generally, however, older (1920s – 1960s) States' legislation concentrates on development rather than the environment.

Pipeline construction is also subject to environmental considerations under various Pipelines Acts. Proposed pipeline routes require EIAs and sometimes public appraisal. However, the environmental element is generally secondary to operator rights specified in the legislation. In some States (WA, Vic.) the law states that development works should not 'unnecessarily' interfere with flora, fauna, scenic attraction, culture or heritage values. In other States, environmental assessment procedures are either weak or non-existent.

The focus of environmental management in relation to energy tends to be on two sectors:

1   *atmospheric*—the control of greenhouse gases (especially CO_2), acid deposition (SO_2) and other air pollutants arising from combustion processes (industry, transport, electricity). These are managed through 'clean' production, more efficient transport and industries, increased public transport and improved urban planning, with measures including emission controls, licensing systems and (controversial) tradeable permits, fines and self-regulation. Arguments in favour of nuclear power arise from time to time in this context, as it is seen as non-polluting but with high risks attached. Australia has *Ozone* protection laws, *Greenhouse* and *Ecologically Sustainable Strategies* which are relevant to this area of management, and a national Greenhouse office has been set up in Canberra (refer Appendix 3); and

2   *renewable energy*—as an alternative to (or complementary with) reducing the emission of pollutants into the atmosphere. There are various strategies, including developing solar, wind, biomass and tidal energy sources as well as recycling (for example, the use of waste heat in industry). A peak national body, the Sustainable Energy Forum, was formed in 1997 to promote sustainable energy. Mechanisms are to be made available to assist with technological development and to interact with governments and industry.

In Tasmania, the Sustainable Development Advisory Council has recommended: the ecologically sustainable development of energy; that alternative energy sources be promoted in the community and industry; that economic instruments be employed to encourage the use of alternative energy, and disincentives removed for under-used energy sources (State of the Environment Unit 1997). In New South Wales, the Sustainable Energy Development Authority and a number of electricity suppliers offer a scheme (*Greenpower*) whereby power can be purchased from 'green' sources (solar, wind, biomass or hydro) in exchange for a tariff to users. Funds raised are then used for further investment in renewable energy. In 1998 there were 17 000 users in NSW and it was hoped that eventually 13 million Australians would have that choice (Lee 1998). In a similar move, the WA Government is to open up its energy markets to alternative sources in a deregulated market. The South Australian Government is also pursuing a sustainable energy policy. However, progress in renewable energy developments has been slow. Less than 0.1% of South Australia's energy is supplied from renewable sources compared with 6% for all Australia (EPA SA 1998:Chapter 14).

Constraints to the development of alternative energy sources include the lack of competition for conventional suppliers, encouragement of the traditional electricity sector, falling petrol prices, insufficient support for research with research grants reduced and small centres closed by the federal government in the latter half of the 1990s, and different priorities. For example, BHP spent $65 million over five years to 1994 on coal research, whereas the Australian National University had spent only $2.25 million on long-term solar power projects (EPA NSW 1995a). In 1998, at the Kyoto conference, Australia argued successfully that its international greenhouse gas emission targets should be *increased*, pleading a 'special case' whereby lowering greenhouse gas emissions would harm industry and employment, even though the nation has one of the world's worst *per capita* emission records. This episode—with implications felt far beyond Australia's shores as some other countries felt less constrained to abide by internationally agreed emission reductions—demonstrated the political power of the industry lobby over that of conservationists. Reliance on voluntary reductions and massive tree-planting programs become futile and are flawed arguments.

INDUSTRY AGREEMENT ACTS

In general, then, each resource tends to be considered primarily in terms of its own point of reference. But in some States (notably Queensland and Western Australia, and in the Northern Territory), it has long been the practice for large-scale, resource-based developments to be approved by governments through the medium of specific Agreement Acts (see Fig. 15.1). In recent years this mechanism has been used to impose environmental protection measures on the developer, by incorporating those measures in the Agreement Act. Failure to conform can incur penalties. This change is closely related to the introduction since 1970 of environmental impact assessment as an environmental management tool. Thus the 1973 *Woodchip Agreement Amendment Act* of WA included the following provision (Clause 19):

> Nothing in this Agreement shall be construed to exempt the Company [WA Chip & Pulp Co.] from compliance with any requirement in connection with the protection of the environment arising out of or incidental to the operations of the Company hereunder that may be made by the State or any State agency or instrumentality or any local or other authority or statutory body of the State pursuant to any Act for the time being in force.

The failure to refer to the possible requirements of any *future* environmental legislation was a significant omission, although other provisions require the Company to avoid water pollution and comply with forestry officers' requirements in relation to the spread of the dieback disease.

More recently, Western Australia's *Yeelirrie Act 1978* required uranium mining to conform to relevant codes of practice, the submission of an environmental management program, and the protection of Aboriginal and historic sites. Mineral sands and alumina Agreement Acts were modified to incorporate new environmental requirements.

Other examples include the mining of coal and the development of the North West Shelf in Western Australia, and the Alligator Rivers mining agreement in the Northern Territory. The Commonwealth's *Environmental Protection (Alligator Rivers Region) Act 1978* covers uranium mining at Ranger and Kakadu. Environmental provisions introduced in this legislation include requirements for rehabilitation work and the employment of an environmental protection officer. There are also provisions (some reflecting concerns over the use of uranium for non-peaceful purposes) under Northern Territory and various Commonwealth atomic energy and nuclear codes Acts. The Supervising Scientist's role under the Alligator Rivers Region Act is to investigate the effects of uranium mining on the environment and to develop standards to protect that environment. Further provisions apply to Aboriginal lands and rights in the region. Expansion into Kakadu National Park was rejected in deference to Aboriginal occupation and beliefs; but the rejection has subsequently been overturned.

Industry Agreements can be used to override local planning powers of environmental controls for the benefit of the developer. Safeguards are not always seen as adequate or upheld (for other examples, refer the Hinchinbrook Canal Estate Development discussed in Chapter 11; and Chapter 4 in Bates 1995).

COMMENT

The legislative and administrative responses discussed in this Chapter could be applied collectively to all four categories of environmental problems listed in the Introduction to Part 2. But they are subject to some considerable failings. These include the fact that environmental protection in virtually all cases is secondary to the primary aim of either the legislation or, especially, the administering body; and the multitude of Acts and agencies created by them. The response to this was the creation of legislation directed specifically and primarily at environment protection as distinct from resource development (Table 4.0). That legislation (and related agencies) is discussed in Chapter 7. But the development of environmental legislation was also a response to international agreements; and these, and the associated evolution of 'environmental law' in Australia, are discussed first, in the following chapter.

CHAPTER 6

NATIONAL AND INTERNATIONAL ENVIRONMENTAL LAW AND POLICIES

Partly in response to the problems discussed at the end of Chapter 5, and also reflecting increasing environmental activism by some Australians, which in turn reflected overseas developments and the increasing number of international environment treaties and conventions, there emerged a distinct body of law which may be termed 'environmental law' in Australia (Fisher 1993; Bates 1995). The upsurge between 1965 and 1986 was detailed in a series of graphs and accompanying text by Grinlinton (1990). Legislation setting up environmental protection agencies is considered in the following chapter.

GROWTH IN ENVIRONMENTAL LEGISLATION

In his first graph, Grinlinton (1990) showed that the growth in the membership of environmental organisations, and (to a lesser extent) the number of those organisations, far outstripped population growth over the two decades. Grinlinton commented that

> with their wide membership, high media profile, political lobbying, and active participation in public inquiries and government consultation, environmental organisations have clearly contributed to the formulation and modification of government policy, and thus to legislative activity (p. 75).

However, other factors such as international obligations, scientific concern, the increasing adverse impacts on the environment, and government policy considerations also contributed to this development. Environmental enactments totalled 199 between 1967 and 1986. However, the level of environmental legislative activity as a proportion of all legislation was still low, ranging from 0.7% in 1969–71 to 1.3% in 1981–83. The Commonwealth passed the largest number of Acts (37) followed by South Australia (29), Victoria (27) and the Northern Territory (26); the least active jurisdictions were the ACT (eight environment Acts) and Tasmania (11).

Grinlinton presented an interesting graph extending from 1986 to before 1910, in which 'environmental' laws are grouped into three categories:

environmental planning and protection; conservation, and resource allocation and development, presented in seven-year groupings. The resource allocation and development group increased only slightly over the entire period, from four or five enactments to a maximum of 12 in each period. In contrast, there were fewer than five 'conservation' Acts in each period until 1951, after which the numbers increased dramatically to 30 in each of the 1973–79 and 1980–86 periods. Environmental planning and protection exhibited the biggest increase, however, from between two and seven in each of the seven-year periods to 1951, to a huge 50 in the final period 1980–86.

Conservation, and environmental planning and protection, were analysed in more detail by Grinlinton (1990) in four-year periods between 1967 and 1986. 'Pollution' and 'conservation' were the largest categories, followed by roughly similar numbers of enactments in 'environmental planning' and 'hazardous substances'. The remaining group, 'waste disposal', had a small and fairly constant number of enactments over the period; whereas 'conservation' rose to a peak in the 1970s, and 'environmental planning' laws clustered in the early and late 1970s and early 1980s.

Other sources encapsulate the change through to the 1990s. In 1980, Fisher listed 130 pieces of environment-related legislation in Australia. By 1990, Bates (1992) recorded over 500, which included more than 100 pieces of Commonwealth legislation. For New South Wales alone, Aplin (1998) found 172 consolidated Acts in 1996 and more than 100 regulations of environmental significance. In 1997, the Industry Commission's inquiry into land management listed the number of land management Acts administered by States' resource management agencies alone—excluding the myriad laws covered by other legislation and Departments (Table 6.1). Appendix 1 shows South Australia's environmental legislation before and after 1970, illustrating the marked growth in legislative activity. More than 140 pieces of environmental legislation were listed for Victoria in 1998 (www.dms.dpc.vic.gov.au). For the Commonwealth, Munchenberg (1994, citing ANZECC 1993) noted that in the entire period prior to 1980 there were only 29 Acts relating to environment protection. This was followed by 35 in the next decade and a further 10 in the 1990–93 period.

In parallel with the increased activity in environmental legislation and land conservation actions, a WA survey found a very high (96%) degree of concern

Table 6.1   **Environmental and land management legislation (excluding regulations and by-laws) administered by States' 'mega' resource agencies. Compiled from the Industry Commission (1998:93).**

Agency	Approx. Number of Acts
DNRE (Vic.)	109
DLWC (NSW)	80
DELM (Tas.)*	48
DNR (Qld)	40
DLPE (NT)	27
Dept Urban Services (ACT)	20

* Now incorporated in DPIWE.

over environmental problems across all levels of society during the early 1980s—although it is difficult to distinguish between cause and effect (Newman and Cameron 1982). This concern fluctuated around slightly lower levels during the 1990s, displaced by more pressing social and economic priorities (Chapter 14) (ABS 1997a). The decline could also be due to the comfort people take from the perception that environmental problems are being addressed.

ENVIRONMENTAL LITIGATION

Not surprisingly, environmental litigation also increased over the 1967–86 period, with peaks of about 40 cases in the mid-1970s and 1980s (Grinlinton 1990). Litigation was grouped into the following six areas: pollution control; conservation and protection of the natural environment; town and country planning; environmental planning and impact assessment; resource allocation and development, and miscellaneous environmental litigation. The latter included diverse areas such as the tort actions of nuisance and trespass, trade practices litigation, wills and trusts cases, freedom of information applications, and cases involving constitutional issues with a strong environmental planning or protection component. Only six cases were recorded by each of the Commonwealth, Western Australia and the ACT over the entire period; at the other end of the scale, Victoria had the largest number of cases (88) followed by New South Wales (63) and Queensland (30). Victoria dominated to the end of the 1970s, as a response to Australia's first comprehensive *Environment Protection Act* (Vic. 1970) which contained elements of substantive environmental protection and environmental planning and impact assessment. But Victoria was then overtaken by New South Wales, reflecting the creation of the Land and Environment Court in 1979. During the 1990s, revisions to environmental protection legislation have increased punitive powers and seen severe penalties applied for the first time (Chapter 7). On the other hand, third party challenges and other rights have been diminished (Chapter 14).

INTERNATIONAL CONVENTIONS, TREATIES AND AGREEMENTS

As indicated above, international conventions and treaties were and are important in influencing environmental legislation and policies in Australia: a number of key national, States and Territories agreements, policies, laws and other measures are detailed in Appendix 3. In 1996, The State of the Environment Advisory Council listed chronologically (in relation to the year the treaty was ratified) 66 international conservation treaties, conventions and protocols to which Australia is a signatory: refer Appendix 2. Some of the more important ones are considered briefly in the following sections.

As noted by Sand (1992), the international conservation treaties, conventions and protocols fall into a number of categories, which in general similarly classify Australia's environmental legislation over the 1970s to 1990s. These are:

- general environmental concerns;
- nature and wildlife;
- atmosphere;
- marine;
- fresh water;
- hazardous substances;
- nuclear safety;
- working environment;
- liability, and
- environmental disputes.

Most conservation treaties, conventions and protocols (termed 'agreements' henceforth) are orientated to environmental conservation and protection against pollution. They are split almost evenly into those dealing with marine and freshwater environments and other aspects. Most agreements set out objectives in somewhat abstract or general terms with few having measurable objectives. Since the report of the 1987 World Commission on the Environment and Development, known as the Brundtland report (WCED 1987), the concept of ecologically sustainable development (ESD) has figured prominently. More than half the agreements are regional or sub-regional in scope; and some relate to developing countries, others to developed nations, and yet others to special circumstances (such as regional seas).

The formulation of the World, National and State Conservation Strategies illustrates the way in which international agreements work: the global overview becomes increasingly specific as it is adopted progressively at national and State levels.

Conservation Strategies

The *World Conservation Strategy* was developed in 1980 by the International Union for the Conservation of Nature and Natural Resources (IUCN) in conjunction with UNEP and the World Wildlife Fund (now the World Wide Fund for Nature), with the following four major objectives (IUCN/WWF 1980):

1   conservation of living resources;
2   maintenance of essential ecological processes and life support systems;
3   preservation of genetic diversity, and
4   sustainable utilisation of species and ecosystems.

Four Conventions were identified in the Strategy as being the most important global Conventions of an environmental and/or conservation nature:

- the Convention on Wetlands of International Importance especially as Waterfowl Habitat (Ramsar Convention);
- the Convention concerning the Protection of the World Cultural and Natural Heritage (World Heritage Convention);
- the Convention of International Trade in Endangered Species of Wild Fauna and Flora (CITES), and

- the Convention for Conservation of Migratory Species of Wild Animals (Bonn Convention).

Australia has few animal species which migrate and no terrestrial international boundaries and is therefore not a signatory to the Bonn Convention. With regard to CITES, the regulation of international trade in plants and animals is considered to be a matter for policing by customs officials (through Commonwealth wildlife protection laws regulating imports and exports), and is therefore in the sphere of federal government regulation. Migratory birds and heritage agreements are discussed below.

The World Conservation Strategy was accepted by the Australian government and, in accordance with the recommendation that every nation prepare an equivalent document, Australia produced its *National Conservation Strategy* in 1983 (Commonwealth of Australia 1983). The National Strategy lists the same four objectives as the World Strategy, with one addition; that is:

5   to maintain and enhance environmental qualities.

The National Conservation Strategy outlines the concept of sustainable development as well as a range of obstacles to achieving conservation goals. The priorities of the National Strategy were:

- to use the body of existing legislation, administration and skills available at all levels of government and to harmonise their activities, and
- to build on the knowledge, expertise, public awareness and participatory processes available in the community.

The National Strategy, in turn, was followed by *State Conservation Strategies*. To illustrate, the National Strategy was accepted by the Western Australian Parliament in 1985, followed by the production of the *WA State Conservation Strategy* in 1987 (Government of WA 1987). That Strategy contained the same objectives as the World and National Conservation Strategies, plus an additional objective:

6   to optimise the quality of life for Western Australians.

'Quality of life' was used to refer to the sense of well-being and fulfilment experienced by each member of the community, a concept which defies precise measurement.

The WA State Conservation Strategy, as with other State Strategies, also addressed more local issues. These included:

- national parks and nature reserves;
- forest management by sustained yield principles;
- wheatbelt salinisation and soil erosion;
- water resources management;
- eutrophication;
- coastal management, and
- mining issues.

The main strategy outlined was to promote an 'environmental ethic' within the community; that is, to promote recognition of the interactions between conservation and sustainable development. Responsibility for co-ordinating the implementation of the National and State Strategies, and to foster the environmental ethic in the community, was allocated to the Environmental Protection Authority.

The above objectives were developed in more precise terms in later national (Commonwealth of Australia 1992c) and State ESD strategies and State of the Environment (SoE) reports (Chapter 8 and Table 8.3). Essentially, the various conservation strategies were policy documents guiding later, more specific strategies (such as for water, atmosphere, soils, forests, biodiversity and coasts: see examples in Appendix 3).

The Rio Conference and Agenda 21

In 1992 at the UN Conference on Environment and Development, also known as the Rio Conference or Earth Summit, four key documents were agreed to by more than 140 governments, including Australia. The four agreements were: The Rio Declaration; Agenda 21; Climate Change and Biodiversity, and a Declaration of Forest Principles. Agenda 21 (UNCED 1992) is an Action Plan to implement the principles of the Rio Declaration. It directs countries to develop policies which include national, State and local ESD policies and SoE reports, with policy links between all levels, thereby providing opportunities for ecologically sustainable development strategies to be monitored and evaluated regularly through state of the environment reports. Some Australian States, such as Tasmania, have legal requirements to produce regular SoE reports; in Tasmania's case, every five years. Annual SoEs are also required of some local governments. On the other hand, Victoria has shut down the agency responsible for preparing its SoE reports on the grounds of economic stringencies. The 40 chapters of Agenda 21 cover a wide range of issues, including: protection of fresh water resources; land degradation and desertification, and conservation of biological diversity. Member States have a responsibility to report annually on their implementation of Agenda 21: Australia presented its first report in 1993 (*Year Book Australia* 1996:391).

Living Resource Agreements

The International Convention on Whaling

Whaling treaties began as early as 1931, with the hunting of whales from Australia ending in 1980 after the closure of the land-based whaling station at Albany in WA. The Australian Nature Conservation Agency (previously the Australian National Parks and Wildlife Service) is heavily involved with international agreements, and through that agency several Australian States implement a national Contingency Plan for Cetacean Strandings. Advice is also provided by the Organisation for the Rescue and Research of Cetaceans in Australia (ORRCA).

The whaling convention, as with others, emphasises the importance of research into the behaviour and habitats of the target species, in order to plan more effectively for its protection and conservation. Since the term 'whale' is poorly defined in the text of the original treaty, it is possible that the provisions of the treaty could be extended to include dolphins and porpoises. Thus the Convention could be applied to the protection of cetaceans (particularly dolphins) which are killed in tuna fishing operations (Lyster 1985).

Agreement for the Protection of Migratory Birds and Birds in Danger of Extinction, and their Environment

Australia is involved in two specific bilateral treaties derived from this international agreement, these being separate agreements with Japan (Japan–Australia Migratory Birds Agreement, or JAMBA) and China (China–Australia Migratory Birds Agreement, or CAMBA). The Agreement involves listing of all species which migrate between the countries and which are to be protected, as well as birds common to both countries. The latter are included on the understanding that there is no evidence that they do not migrate. The Agreement also provides for non-migratory birds which are endangered and require special protection (Lyster 1985).

The Migratory Birds Agreement overlaps with CITES, as it prohibits trade in individuals of listed species, their eggs, products and specimens (Lyster 1985). It also overlaps with the Ramsar Convention (discussed below), as it touches on the environment of the birds as does Ramsar, which also provides for the habitats of migratory birds.

Migratory birds travel from Siberia, China and Japan, through southeast Asia, into Australia. Species include the Great Knot, Tattler, Turnstone, large Sand Plover, Curlew Sandpiper, Terek Sandpiper and the Red-Necked Stint. In all, there are 34 species of birds from Eurasia and one from New Zealand which migrate to Australia (Migratory Birds Program n.d.).

The Agreement refers to prohibiting the 'taking' of listed species, which carries the assumption of the shooting and trapping of protected, migratory birds being prohibited. The ambiguity of the term 'taking' has been used in the USA to prosecute where chemical spills and pollution have caused the death of protected species (Lyster 1985). Thus it is not only the hunting and trapping of birds that is illegal: any activity which is directly related to the death of protected birds may result in prosecution.

JAMBA and CAMBA contain an indigenous people's traditional lifestyle clause permitting Aboriginal hunting and gathering of protected species. However, this is only the case if the population of each species is maintained in optimum numbers and the adequate preservation of the species is not prejudiced (Lyster 1985).

Terrestrial Agreements

Convention on Wetlands of International Importance especially as Waterfowl Habitat (Ramsar Convention)

Ramsar is not an acronym, but the name of the town in Iran where the convention was signed in 1971, Australia being one of the 18 signatories (92 in the mid-1990s). The main goal of the Convention is to protect wetlands, especially where they are important for the continued existence of waterfowl. This may refer to wetlands being sites where rare and/or endangered species are found, where significant populations of avifauna congregate, or where there are migratory species. All wetlands are included, as well as mangrove swamps, peat bogs, water meadows, coastal beaches, coastal waters, tidal flats, mountain lakes and tropical river systems (Lyster 1985; Comino 1997).

The objectives and obligations of the Convention are:
- research and training in wetland management;
- definition and delineation of wetlands;
- development of national inventories of wetlands;
- listing of sites and establishment of reserves;
- formulating and implementing 'wise' land-use planning, and
- creating national policies on wetlands.

The Convention involves the nomination, acceptance and placing of wetland areas on a List of Wetlands of International Importance (the List). Wetlands are selected for their ecology, botany, zoology, limnology or hydrology, and those of international importance to waterfowl at any season. The last criterion is particularly significant in the semi-arid and arid parts of Australia, where the seasonal drying out of wetlands leads remaining water bodies to become especially important to waterfowl as drought refuge areas.

In Australia, the Ramsar Convention is co-ordinated by the Council of Nature Conservation Ministers (CONCOM), with first the Australian Nature Conservation Agency and then Environment Australia being responsible for listings and proposals for listings; although many of the Ramsar obligations are now met under State legislation. Australia has 42 Ramsar sites listed (Fig. 6.1). Ramsar listing has the potential for national protection of any site, and listing is usually followed by statutory protection if the site is not already protected. Most wetlands on the Australian List already have national park or reserve status.

The Convention incorporates a general obligation for parties to include wetland conservation within their planning policies, and several Australian States have produced Wetland Conservation Policies. Other Australian initiatives, such as the actions of the Murray–Darling Basin Commission, heritage and endangered species Acts, national water resource policies and wetland inventories, bear on Ramsar with respect to the definition of important wetland sites or inclusion of Ramsar sites.

In 1996 the Australian government (hosting a Ramsar conference in Brisbane) released its *Draft National Wetlands Policy*, which seeks to harmonise poli-

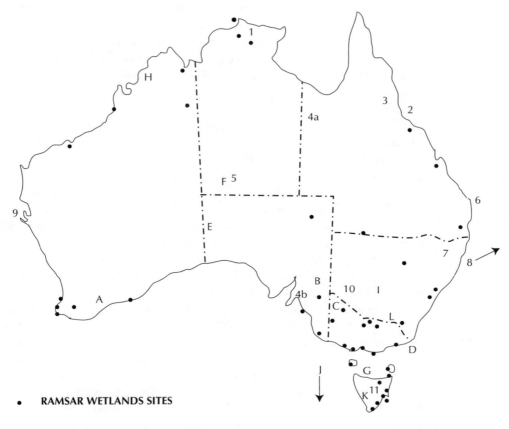

Fig. 6.1   Ramsar sites, International Biosphere reserves and World Heritage areas in Australia. From Aplin (1998, Figs. 7.2 and 9.2).

- **RAMSAR WETLANDS SITES**

INTERNATIONAL BIOSPHERE RESERVES

A   Fitzgerald River National Park, WA (1978)
B   Danggali Conservation Park, SA (1977)
C   Hattah-Kulkyne National Park &
     Murray-Kulkyne Park, Vic. (1982)
D   Croajingolong National Park, Vic. &
     Nadgee Nature Reserve, NSW (1977)
E   Unnamed conservation park, SA (1977)
F   Uluru-Kata Tjuta National Park, NT (1977)
G   Wilson's Promontory National Park, Vic. (1982)
H   Prince Regent Nature Reserve, WA (1977)
I   Yathong Nature Reserve, NSW (1977)
J   Macquarie Island Nature Reserve (1977)
K   Southwest National Park, Tas. (1977)
L   Kosciuszko National Park, NSW (1977)

WORLD HERITAGE AREAS

1   Kakadu National Park
2   Great Barrier Reef
3   Wet Tropics of Queensland
4   Australian Fossil Mammal Sites
    a Riversleigh
    b Naracoorte
5   Uluru-Kata Tjuta National Park
6   Fraser Island
7   Central Eastern Rainforest
    Reserves (50 sites)
8   Lord Howe Island
9   Shark Bay
10 Willandra Lakes Region
11 Tasmanian Wilderness

cies, legislation and programs and promote the ecologically sustainable management of wetlands. Under the Natural Heritage Trust (NHT) umbrella, the *National Wetlands Program* delivers funds and develops objectives which include wetlands listings, good management, research and community involvement.

Over 500 nationally important sites were identified (see also ANCA 1996). Unfortunately, Ramsar guidelines for their listing and 'wise use' (as well as Commonwealth and State legislation) are not always clear, adequate or enforceable, resulting in some very public land-use conflicts. Such conflicts have included the Creery wetlands in the Peel–Yalgorup system in WA, the Hindmarsh bridge proposal in SA and the Macquarie Marshes in NSW (Comino 1997).

Two years after the Brisbane Ramsar conference, the World Wildlife Fund pointed out Australia's poor track record. Despite being one of the first Ramsar signatories 27 years previously, only three of the nine Commonwealth/State/Territory jurisdictions have a wetlands policy. None has an implementation plan and most Australian wetlands have no active conservation management. Most of the 698 sites identified as being of national importance are not protected by conservation reserves or programs, and only 14 are listed under Ramsar. Thus only 5 million hectares of more than 24 million hectares of nationally important wetlands are conserved under the treaty (Wildflower Society of Western Australia *Newsletter*, Nov. 1998:16, citing the Newsletter of the Threatened Species Network (WA), August 1998).

Jackson and Crough (1995) have pointed out other weaknesses in the Convention. Australia appears to have failed to implement effectively its international obligations with respect to indigenous peoples. For example, Roebuck Bay (in WA) was designated in 1990 under the Ramsar convention. Since 1993, much of the area has been subject to native title claim, although Aboriginal interests had been documented in earlier environmental studies. The traditional owners have strongly defended their right to continue their traditional practices, and are keenly interested in management issues, especially visitor use and fisheries. Yet the Australian Nature Conservation Agency listing of the Ramsar area reads: 'the principal social value of the wetland is recreational fishing'. Jackson and Crough (1995) go on to comment that consultative mechanisms between government agencies and indigenous people are inadequate, with policy responses to the implementation of international protocols being fragmented and dispersed across a range of agencies.

Convention Concerning the Protection of the World Cultural and Natural Heritage (World Heritage Convention)

The objective of this convention is to protect natural and cultural areas of outstanding universal value, by use of the World Heritage List, World Heritage in Danger List, and a World Heritage Fund. Twelve world heritage sites in Australia (to 1995) are listed in Table 6.2 and located in Fig. 6.1.

World heritage listing imposes a legal duty on signatory parties to protect listed sites, as the country in which a site is located is responsible for its nomination. In Australia, the protection of world heritage sites is undertaken through the *Heritage Commission Act 1975* and the *World Heritage Property Conservation Act 1983*, with relevant State legislation including those dealing with environmental protection, fauna and flora conservation, forests and fisheries.

Table 6.2   **Australia's World Heritage sites (located in Fig. 6.1). Source: State of the Environment Advisory Council (1996:Table 9.4) and Lane *et al.* (1996).**

World Heritage Property	Date Inscribed	Values Recognised
Great Barrier Reef, Qld	1981	Natural
Willandra Lakes region, NSW	1981	Natural, cultural
Kakadu National Park, NT		Natural, cultural
– stage 1	1981	
– stage 2	1987	
– stage 3	1992	
Tasmanian wilderness World Heritage area		Natural, cultural
– stage 1	1982	
– stages 2, 3	1989	
Lord Howe Island, NSW	1982	Natural
Central eastern rainforest reserves,NSW & Qld		Natural
– original	1986	
– extensions	1994	
Uluru–Kata Tjuta National Park, NT		Natural, cultural
– inscribed for natural values	1987	
– inscribed for cultural values	1994	
Wet tropics World Heritage area, Qld	1988	Natural
Shark Bay, WA	1991	Natural
Fraser Island World Heritage area, Qld	1992	Natural
Fossil mammal site, Riversleigh, Qld	1994	Natural
Fossil mammal site, Naracoorte, SA	1994	Natural

The objectives of the World Heritage Convention are to:

1   integrate protection of heritage into comprehensive planning programs;
2   protect and conserve the area through the provision of appropriate staff;
3   encourage scientific research to counteract dangers to listed sites;
4   identify appropriate measures to identify, protect, conserve and present heritage, and
5   provide centres for training and education.

There are four criteria for World Heritage Listing. They are that a proposed site:

•   is an example of major stages of the earth's evolutionary history;
•   is an example of significant ongoing geological processes, biological evolution and human interaction with the natural environment;
•   contains unique, rare or superlative natural phenomena, formations or features, or areas of exceptional natural beauty, or
•   is a habitat where populations of rare or endangered species of plants or animals still survive.

The World Heritage Convention is one of the most recognised of all international agreements, and in recognition of international pressure for conservation of listed sites, the previous (Labor) Federal Government was particularly active in preserving its nominated sites. This was especially evident from the Franklin dam case, following which State governments have tended to resist the inclusion of any part of 'their' territory on the World Heritage List. Essentially this is due to a State mistrust of the exercise of Federal power within the State (Davis 1989). This is also reflected in the dramatic variation between sites in terms of funding and management approaches. For a variety of political and historical reasons, some lack dedicated strategy or management plans and are world heritage areas in name only; not all are equal (Lane *et al.* 1996).

National Heritage

Nevertheless, Australia proposes to develop national heritage legislation on completion of its National Heritage Strategy. The new legislation will replace the *Heritage Commission Act 1975*. While protection of heritage places will be primarily the responsibility of the States and Territories, those of outstanding or national importance will remain the responsibility of the Commonwealth (Commonwealth of Australia 1998). A *Register of the National Estate* covers natural, built and cultural places of special value, although listing does not ensure their protection. In 1998, more than 12 000 places were on the Register—77% were historical, 16% natural and 7% indigenous (State of the Environment Advisory Council 1996:9.21; AHC 1998).

International Biosphere Reserves

International Biosphere Reserves are part of the UNESCO/UNEP Man and the Biosphere (MAB) program, involving 112 countries, and concerned with:

1   conservation of genetic resources and ecosystems;
2   establishment of an international network of areas directly related to MAB research, monitoring, training and information exchange, and
3   concern for associating environment and development in research and education (UNESCO 1986).

The biosphere reserve concept was devised to respond to the problem of increasing losses of living species, exacerbated by the absence of scientific knowledge and the inadequacies of traditional means of conserving species and ecosystems. Areas nominated must be representative of:

- species composition;
- physical structure;
- parameters of certain phenomena; for example species, diversity, and
- human land-use patterns and intensities.

The aim is to reserve examples of every major type of biological community, and to establish a concentric zoning system comprising: a core area, which is strictly protected; a buffer zone, to protect the core, for which some human management is permitted, and an outer transition area, wherein some management, research stations and human activities may be located. The twelve declared International Biosphere Reserves in Australia are located in Fig. 6.1.

The Convention on Biological Diversity

This is arguably one of the most important of the international environment agreements. The *Convention on Biological Diversity* was negotiated in the lead up to the 1992 UN Conference on Environment and Development (UNCED) in Rio de Janeiro, and has since been signed by 156 nations, including the European Community. Australia ratified the Convention in June 1993, and it came into force on 29 December of the same year. A wide range of technical matters has been considered under the Convention at subsequent meetings, including conservation and sustainable use, and the range of national activities to be undertaken to reduce the loss of biological diversity. It is also intended that the Convention will provide a mechanism to spread the costs of carrying out these activities.

The Convention grew out of a number of concerns, starting with the obvious destruction of habitats and extending to the consequential extinction of species and the vulnerability of key agricultural crops. According to Cave (1993), one catalyst to the drafting of the Convention was the resentment of many countries regarding the exploitation of genetic resources, particularly plant species, which came from their country but for which they received no benefit. In some cases where the plants were modified, the countries were even charged for the use of their products. This meant that there was no recognition of the value of the habitats which had produced these valuable species, although habitat preservation is a costly process. As these habitats have become degraded, so concern has grown over the reducing gene pool of commercially useful species. Propagation of the most productive species of potatoes, for example, means that if a disease strikes those few species, huge damage is caused, despite the fact that there may be wild or little-used varieties of potato resistant to the disease. Preserving habitats for unknown future benefits, often medical in nature, is also costly.

The 1996 *National Strategy for the Conservation of Australia's Biological Diversity* acknowledges the core objectives set out in Australia's National ESD Policy in 1992. It recognises:

- that conservation of biological diversity (genetic, species and ecosystems) provides significant cultural, economic, educational, scientific and social benefits to all Australians;
- the need for more knowledge and a better understanding of biodiversity;
- the need to strengthen current policies, practices and attitudes to achieve conservation and sustainable use, and
- intrinsic values, regardless of whether they are to our benefit (Commonwealth of Australia 1996a:5).

Benefits arising from conserving biodiversity include not solely species and habitat protection, but also the maintenance of a wide array of ecological 'services' such as the hydrological cycle (groundwater recharge, watershed protection and buffering against extreme events), climate amelioration, soil fertility, protection from erosion, nutrient storing and recycling, and pollutant breakdown and absorption.

A more recent concern has been the possible damage from the release of genetically engineered plants and animals into natural environments. There are also many well-known examples of damage from the transfer of naturally occurring species to foreign environments. Therefore it is most important that organisms must not be released until it is clear they pose no threat to existing species and ecosystems. The accidental release in Australia of the rabbit *calicivirus* from its research environment was an example of precisely such a situation, which generated considerable concern amongst a number of scientists in Australia and internationally.

A new Commonwealth *Environment Protection and Biodiversity Conservation Act* was passed in 1999. It amalgamates several pieces of Commonwealth legislation on environment protection, conservation and wildlife, and acknowledges the National Strategy on Biodiversity (discussed in Chapter 7).

The International Convention to Combat Desertification

Australia is a signatory (1994) to this Convention, which includes the establishment of National Action Plans (NAP). However, Australia has not adopted a NAP since its provisions are covered by a range of national and State programs (Decade of Landcare, National Drought Policy, Rural Adjustment Schemes) and the National Rangelands Strategy, which particularly focuses on the management of arid lands. Australia also offers technical and scientific aid in arid lands management to other countries (Williams *et al.* 1995).

Atmospheric Agreements

The Protocol on Substances that Deplete the Ozone Layer (Montreal Protocol)

Internationally, this agreement came into force in 1987. It is concerned with preventing the emission of ozone-depleting substances into the ozonosphere or upper atmosphere. Western Australia's experience with implementation of the Protocol is instructive—particularly in comparison with the Greenhouse agreement.

The Western Australian Government accepted the Montreal Protocol and in 1988 developed a State Strategy in response. That Strategy took three steps:
1 change government purchasing policy to avoid, where possible, products containing chlorofluorocarbons (CFCs);
2 legislate to phase out the use of CFCs by 1989, and
3 formulate an Environmental Protection Policy (EPP) to improve the control of CFCs used as refrigerants (EPA 1990a).

This strategy was followed up by the development of the Environmental Protection (Ozone Depleting Substances) Policy in 1989. As an EPP, the policy has statutory authority through the *Environmental Protection Act 1986* (WA). The Policy was one of the first of its kind in the world. Its implementation was facilitated by technology developed in Western Australia making possible the collection and recycling of CFCs, which meant that they no longer need to be released to the atmosphere (EPA 1990b). The EPA can license distributors of CFCs and has the authority to both monitor recycling procedures and to inspect any premises containing significant quantities of CFCs. The policy makes it an offence to fill leaking equipment with CFCs or to release them directly to the atmosphere, with fines of $5000 for individuals and $10 000 for companies found guilty of infringing the policy.

This particular example is the strongest illustration of Western Australian implementation of an international agreement relating to an environmental issue. However, it is a considerable drain on the EPA's resources, as the agency is solely responsible for implementing the policy. This may inhibit further innovations of this kind and also discourage the EPA from complying with the State government's commitment to achieving a uniform national strategy.

Greenhouse/Climate Change

The Toronto Agreement of 1988 and Noorduijlk (The Hague) Declaration of 1989 are both concerned with the reduction of carbon dioxide emissions by 20% of 1988 levels, by the year 2005. More recently, the United Nations *Framework Convention on Climate Change* was signed by more than 150 governments at the June 1992 'Earth Summit' held in Rio de Janeiro, and ratified by Australia on 12 December of the same year. The Convention came into force in March 1994, and most developed countries indicated their intention to bring emissions of greenhouse gases back to 1990 levels by the year 2000. Developing countries have been funded by the United Nations' Global Environment Facility for projects and activities related to the convention (Williams 1993); but it is unlikely that the above target and funding are sufficient to prevent greenhouse warming.

The international trend has been towards:

1   reducing greenhouse gas emissions;
2   reforestation, and
3   promoting energy efficiency.

A *National Greenhouse Response Strategy* was developed in 1992 (Commonwealth of Australia 1992a). There has been some promotion of energy efficiency at federal and State levels, including the use of energy-efficiency ratings on household appliances and development of renewable energy alternatives. A national Greenhouse Office has been set up. One of its objectives is to develop a national Carbon Accounting System for land-based sources and sinks. Integrated research programs, overseen by the CSIRO, have been instigated. Amongst the research areas has been the removal of methane (a greenhouse gas) from landfill sites, to serve as an alternative energy resource and to prevent the release of methane to the atmosphere. Other proposed actions have included tree-planting programs and the reduction of sulphur emissions to the atmosphere. In Western Australia, a Greenhouse Gas Co-ordinating Committee was established and a State Greenhouse Gas Audit was carried out in 1988 to document the history of emissions at the State level, and to contribute to a national audit (WAGCC 1989, and revised, DEP 1994).

However, in the early 1990s there developed increasing industrial pressure, particularly from the coal mining lobby, for Australia to be exempted from the climate change targets. This pressure resulted in Australia arguing at an international level that it be given special treatment, because the country relies heavily on the export of primary sources of energy (particularly coal and gas). In 1997 Australia failed in an attempt to enlist the support of the Japanese, British and EU leaders, on the grounds that special pleading for one country would very rapidly unravel the whole agreement; but gained some support from the USA. At the Tokyo meeting Australia argued successfully that it be permitted to *increase* its greenhouse gas emissions by 8% on 1990 levels to 2012, compared with a 5% *reduction* by other developed countries. Australia has not made too many international friends in this regard and indeed is making it very difficult for effective international action on global warming to be implemented. However, others have argued that few other countries are expected to meet their Convention targets.

COMMENT

Sand (1992) has drawn attention to some of the problems with international conventions. It is difficult to measure their success—some are self-evaluating and some incorporate regular monitoring, but evaluations eventually falter. There is a huge volume of agreements—2000 for fisheries alone. The scope of some agreements is too narrow for them to be of global significance. Exemptions given to some countries, to assist their development, undermine the global effectiveness of the agreement. Many developing countries are not participating actively (partly due to lack of resources and expertise); binding agreements are being signed predominantly by developed countries. There are long delays in implementation, often with endless ratifications (such as the Law of the Sea Convention). National reports differ widely and are not always written to an agreed format. There are poor levels of compliance in reporting activities in some areas, such as air pollution. Reports are often late, incomplete or not produced at all, with difficulties including inadequate financial resources, technical assistance and public information. In response to some of these difficulties, new procedures are being considered to assist reporting and compliance, through 'expert' groups.

With specific reference to commitments made in Rio, a five-year stocktake by the World Bank of governments' performance found that:

> the high rhetoric of Rio has been followed by an often painful silence as policy makers face the challenge of converting rhetoric to action (World Bank 1997:5).

Many countries lack mature institutional bases for environmental management and find it difficult to create institutions to implement changes. About

100 (of 178) countries had prepared ESD strategies or national environmental action plans by 1997 but were not always successful in implementing or changing policies. The World Bank (1997) document set out a range of recommendations for achieving change, with market approaches and public participation seen as some effective methods, and included some case studies.

It should also be noted that there is an increasing level of internal concern and opposition to Australia's participation in the above (and many other) international agreements. In particular, it seems that Australian ratification is carried out without reference to Parliament; and it has been argued that, consequently, Australia has been undemocratically handing over its sovereign powers to largely anonymous and bureaucratic international bodies. This view does have some validity, even though international agreements are essential if global degradation is to be dealt with effectively. The lesson here, as at all scales, is that for environmental planning and management to succeed, the population for which the management is being done must be included as a full, equal and active partner in the process. This is discussed further in Part 4, Public Participation in Environmental Decision-making. There also needs to be a genuine commitment to implement the spirit of the agreements.

One of the key means of implementing environmental protection was the development of legislation, policies and agencies designed specifically for that purpose, and not as an attachment to prior resource development Acts (Table 4.0). This is discussed in the following chapter.

CHAPTER 7

ENVIRONMENTAL PROTECTION LEGISLATION AND AGENCIES, OTHER MEASURES AND POLICIES

Environmental Impact Assessment (EIA) and the creation of new laws and agencies whose sole or prime objective and responsibility was environmental protection, were seen as the answer to the proliferation of 'environmental' and resource legislation and other problems discussed in previous chapters. But, as will be seen, the EIA approach was almost entirely restricted to only one of the four categories of environmental stresses listed in the Introduction to Part 2— project or resource-specific problems. Further, the EIA method, whilst revolutionary at the time of its introduction 30 years ago, is increasingly being recognised as having important deficiencies. This is discussed in Chapter 11.

INTRODUCTION OF ENVIRONMENTAL PROTECTION AND IMPACT ASSESSMENT LEGISLATION

Federally, and in every State and Territory, there is now generally one agency which can be identified as exercising primary responsibility for environmental protection and pollution matters, and one Act which provides the statutory basis for those agencies. Table 7.1 summarises the position; but it is stressed that (frustratingly for the authors) the situation is in a state of constant flux, particularly with moves towards structural and legislative reform and integration in planning and resource/environmental management (Part 5). In some instances EIA operates under other laws, notably planning.

THE COMMONWEALTH'S ROLE IN NATIONAL ENVIRONMENTAL PROTECTION

The earliest environmental protection Acts in Australia were passed in 1970 (Vic.) and 1971 (WA). But to 1978, when Victoria's *Environmental Effects Act* came into force, only the Australian Commonwealth had created detailed statutory procedures, in 1974, for environmental impact assessment (EIA) and public inquiries. However, the 1974 *Environment Protection (Impact of Proposals) Act* (Commonwealth) did not create an administering body. Instead, the

Table 7.1 **Key environment protection, planning and impact assessment agencies and legislation in Australia. Updated from Harvey (1998), various Websites (March 1999) and personnel in environment protection and planning agencies (see Acknowledgments).**

Jurisdiction	Administering Agency	Legislation
Commonwealth	Department of the Environment and Heritage Environment Australia (see Appendix 4)	Environment Protection (Impact of Proposals) Act 1974–87 Environment Protection & Biodiversity Conservation Act (1999)
New South Wales	Department of Urban Affairs & Planning Environment Protection Authority	Environmental Planning & Assessment Act 1979 – 1997 Environmental Offences & Penalties Act 1989 Protection of the Environment Administration Act 1991 Protection of the Environment Operations Act 1997
Victoria	Department of Infrastructure (Division of Planning, Heritage & Market Information) Environment Protection Authority	Environment Protection Act 1970–96 Environmental Effects Act 1978–95 Planning and Environment Act 1987–97
Queensland	Environmental Protection Agency Department of Communication and Information, Local Government and Planning	State Development & Public Works Organisation Act 1971–85 Environmental Protection Act 1994–97 Integrated Planning Act 1997
South Australia	Department for Environment, Heritage & Aboriginal Affairs Department of Transport, Urban Planning and the Arts (Planning South Australia) Environment Protection Authority	Development Act 1993 Development (Major Development Assessment) Act 1996 Environment Protection Act 1993–97
Western Australia	Environmental Protection Agency Department of Environmental Protection Ministry for Planning	Environmental Protection Act 1971–98 Planning Legislation Amendment Act 1996
Tasmania	Department of Primary Industries, Water and Environment (Environment & Planning Division)	Environmental Management & Pollution Control Act 1994 (replaces EP Act 1973–85) Land use Planning & Approvals Act 1993–95
Northern Territory	Department of Lands, Planning & Environment (Environment & Heritage; and Planning Division) Department of Housing & Local Government	State Policies & Projects Act 1993 Environmental Assessment Act 1982–94 Planning Act 1979 – 1993 Environmental Offences & Penalties Act 1996 Waste Management & Pollution Control Act 1999 (equivalent of EP Act)
Australian Capital Territory	Department of Urban Services (Planning & Land Management Group) Environment ACT (Environment Management Authority)	Land (Planning and Environment) Act 1991–97 Environment Protection Act 1997

control structures revolved around existing federal institutions—what Fisher (1980) termed 'structural legislation'. Subsequently, however, there have been significant changes.

Until the passage of the 1974 Commonwealth Act, it could be said that there had been no Commonwealth legislation to address environmental issues. As noted previously, the Australian constitution makes no reference to environmental protection: any environmental powers which do exist are incidental to other powers such as those related to export approval, heritage or foreign investment. These may provide the Commonwealth with the ability to control the environmental impact of large-scale developments in mining and forestry, for example. The same powers have been invoked periodically from the early 1970s by environmental groups challenging State development projects seen as having undue adverse effects on the environment. Such challenges, however, have generally proven to be lengthy and costly to proponents and with few successes (Munchenberg 1994; Bates 1995). Significantly, the above triggers of Commonwealth involvement in EIA were removed in the new *Environment Protection and Biodiversity Conservation Act 1999*, which nevertheless appears unlikely to come into force until well into 2000.

Following the passage of the *Environment Protection (Impact of Proposals) Act 1974*, and as the previous chapter has shown, there was a marked surge in other Commonwealth environmental legislation. These included several specific environmental protection Acts relating to nuclear activities and uranium mining (1978), sea dumping (1981) and import/export of hazardous wastes (1989–96). The increased Commonwealth activity also reflected legislation supporting Australia's international obligations covering areas such as marine protection, nature conservation and hazardous wastes (Chapter 6 and Appendix 2). However, there are no general Commonwealth laws covering water pollution, waste management or air pollution; such powers reside with the States.

Additional, non-statutory Commonwealth powers in environmental protection and management extend rather more indirectly to areas such as funding for environmental research through agencies and institutions or other arrangements, including CSIRO, LWRRDC, CRCs, ARC and the NHT. As landowner, the Commonwealth also has certain responsibilities relating to airports, telecommunications and defence facilities, noting however a shift of responsibilities with increasing corporatisation and privatisation of some of these functions. The Commonwealth also has exclusive control over the Federal Territories (Appendix E in Senate Standing Committee on Science, Technology and the Environment 1984).

Commonwealth Environmental Administration

Commonwealth environmental administration takes place within the Department of Environment and Heritage (previously Department of Environment, Sport and Territories (DEST)), with its environment department being known as Environment Australia (Table 7.1: see also Appendix 4). With a staff of over

1000, this agency is responsible for six environment groups, with the Environment Protection Group being responsible for EIA. However, other Commonwealth departments may also exercise certain environmental responsibilities: Agriculture, Fisheries and Forestry Australia (AFFA) (formerly part of the Department of Primary Industries and Energy (DPIE))—responsible for natural resources and the Murray–Darling Basin; Department of Transport and Regional Services (DTRS)—marine spills, aircraft/vehicle noise, and the Department of Foreign Affairs and Trade (DFAT)—international matters and exports.

Commonwealth Environment Protection Agreements, Policies and Initiatives

By the mid-1990s, the Commonwealth government had set up a series of national policies and strategies to support environmental protection and better environmental management, incorporating national ecologically sustainable development (ESD) goals and improved environmental reporting. The various strategies and policies cover a wide range of matters, from natural resource management and land degradation to pollution control and standards (Appendix 3). In turn, the Commonwealth initiatives strongly influenced the framing of State and Territory environmental legislation and policies in natural resource management and regional planning, resting on improved, integrated co-operation and structural and legislative reform.

The main initiatives included formation of: the Australia and New Zealand Environment and Conservation Council (ANZECC); the Council of Australian Governments (COAG); the Intergovernmental Agreement on the Environment (IGAE); the Intergovernmental Committee on Ecologically Sustainable Development (ICESD); Environment Australia, and the National Environment Protection Council (NEPC) (Appendix 4).

In brief, the IGAE outlines the roles and responsibilities of all levels of government in Australia and provides for a co-operative federalism in environmental decision-making. This is implemented through a set of schedules which cover major environmental issues. ANZECC is the main inter-governmental ministerial co-ordinating committee on the environment. It is supported and advised by other ministerial committees covering a wide range of environmental matters. ICESD is an inter-governmental committee which is responsible for reviewing and implementing the national ESD strategy. It reported to COAG in 1993 and 1996 on implementation, focusing on processes. A review of ESD implementation was under way in 1998. Finally, COAG is a peak council of all Commonwealth, State, Territory and local government leaders to oversee co-operative arrangements, roles and responsibilities amongst governments.

The main thrust of these initiatives was to seek greater national co-operation and sharing of responsibility in environmental matters amongst governments at all levels (Commonwealth to local) and implementation of better environmental protection around Australia. A clearer definition of roles and responsibilities, and development of consistent sets of national environment protection measures (NEPMs—Appendix 4 and below) which rely on complementary legislation in all jurisdictions, are critical to the achievement of these goals. Thus, the National Environment Protection Council (NEPC; see below) supports the latter goal, and a number of State and Commonwealth environmental protection reforms, amendments and new laws are also an acknowledgment of the reform goals. Paralleling the above initiatives was an influential economic rationalist ideology contained in a key document, the *National Competition Policy*, which drove much public policy and reform in the 1990s and was endorsed by COAG (Hilmer 1993). Essentially, the policy relies on market forces to determine the most 'efficient' outcomes in resource allocation. However, the environment is not necessarily ignored. COAG's water reform policy, for example, recognises the environment as a legitimate user of water (COAG 1994, 1995).

Thus, much happened at administrative and legislative levels of Australian governments during the 1990s. While it can be argued that the changes were often positive, confusion, uncertainty and destabilisation accompanied the changes. Government agencies were named, renamed, amalgamated or split, often several times, or simply disappeared altogether. In many instances bigger and more costly monolithic agencies created yet another layer of bureaucracy, with their new, regional offices complaining about being out of touch with head office. In Powell's words:

> throughout the nation, government agencies suffered epidemics of merging, scrapping, re-constituting, re-skilling, re-briefing and re-ranking. Hallowed conceptions of the public service vocation were ignored or gracelessly devalued (Powell 1998:67).

The National Environment Protection Council (NEPC)

In terms of seeking nationally 'better environmental protection' (COAG 1992), the establishment of NEPC was an important move, enacted in Commonwealth law in 1994. Details are presented in Appendix 4. NEPC's role is to oversee the development and implementation of sets of National Environment Protection Measures (NEPMs), including the establishment of a National Pollutant Inventory. Two other measures on ambient air quality and the movement of controlled wastes between States have been finalised, whilst a further two on used packaging materials and assessment of contaminated sites were under development in the first half of 1999. Other ambient standards may be developed for marine, estuarine and fresh waters, and noise. Under the NEPC agreement, States and Territories were to adopt NEPMs into State/Territory policies covering areas such as air, water, waste and noise. The measures take the form of Environment Protection Policies (EPPs) or equivalents, or they may be absorbed into various State and Territory environment protection, planning and resource management legislation. In reviewing NEPMs, Streets and di Carlo (1999) highlighted some possible weaknesses in relation to reporting and enforcement procedures, and the discretionary nature of their implementation. However, the use of NEPMs is still in its infancy. Their establishment is seen as a step in the right direction in pollution control in particular and environmental protection more generally.

As a point of interest, when the NEPC legislation was mooted in 1994, it was vigorously opposed by the WA Coalition Government (Minson 1994). Amongst a range of concerns, NEPC was viewed as an unwarranted intrusion into the State's affairs, the more so for Western Australia which has long been a proponent of 'States' rights' (versus those of the Commonwealth), and because of its resource-rich economic base. It was considered that unnecessary costs and uncertainty would be imposed on developers, and the proposed uniform environmental protection standards were seen as too rigid and inappropriate for the wide range of environments in that State. Some of the areas of concern, such as reference to 'uniform' standards, have since been modified, and WA is now party to NEPC legislation, as are most other States.

The concerns raised by WA are not uncommon and have appeared around the country at various times and in different contexts. In general, they reflect strong business, industry and mining interests concerned over potential constraints, costs and delays to developments. The same arguments have appeared in various environmental reform agendas put forward by pro-development governments. Perhaps WA showed its hand rather too obviously in this instance when stating in its NEPC discussion paper (previous paragraph) that economic growth was an 'essential forerunner of the rational protection of the environment'—as did Queensland's Labor Premier Goss, who in the early 1990s frequently emphasised the need to develop natural resources and castigated environmentalists for retarding economic growth (Briody and Prenzler 1998). Neither was Kennett's Victorian Government blameless in this regard (Raff 1995a; Christoff 1998); and Stein (1998) has expressed his deep concern over the watering down of planning and environmental legislation in NSW in favour of development processes (Chapter 14).

Reforms to Commonwealth Environment Protection Legislation

The Commonwealth parliament passed the new *Environment Protection and Biodiversity Conservation Act* in mid-1999. It replaces the 1974 environmental protection Act and previous conservation laws and introduces new biodiversity legislation.

However, the passage of the Act did not take place without controversy. A major concern with the new Act has been the transference of Commonwealth approval and environmental assessment powers for major developments to the States. The Commonwealth has thereby withdrawn from triggering mechanisms whereby under the 1974 Act it could involve itself in major and protracted environmental disputes. Bilateral agreements will now take place between the Commonwealth and the States under an accreditation scheme. While public comment on draft bilateral arrangements is possible, governments will make final approvals with no requirement for parliamentary review. Some see these new arrangements as placing no impediment in the way of States' economic development programs: indeed, a recent Environment Institute of Australia *Newsletter* (No. 21, June 1999, pp. 8–9) comments on a line-up of development projects even before the Act is promulgated in 2000.

The new Act has also been criticised for its failure to embrace serious environmental concerns such as vegetation loss, aspects of climate change and forestry issues. However, it is more clear than the 1974 legislation on the Commonwealth's environmental responsibilities concerning matters of national environmental significance, including international environmental obligations relating to wetlands, migratory birds, biological diversity, threatened species and nuclear matters—although the possibility remains that even these responsibilities can be devolved to the States (Connor 1999; Fowler 1999).

National Agreement on Environmental Impact Assessment

In 1991, ANZECC set up a working group to develop an integrated and co-ordinated approach to EIA in response to calls from industry and the community for greater uniformity amongst jurisdictions and a speedier process. The agreement was endorsed by ANZECC ministers and has been used to satisfy the requirements of Schedule 3 of the 1992 IGAE. In 1996, national EIA guidelines and criteria were issued, with some revisions as a result of public input. Some of the public concerns raised were in relation to: assurances of public participation in the process; procedural differences amongst jurisdictions; fast-tracking procedures; insufficient recognition of ESD, and standards of accreditation for EIA practitioners (ANZECC 1997). The intent of the agreement has been reflected in amendments to environment protection Acts in the States and Territories during the 1990s, particularly in relation to national uniformity, fast-tracking procedures and the diminution of public participation.

STATES AND TERRITORIES

Environmental Administrative Arrangements

As noted at the start of this chapter, all States now have environmental protection Acts or equivalents and their administering departments (usually EPAs) (Table 7.1). Despite the reforms, institutional arrangements, assessment conditions and legislative requirements still vary in the different jurisdictions. Several States and Territories still keep their planning departments as separate agencies (WA, Qld, NSW) but some (NT, SA, ACT, Vic., Tas.) group them within mega-natural resource or other agencies. A few jurisdictions have a distinct Department of Environment (WA, Commonwealth) or the equivalent nested within larger agencies (SA, NT, Tas.). Most States keep parks and wildlife separate, though in WA, CALM includes forestry. Some States keep natural resource management agencies in super-departments (SA, Qld, Tas., NSW) or retain traditional and separate arrangements (agriculture, water, rivers, land administration—WA). Most jurisdictions now separate their managerial/corporate arm from environmental or resource management, perhaps inadvertently formalising the potential for conflict of interest in environmental management.

Changes to Environmental Protection Legislation

Stepping aside from the Commonwealth's roles and responsibilities, a number of observations can be made in relation to the progress of environmental protection legislation around Australia since the 1970s. More recent legislation (as amended or new environment protection and planning Acts) in several jurisdictions suggests that the means of maintaining and improving environmental quality have moved beyond EIA and its equivalents. On the one hand there has been a narrowing of focus to strengthen pollution control, incorporating air, water and noise pollution legislation under one Environment Protection Act; on the other hand there are more holistic approaches which inevitably merge with the planning process, with increasing statutory powers being given to regional and local government levels. Much of this is reflected in new integrated planning or resource management legislation or policies; and it also responds to various Commonwealth environmental protection initiatives (Part 5).

Whether as existing or new laws or amendments, environmental protection legislation in most jurisdictions has several features in common. Although statutes are set out differently under individual pieces of legislation (Table 7.1), the underlying aim is to cover anything which may cause damage to the quality of people's environment or affect people's comfort or health. In some cases (NSW, Tas.), definitions are broad enough to exclude nothing. WA includes direct and indirect effects.

In general, the environmental protection laws of the 1990s incorporate: ESD principles; protection of economic, environmental and social values in planning and development; a broadly consistent EIA process; powers to develop environmental protection policies; increased penalties; tiered systems of offences; recognition of due diligence in avoidance of environmental harm; support for self-regulation; and licensing and other regulatory arrangements. Most Acts include EIA but increasingly this is also being included in planning legislation (such as Queensland's *Integrated Planning Act 1997* and WA's *Planning Legislation Amendment Act 1996*).

Victoria

Victoria was the first State to legislate for environmental impact assessment (EIA) in an environmental protection Act, in its *Environmental Effects Act 1978*, which applies to public works but not municipalities. Any Minister may seek an Environmental Effects Statement (EES) for assessment by the Minister for Environment. The EPA or planning bodies may also seek ministerial advice—or be directed to do so—under the *Local Government Act 1989* and the

Planning and Environment Act 1987. Again, the Minister for Environment may request an EES. Under the latter Act, State, regional and local planning schemes are seen as the mechanisms for achieving sustainable development and environmental and heritage protection, and the EPA has the power to review such schemes. Amongst the more usual planning objectives, the Act is intended 'to provide for the protection of natural and man-made resources and the maintenance of ecological processes and genetic diversity'. The Act has undergone several amendments since its passage, to: speed up the approvals process; limit objections, and incorporate inter-governmental agreement guidelines on EIA.

Victoria's *Environment Protection Act*, introduced in 1970 with subsequent amendments, attends to pollution control—as in the other States. This Act set up an independent, statutory Environment Protection Authority (EPA). In 1992, the EPA became the first environmental regulator in Australia to start auditing companies—checking on industry's compliance with regulations as well as pin-pointing areas of potential waste minimisation.

New South Wales

The *Environmental Planning and Assessment Act 1979* (NSW) was Australia's most comprehensive at the time and was something of a model for the other States in its acknowledgment of biophysical, economic and social environmental values. In particular, the Act set out to combine environmental protection with planning in one agency, using State Planning Instruments (SPIs)—SEPPs (State Environmental Planning Policies), REPs (Regional Environment Plans), LEPs (Local Environment Plans) and DCPs (Development Control Plans). Environmental planning and impact assessment are implemented through the legislation with regional plans playing a particularly important role. 'Significant' developments require the preparation of an EIS which, under more recent amendments, are to include ESD principles. Notably, local governments were given considerable powers in relation to pollution control as well as in the preparation of local plans. The Act also set up a Land and Environment Court to hear appeals under the legislation. The Court functioned effectively for more than a decade, but its role has been progressively reduced in recent years. Also, the combination of planning and environmental protection in the single agency was unsuccessful and the agency was split in the late 1980s. The main responsibility for EIA is now with the Department of Urban Affairs and Planning via the Director General of the Department or the Minister for Environment. Later amendments to the *Environmental Planning and Assessment Act* in 1997 saw some further changes, altering the balance between State and local government functions. The amendments were particularly aimed at streamlining development approval processes, but local powers were wound back, ministerial discretion increased and opportunities for public participation reduced (Chapter 14 and Part 5). Other changes included a requirement for Species Impact Statements under the *National Parks and Wildlife Act 1974* (NSW), and the introduction of 'thresholds' and criteria for designated developments.

The 1991 New South Wales *Protection of the Environment Administration Act* set up a separate Environment Protection Authority. The EPA is required 'to have regard to' maintaining ecologically sustainable development (ESD) and to reduce risks to human health as well as preventing environmental degradation. The following principles and programs are included: the precautionary principle; inter-generation equity; conservation of biological diversity and ecological integrity, and improved valuation and pricing of environmental resources (Woolf 1992). As in other jurisdictions, the more recent *Protection of the Environment Operations Act 1997* draws together disparate laws dealing with air, water, noise and waste pollution, and applies stronger penalties and licensing systems. The new Act streamlines the approvals system which had been regarded as over-regulated, lengthy and full of duplication. The Act also provides for Protection of the Environment Policies (PEPs) which detail environmental standards and goals. Contravening a PEP does not constitute an offence; rather, the Policies provide guidance to regulatory bodies with reference to the licensing of an activity or development. Additional powers are available to local councils to issue licences for minor developments, or clean-up notices.

As in other States, there has been a shift towards developing partnerships between regulators and developers. This is supported by a range of policies and strategies including self-regulation schemes, rather than a complete reliance on licensing or penalties. However, concerns have been raised over the effectiveness of voluntary schemes (refer, for example, EMS in Chapter 8).

Northern Territory

The Northern Territory's *Environmental Assessment Act 1982* (amended) parallels environmental protection legislation in other jurisdictions and incorporates EIA. In line with the States, there are tough new penalties under its *Environmental Offences and Penalties Act 1996*, which consolidates environmental protection legislation. The *Waste Management and Pollution Control Act 1999* encourages 'best management practices' and adds to the former pieces of legislation, which are administered by the Division of Environment and Heritage within the Department of Lands, Planning and Environment (DLPE). The *Planning Act 1979* (amended 1993) requires consideration of environmental impacts in planning procedures; the Act is administered by the Planning Division of DLPE.

Australian Capital Territory

The ACT Government passed its *Environment Protection Act* in 1997, which integrated five different Acts and encouraged industry to adopt 'best practice' in their manufacturing processes to assist in preventing environmental detriment. The Minister can determine whether an EIA is required. The Act provided for the establishment of an Environment Management Authority (EMA) to administer environment protection policies and provide guidance on environmental impacts. Further, the ACT's *Land (Planning and Environment) Act 1991–97* includes in its objectives the need to ensure an 'attractive, safe and efficient environment'. The Act is administered by the Planning and Land Management Group within the Department of Urban Services.

Queensland

Queensland was actually the first State to make statutory provision for EIA, in 1973 (and there were provisions for the process even earlier); however, this was in an amendment to the *Local Government Act* 1936, in the context of local Councils possibly being required to exercise development control powers (Fowler 1982). Unlike Victoria's 1978 legislation, it was not part of an environment protection Act and it did not set up an environmental protection agency. Somewhat belatedly, Queensland passed its *Environmental Protection Act* in 1994 (amended 1997). The Act was designed to establish the state of the environment, define environmental objectives and develop and implement 'effect management strategies'; it consolidated various pollution laws and placed significant obligations on both public and private sectors. Fines of up to $1.2 million can be imposed on companies which wilfully or negligently cause serious environmental harm, and they can also be levied for the costs of any clean-up and the State's investigation and enforcement expenses.

An Environmental Protection Agency was set up in 1999 (replacing the former Department of Environment and Heritage), with its first Director being the previous head of both WA's and the Commonwealth's environmental protection agencies. The new EPA is responsible for urban environmental issues, and it has five Divisions: sustainable industries; environmental planning (including responsibility for the Integrated Development Assessment System, IDAS—set up under the 1997 *Integrated Planning Act*); regulatory and operations (with responsibility for the EPA Act); policy and economics, and environment and technical services. The Department of Natural Resources deals with rural impacts and the environment, and the Department of Parks and Wildlife (now the Parks and Wildlife Service) has responsibility for preservation of 'pristine' regions (Environment Institute of Australia *Newsletter* No. 18, February 1999).

Prior to this, under Queensland's 1990–92 *Local Government (Planning and Environment) Act* (subsequently replaced by the *Integrated Planning Act 1997*, administered by the Department of Communication and Information, Local Government and Planning), responsibility for planning and environmental matters was attributed directly to local authorities, with the (former) Department of Local Government acting when necessary. EIA was required for 'designated developments', or as decided by the Chief Executive of the then Department of Environment and Heritage, but the legislation had significant deficiencies with regard to public participation and appeal rights. (At the central government level an EIA could be initiated under the *Public Works Organisation Act 1971*). On the other hand a Planning and Environment Court was established. Under the *Integrated Planning Act 1997*, comprehensive EIA is included, the development approvals process is speeded up and voluntary environmental protection measures are encouraged. The State has control over major developments and local authorities have responsibility for local matters, which are expected to harmonise with regional plans (Chapter 19).

South Australia

South Australia passed its *Environment Protection Act* in 1993 (with later amendments). This Act also replaced a string of specific environment Acts, dealing with clean air, the marine environment, noise, waste management, beverage containers, water quality, and the Environment Protection Council. An Environment Protection Authority was established and vested with a broad range of powers for licensing and approving activities of environmental significance, with penalties of up to $1 million for non-compliance. It does not seem that EIA has a role in the agency's activities.

It is under the *Development Act 1993*, amended by the *Development (Major Development Assessment) Amendment Act 1996*, that EIA is required for major developments or projects; the Development Act has appropriate linkages with the *Environment Protection Act*. The EIA process is administered by the EIA branch within the Department of Transport, Urban Planning and the Arts, although the initial stages are controlled by an independent Major Developments Panel (Harvey 1998:31). Environmental planning policies can be developed under the *Development Act* (amended) and can apply from the State to local levels. Natural resource management takes place either through the Department of Environment, Heritage and Aboriginal Affairs (DEHAA) or the Department of Primary Industry and Resources (PIRSA), although DEHAA is the main environmental management and protection agency. As in NSW, it seems that planning and environmental protection have gone their separate ways.

An Environment, Resources and Development Court was established (succeeding the Planning Appeals Tribunal) and is authorised by the *Environment Protection Act* to enforce the Act's provisions; it is also possible for a third party, that is 'any person', to seek an order of the Court to enforce the provisions of the Act. But although third party appeals have been a feature of SA planning legislation since 1972 (under the *Planning and Development Act 1972*), there was a progressive reduction in the availability of such rights over the subsequent two decades (Hodgson 1996).

Tasmania

In Tasmania, the *Environmental Management and Pollution Control Act 1994* replaced the 1973 *Environment Protection Act*. It provides for EIA (which had not been referred to explicitly in the previous legislation) in both the public and private sectors, although the sections addressing EIA were not enacted until January 1996. The Act is designed to complement the 'framework' legislation for the State's Resource Management and Planning System (RMPS), which embraces a number of environment-related Acts, State policies, regional and local planning, natural resource management and pollution control. The RMPS has design similarities with the provisions of New Zealand's *Resource Management Act 1991* (discussed in Chapter 20). Issues and objectives of the Tasmanian *Environmental Management and Pollution Control Act* incorporate principles of sustainable development and the precautionary principle. The EIA process is closely integrated with planning processes under the *Land Use Planning and Approvals Act 1993–95* and it is also linked with the *State Policies and Projects Act 1993*, which integrates the RMPS. A Resource Planning Development Commission amalgamates the functions of former development assessment, planning and resource management advisory bodies. The Environment and Planning Division within the Department of Primary Industry, Water and the Environment administers the legislation, and a Resource Management and Planning Appeal Tribunal hears appeals on environmental management matters.

Western Australia

Although Western Australia had one of the earliest environment protection Acts, in 1971, it was not until 1986 that the State legislated for the equivalent of EIA, even though the process had been used since the early 1970s by the then Department of Conservation and Environment. The 1986 Act has particular strengths in three areas: pollution control; environmental impact assessment (or environmental review and management programs—ERMP), and the formulation of environment protection policies (EPPs) which, once accepted by Parliament, have the force of law. Both private and public development proposals may be referred to the EPA for assessment. Major proposals may require a full ERMP with lesser levels of assessment being applied to less important projects. Other government agencies and local authorities with responsibility for environmental and resource management are required to operate within the terms of the environment protection legislation. There is provision for appeals to the Minister, whose decision is final.

Amendments to the *Environmental Protection Act* were passed in 1998, considerably strengthening the legislation in terms of its 'teeth' but reducing the provision for public participation. The changes broadly mirrored the provisions in equivalent legislation in other States and national EIA policy. The roles of the Department of Environmental Protection (DEP) and the EPA were clearly separated (already achieved in 1993), the notion of 'environmental harm' was clarified, more rapid environmental assessment and approvals were made possible, tiered offences and increased penalties (to $1 million) were introduced and appeals processes were reviewed (Edwardes 1997).

In 1996, new planning legislation was introduced (*Planning Legislation Amendment Act*) (see Chapter 17) which incorporates environmental assessment as an integral part of the planning process and enables early environmental assessment of town or regional planning schemes to be carried out by the EPA. However, there is still a lack of clarity concerning the relative roles of the

Department of Environmental Protection (DEP) and the Ministry for Planning's Environment Section in relation to environmental matters.

A Comment on Definition

Despite Queensland's poor environmental and planning record of the previous two decades, the State's definition of 'environment' in the 1990–92 *Local Government (Planning and Environment) Act* was arguably the most enlightened in Australia at the time, with reference to its unusual breadth. Environment was defined to include:

a)   ecosystems and their constituent parts including people and communities;

b)   all natural and physical resources;

c)   those qualities and characteristics of locations, places and area, however large or small, which contribute to their biological diversity and integrity, intrinsic or attributed scientific value or interest, amenity, harmony, and sense of community, and

d)   the social, economic, aesthetic and cultural conditions which affect the matters referred to in paragraphs a), b) and c) or which are affected by those matters (Leadbetter 1991).

This explicit incorporation of social, economic and community considerations in the definition of 'environment' is both philosophically and technically correct (Chapter 1) and widely adopted in the USA and Canada. In Australia, the Commonwealth, Victoria and Tasmania similarly adopt a broad definition. New South Wales previously used an enlightened definition of 'environment' in its 1979 *Environmental Planning and Assessment Act*, but its 1991 *Protection of the Environment Administration Act* redefined the term much more narrowly (although 'human-made or modified structures and areas' are included).

Some of the other Australian jurisdictions legislatively restrict 'environment' narrowly to the biophysical environment and aesthetics, including the somewhat ambiguous definition in the Western Australian *Environmental Protection Act 1986*:

'Environment', subject to subsection (2), means living things, their physical, biological and social surroundings, and interactions between all of these.

So far so good; the sting is in subsection (2), which reads:

for the purposes of the definition of 'environment' in subsection (1), the social surroundings of man (sic) are his aesthetic, cultural, economic and social surroundings **to the extent that those surroundings directly affect or are affected by his physical or biological surroundings** (emphasis added).

This and other restrictive definitions of 'environment' have serious, adverse consequences. In Western Australia, for example, the Social Impact Assessment

Unit (which in any event should have been an integral part of the EIA process, but which was located within the Department of Premier and Cabinet) was disbanded in 1993, following a change of government. The mining lobby in particular, and the Chamber of Commerce, had insisted (at a public meeting held as part of a process of reviewing the 1986 environment Act) that EIA be restricted to what was quaintly termed 'technical matters', meaning physical aspects of the environment. Later, the problem of interpretation and definition also arose in the *Cockburn Cement Ltd v. Coastal Waters Alliance* (representing various community interests) case where the EPA was ruled to have stepped outside the intent of the Act by including economic considerations (in support of its recommendation that a proposal to continue the mining of limestone from the seabed be approved) (Bache *et al.* 1996).

Compliance and Enforcement

As the foregoing has shown, heavy penalties for breaches of environmental law now apply around Australia. At the same time, there is greater fairness under tiered systems, where the seriousness of the offence is matched by the penalty—ranging from infringement notices to $1 million fines or imprisonment. Until recently, there has been a general reluctance by agencies to prosecute offenders. However, following a strengthening of the legislation, Australia's first gaol sentence was handed down in WA to a company director in 1995 (Brunton 1995). A similar case occurred in NSW in 1997 (Lipman 1998). Both penalties were accompanied by fines.

There is now some discussion as to whether punitive measures are better than voluntary or other approaches to secure compliance with environmental protection laws and standards (for example, Cunningham and Sinclair 1999). Smith (1995a, b), citing US successes, believed that large fines have a deterrent effect. But Australia's environment protection agencies prefer to concentrate on outcomes rather than penalties. Deegan (1998) partly supported this approach, noting improvements in Australian corporate environmental reporting practices in recent years. Nevertheless, there was room for further improvement.

Briody and Prenzler (1998) seemed less than impressed with Queensland's record in environmental law enforcement. They believed that under-enforcement by the Department of Environment was partly a result of indirect, undue influence by government and industry, and the Government's political *milieu*. Civil action and an independent EPA could help to improve these shortcomings. However, Kelleher (1998) found that the Government's reluctance to prosecute appeared to have diminished following the passage of the *Environmental Protection Act 1994*.

Neither has WA's environmental prosecution record been impressive. Between 1988 and 1997, the WA Department of Environment mounted 54 prosecutions under the *Environmental Protection Act*. More than a quarter failed in court or were withdrawn by the Department. The total penalties imposed amounted to a mere $215 000 (Spellman 1998).

EXPENDITURE ON ENVIRONMENTAL PROTECTION

Environmental expenditure can be used as an indication of public and private commitment to environmental protection, and to some extent it reflects the effectiveness of environmental policies and laws. The level of expenditure on environmental protection in Australia is still low: at $6.5 billion in 1993/4 it represented an estimated 1.5% of GDP (Table 7.2). The bulk of this was spent on water and waste management. Approximately half of the total environmental expenditure occurred in the government sector (primarily water, waste and pollution abatement), with private sector expenditure incurred mainly on household waste and sanitation. Thus the total expenditure figure is misleading and indeed a distortion. The above areas of expenditure have always been seen as providing essential public services, rather than being considered as fulfilling environmental management programs and policies in a broader and more contemporary sense. Total *per capita* government purpose expenditure on the environment in 1993/4 ranged from $124 in Victoria to $39 in WA. Other purpose *per capita* expenditure (such as soil and water protection, biodiversity and landscape) ranged from $240 in the Northern Territory to $25 in Victoria (ABS 1997b).

Table 7.2   **Australian expenditure on environmental protection 1993–94. Source: ABS 1997b.**

Total Australian Expenditure:	$6.5 billion = 1.5% of GDP
public sector (includes sanitation, rubbish pollution abatement, environmental protection)	$3.1 billion
private sector	$3.4 billion

$ per capita expenditure by State/Territory 1993/4

	Govt Purpose	*Other Purpose*
Vic.	124	25
SA	121	38
NSW	104	76
NT	82	240
Tas.	74	96
Qld	72	52
ACT	70	37
WA	39	42
Total	61	66

Both the agricultural and the mining sectors spend less than 1% of their total expenditure on environmental protection. The greatest expenditures in the agricultural sector are in beef, sheep and grain—that is, broadacre agriculture—and the least in poultry. Within individual sectors, the sugar industry outlays the greatest proportion at 8% of its total expenditure, mainly on earthworks for erosion and salinity control, followed by sheep and beef expenditure of 1.4% on animal, insect and weed eradication (ABS 1997b).

COMMENT

Australia introduced EIA gradually from the 1970s onwards, mostly in environmental legislation in the late 1970s and 1980s, but not until the 1990s in some jurisdictions. However, this was not the only or even the main legislative response to the increasing problem of environmental degradation. Pollution control (broadly defined) and the integration of environmental protection with land use planning, with a more holistic approach to environmental protection, were also introduced in some jurisdictions, although with mixed results. The current situation still exhibits a number of deficiencies which are discussed further in Part 5.

One requirement which is common to most approaches is an evaluation of the 'state of the environment', with reports being required every 3–5 years in some jurisdictions. The next Part examines some of the methods used to evaluate or assess the environment. Of particular interest is the prospect that these techniques have the potential to provide a bridge between planners (with disciplinary expertise in architecture, engineering or human geography) and environmental scientists (botanists, zoologists, physical geographers). In the past, even in agencies combining 'environmental protection' and 'planning', there has been a marked lack of effective communication between the two groups. This has undoubtedly contributed to some of the difficulties in resolving environmental problems and issues.

Part 3

Environmental and Land Assessment Methods

Environmental assessment, measurement or evaluation has three main roles: ambient or base-line studies; ongoing environmental monitoring, and project- or activity-specific studies. Chapter 8 considers some of the methods which are appropriate to one or more of these roles: namely, toxicological/bioassay approaches; physical/chemical monitoring; biomonitoring; ecological approaches, and environmental quality indicators. Subsequent chapters discuss methods more appropriate to the evaluation of land, largely for planning purposes, and environmental impact assessment is examined in the final chapter of this Part.

CHAPTER 8

METHODS OF ENVIRONMENTAL ASSESSMENT, MEASUREMENT AND REPORTING

This chapter discusses several assessment and measurement techniques used to describe environmental condition and change. The methods range from relatively straightforward laboratory and field techniques to more complex assessments of ecosystem dynamics and environmental indicators. Each has an important role to play as an aid to understanding environmental behaviour, and to support natural resource and environmental planning and management processes.

Several publications provide further details relating to the early sections of this chapter. They include Holling (1978), Best and Haeck (1982), Westman (1985), WHO (1993), Schmitt and Osenberg (1996), Grant *et al.* (1997), and McKinney and Schoch (1998).

TOXICOLOGICAL/BIOASSAY APPROACHES

Toxicological and bioassay approaches are concerned with investigating the direct and indirect impacts of environmental substances—anthropogenic or natural—on organisms or environmental systems (terrestrial, aquatic or atmospheric). Such studies can be specific (to a species or toxin) or short term (such as investigating acute impacts in lethal dose studies). Conversely, studies may be broader, in time and space. These look at whole systems and complex responses—such as epidemiological approaches which use statistically determined populations to identify long-term health outcomes to environmental exposures.

The techniques used to assess environmental effects on systems and organisms are numerous, covering instrumental, pharmacological or biochemical methods. The approaches may be field- or laboratory-based. In general, toxicological and bioassay methods are concerned with setting human and environmental tolerance levels via experimental studies on living materials. They are generally specific to particular species or toxins; such as investigating the tolerance of species to salinity, acid rain or synthetic chemicals.

The toxicity of pesticides, for example, can be measured in various ways. Acute toxicity is evaluated in terms of oral (ingested) or dermal (absorbed through the skin) Lethal Doses (LD). Doses are determined by the concentration of active ingredient in milligrams per kilogram required to kill 50% of test animals in a laboratory (LD50). The lower the figure, the more toxic the chemical. For instance, the highly toxic chemical aldicarb has an oral LD50 of 0.93 (and a dermal LD50 of 5), while the oral LD50 of the mild herbicide atrazine is 2000 (and the dermal LD50 is 7500) (DPI 1980). Other measures of toxicity, under controlled conditions, involve observation of the long-term effects of dosages on test animals (including behavioural and immune system responses), and the influence of the chemicals as mutagens (cell damaging), teratogens (foetal damaging) and carcinogens (cancer forming). More recently, alternative tests on living materials such as micro-organisms, but for humane reasons avoiding the use of laboratory animals, have been developed. Other methods include molecular toxicology and computer modelling. But such methods also have their limitations in terms of accuracy or the amount of information provided.

PHYSICAL/CHEMICAL MONITORING

These are field-based environmental programs designed to sample the occurrence of particular substances in space and time. Measurements of odours and noise are also included under this heading. Sampling techniques range widely and include use of dye tracing (as for water and air movements), traps (animals, water or air-borne sediment), bores (groundwater movements) and automated collection devices (rainfall and a wide range of air- and water-borne pollutants). Field monitoring is often conducted at a regional scale, and is usually compound-specific; for example, monitoring the concentrations of atmospheric SO_2 over or adjacent to industrial areas. Sampling and monitoring may take place over relatively short periods of time or continue for many years, depending on the objectives. Selection of appropriate sampling frequencies and spatial distributions is crucial, and needs to consider the nature of the medium in which the substance occurs. For example, monitoring of SO_2 concentrations in the atmosphere needs to take into account wind strengths and frequencies and distance downwind from the emitting source(s).

In general, Australia is poorly served with air-quality monitoring stations. They are primarily based in major cities or selected industrial areas. Some 95% of Australia's area is not routinely monitored (State of the Environment Advisory Council 1996:5.21), which neglects several areas of concern in non-metropolitan Australia. These include pollutants from coal-fired power stations (SO_2 and fly ash), fallout from ore processing (heavy metals) and pollution from primary industries (such as agricultural chemicals and prescribed burns of naturally vegetated areas).

Sampling design for stream salinity monitoring must incorporate not only seasonal flow regimes but variations in discharge associated with individual rainfall events. Salt concentrations will differ between rising and falling limbs of the stream hydrograph (Fig. 8.1), and prior rainfall events. The first flow after a long, dry period will flush salts from the system, yielding high concentrations. A similar discharge coming towards the end of a long period of high rainfall will contain relatively little salt.

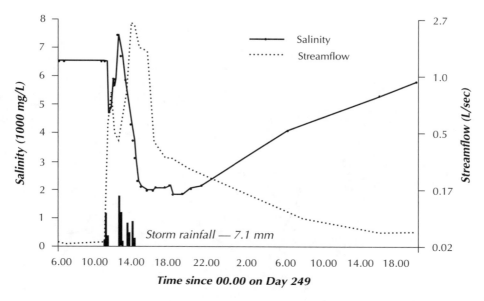

Fig. 8.1   Salt concentrations associated with the rising and falling limb of a flow hydrograph in relation to one rainfall event. From George (1991a).

Similar arguments apply to most pollutants in water, indicating that continuous monitoring is necessary in order to obtain accurate results. But for practical and cost reasons this is not always possible, and periodic sampling is often the only practicable alternative. Possible deficiencies in the data arising from 'second best' methods need to be recognised and acknowledged.

Spatial sampling designs also need to be considered carefully in relation to the objectives of the study. For example, sampling in a drainage basin where the sources of the pollutants are not known, should take drainage basin morphometry into account so that the culprit sub-catchment(s) can be identified (Fig. 8.2). Green (1979) set out ten useful principles of sampling design for environmental biologists (Table 8.1); they are also relevant to other environmental scientists.

Targets for monitoring may be derived from toxicological studies, such as lakes for pH or streams and soils for salinity or pesticide residues. The output is often in the form of maps produced with the aid of a geographical information system (GIS), from which assessments may be made concerning the quality or state of the environment (see 'Quality Indicators' below). The data base so generated may also be 'overlaid' in the GIS with other environmental variables and/or subjected to statistical analyses with the objective of identifying cause–effect relationships. However, if this is to be done, then this should be known from the outset (Green's first principle—Table 8.1) as it would probably influence the sampling design.

Another form of regional and spatial environmental monitoring is possible at a vastly different scale through the use of satellite imagery. Data from satellites (such as Landsat, NOAA or SPOT) are captured and manipulated or combined with ancillary land information, such as ground-derived data on geology, soils,

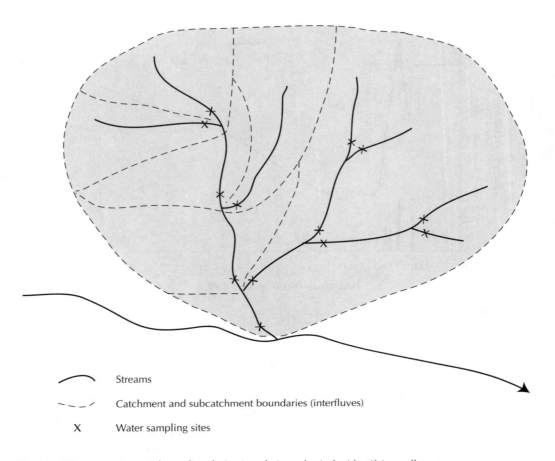

Streams	
Catchment and subcatchment boundaries (interfluves)	
X	Water sampling sites

Fig. 8.2 Diagrammatic spatial sampling design in a drainage basin for identifying pollutant sources.

landform and hydrology, to obtain a time/space 'picture' of land condition. Examples include changes to land cover and land use or in land management practices, crop diseases, fires and floods. These methods have been particularly useful in building up a base of information on Australia's rangelands (Graetz *et al.* 1982, 1992) and on land degradation, land resources and land management (Thorburn and Till 1988; Johnston and Barson 1990; Kelly *et al.* 1990). Nevertheless there are several problems with the technique, including: access to and cost of the data; studies being driven by the technology rather than the problem; loss of smaller details (resolution); difficulties with calibrating data to ground conditions, and problems in untangling cause/effect relationships.

In general, monitoring does not involve much understanding of mechanisms, and assessment may be very narrow. It also needs to be clearly recognised that monitoring in itself does not solve anything. Assurances (as in the case of the controversy associated with the woodchip industry) that there is nothing to worry about because the environment is being monitored, miss the point of monitoring. Clearly, if monitoring identifies that a problem has occurred then

Table 8.1   **Roger Green's ten principles of sampling design for environmental biologists (Green 1979).**

1	Be able to state concisely to someone else what question you are asking. Your results will be as coherent and comprehensible as your initial conception of the problem.
2	Take replicate samples within each combination of time, location and any other controlled variable. Differences among can only be demonstrated by comparison with differences within.
3	Take an equal number of randomly allocated replicate samples for each combination of controlled variables. Obtaining samples from 'representative' or 'typical' places is *not* random sampling.
4	To test whether a condition has an effect, collect samples both where the condition is present and where the condition is absent but all else is the same.
5	Carry out some preliminary sampling to provide a basis for evaluation of sampling design and statistical analysis options. Those who skip this step because they do not have enough time usually end up losing time.
6	Verify that your sampling device or method is sampling the population you think you are sampling, and with equal and adequate efficiency over the entire range of sampling conditions to be encountered. Variation in efficiency of sampling from area to area biases among-area comparisons.
7	If the area to be sampled has a large-area environmental pattern, break the area up into relatively homogeneous subareas and allocate sampling sites to each in proportion to the sizes of the subareas. If it is an estimate of total abundance of individuals over the entire area that is desired, make the allocation proportional to the approximate or estimated number of individuals in the subareas.
8	Verify that your sample unit size is appropriate to the size, densities and spatial distributions of the individuals you are sampling. Then estimate the number of replicate samples required to obtain the precision you want.
9	Test your data to determine whether the error variation is homogeneous, normally distributed and independent of the mean. If they are not, as will be the case for most field data, then (a) transform the data appropriately, (b) use a distribution-free (non-parametric) procedure, (c) use an appropriate non-sequential sampling design, or (d) test against simulated H_o data.
10	Having chosen the best statistical method to test your hypothesis, stick with the result. An unexpected or undesired result is *not* a valid reason for rejecting the method and hunting for a 'better' one.

it is too late to prevent it. Although remedial measures may be taken, it is usually much more difficult, and less efficient, to repair a problem after it has developed than to prevent it from occurring in the first place. In this context, application of the precautionary principle through proactive management, or more complex monitoring to detect subtle but critical changes, are needed.

BIOMONITORING (the canary in the coal mine method)

This includes inventory-style methods (refer Chapter 9) including the production of species lists, frequency and spatial distributions. These are generally static in approach and therefore somewhat limited in their application. The species list in the Attachment to the Western Australian woodchip EIS, for example, provides no information on the dynamics or processes of the ecosystem involved and is therefore virtually useless as a management tool. Nevertheless species lists are important in particular contexts, especially in relation to species distribution or the identification of rare or endangered species.

More useful is the identification of **indicator species** which can be used as a measure of environmental quality, or as an indicator of the onset of degradation or change. Indicator species may be classed as sentinels, detectors, exploiters or accumulators as well as bioassay organisms. For example, macro invertebrate assemblages were used as indicators of river health in the Natural Rivers Health

Program, in which more than 6500 reference sites were sampled between 1997 and 1999 (ANZECC 1998:41). In the Western Australian wheatbelt, the appearance of sea barley grass (*Hordeum marinum*) in otherwise healthy pasture is an excellent indicator of incipient secondary soil salinity. Similarly, the presence of South African veld grass (*Ehrharta* sp.) in native vegetation remnants or bushland indicates disturbance as a result of grazing, human activities or inappropriate fire regimes.

However, this approach relies heavily on toxicology, detailed field and species observations, or other techniques in order to identify correctly and unambiguously responses which are indicative of environmental change. The method relies on a high degree of knowledge of how systems work with regard to expected perturbations and the reactions of species to them (for example, Hellawell 1986). One potential weakness is that single programs may focus on only one indicator for a limited range of perturbations. That narrowness stimulated the search for more general indicators of environmental quality, or broader approaches to environmental assessment, some of which are discussed below.

SPECIES GUILDS

In increasing order of complexity, the guild approach incorporates aspects of:
- individuals (health),
- populations (structure and/or dynamics),
- community (structure), and
- ecosystem (function)

into a sampling design used to generate an index of biotic integrity for a particular environment, such as a catchment. The method demands a great deal of expert information and involves detailed research, often over a long period of time. The famous Hubbard Brook ecosystem study, which continued over a period of many years, is a particularly good example (Likens *et al.* 1977).

ECOLOGICAL APPROACHES

The approach can be further structured so as to explicitly incorporate ecological principles of:
1   **organisation** (community structure, including an inventory of the composition and diversity of communities, indicator species and the congruity of flora and fauna);
2   **function** (the physiology, growth and behaviour of organisms or species, and energy transfers, resource partitioning, biomass production, competition, pest–predator relationships and population dynamics within communities or ecosystems);
3   **state** descriptors, including inertia, redundancy, stability, equilibrium or fluctuating condition and recovery potential following disturbance; and

4   **habitat** assessment, both physical and chemical, and including evaluations of quality.

The concept of *habitat site* includes a distinct area of the landsurface with a defined range of slope angles and particular position on the slope, possessing defined morphological limiting factors with regard to agricultural development, and which is also a recognisable, distinct unit of the landsurface. This basis is very similar to the land unit in the land systems technique (Chapter 9). But the starting point is land use/vegetation patterns, with an emphasis on recognising the interactions between ecological and socioeconomic systems, of which land-use patterns are an expression (Fig. 8.3). In other words, and recalling the discussion in Chapter 1, the emphasis is on dynamics rather than static descriptors.

The above is one type of ecological approach, and it arose from recognition of the failure of other methods to take an holistic approach. Beanlands and Duinker (1983) have discussed the use of two forms of scoping to identify valued ecosystem components:

1   social scoping—identifying societal values and perspectives in order to define the valued ecosystem components, and

2   ecological scoping—the methods and degree to which the above components can be assessed.

Under its Sustainable Landscapes Program, CSIRO's Division of Wildlife and Ecology is developing different means of measuring and monitoring biodiversity and ecological sustainability (CSIRO 1998:6–7). The program involves descriptions of the spatial distribution of biodiversity and biogeographic

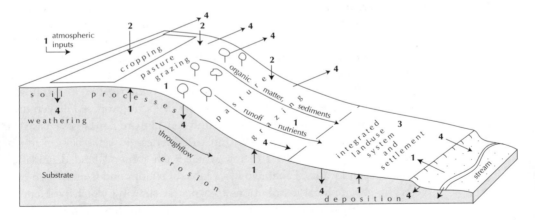

1	Natural nutrient inputs. Dust, alluvial sediments, organic matter, weathering.	**3** Local enrichment — input of manures, crop residues, recycled wastes, slopewash sediments.
2	External (artificial) nutrient inputs.	**4** Nutrient export. Biomass (crop, timber), stock, volatilisation, fire, leaching, runoff, sediment.

Fig. 8.3   Diagrammatic relationship between land use, topography and nutrient flows; illustrating interactions between ecological and socioeconomic systems. Modified from Pieri *et al.* (1995, Fig. 3.1).

characteristics, and development of reference taxonomies. Measurements of changes in distribution, abundance and condition will be used to determine responses to land-use practices and other disturbances. Tools are to be developed to identify those variables representing, or acting as surrogates for, biological diversity. Where necessary, the variables will be used as indicators in planning for conservation and biodiversity, and for monitoring trends.

Ecological approaches have also been used to develop the concept of *strategic assessments*, which form part of an environmental management hierarchy:

> government policies and programs → strategic assessments → project assessments → regulatory procedures.

Strategic assessments incorporate an assessment of the biophysical environment with an assessment of likely cultural demands in the form of (for example) five-year projections of development potential. This places environmental assessment within a feedback framework which demands continued reassessment via environmental monitoring and changing social values. This latter step—incorporating societal values into the assessment process—is critical if environmental management is to be effective. Strategic assessments also enable evaluation of cumulative effects of multiple and continuing developments (Chapter 11). The process is both dynamic and adaptive.

In 1999, Tonk, Sadler and Verheem, members of the International Association for Impact Assessment (IAIA), proposed strategic environmental assessment (SEA) quality principles under two headings: generic principles and process principles (Table 8.2).

More specifically, the role of this kind of ecological analysis in regional planning is to provide information on the environmental implications of alternative plans. The landscape is approached as a complex, functional entity (Simmons 1980; Beanlands and Duinker 1983): refer 'bioregions' in Chapter 16.

For example, agricultural development is considered in the context of the biological system in which it is to take place. In this context, Williams and Hamblin (1991—an unpublished report for a working party on sustainable development in agriculture, cited in Roberts 1995:57) proposed eleven *agroecological zones* for Australia based mainly on climate zones as a basis for resource management analysis. Within these zones, 46 *agroecological regions* were mapped provisionally to show where each form of land degradation occurs in each locality. Each map has a table of regions and issues which converts the geographical distributions to a matrix showing which problems of unsustainability require attention in each region. Maps have also been drawn up from information provided by each State so that each major soil problem (such as erosion or acidity) can be plotted. These maps are to aid planners and managers as geographical databases of degradation types. They make it possible to calculate total areas in need of corrective action and to estimate predicted costs for budgeting.

In general, however, the operation of the ecological system is often independent of much of the data collected during a 'land system' survey (for example,

Table 8.2   **Proposed Strategic Environmental Assessment (SEA) quality principles to apply to the environmental assessment of all strategic decisions (legislation, regulations, policies, plans and programs) with potentially significant environmental consequences. Proposed by J. Tonk, B. Sadler and R. Verheem, International Association for Impact Assessment (IAIA) e-mail discussion paper, 21/1/99.**

1 Generic Principles
SEA is:

- *fit for purpose*: customised to the characteristics of the strategic decision-making process;
- *objectives-led*: undertaken with reference to environmental goals and priorities;
- *sustainability-orientated*: facilitates identification of development options and proposals which are environmentally sustainable;
- *integrated*: related to parallel economic and social appraisals and tiered to project EIA where appropriate;
- *transparent*: has clear, easily understood information requirements including provision for public reporting;
- *cost effective*: achieves its objectives within limits of available information, time and issues;
- *relevant*: focuses on the issues that matter;
- *practicable*: provides information which is required for decision-making.

2 Process Principles
A SEA process ensures:

- *good screening*: an appropriate assessment of all strategic decisions with significant environmental consequences;
- *ownership*: an assessment is carried out by the responsible agencies;
- *good timing*: availability of the assessment results early enough to influence the decision;
- *good scoping*: provision of all relevant information to judge whether an initiative should go ahead, and whether its objectives might be achieved in a more environmentally friendly manner;
- *good reviewing*: safeguarding the quality of process and information by an effective review mechanism;
- *credibility*: appropriate consultation to include the views of the affected public early enough to influence the decision;
- *clear documentation*: assessment results are identifiable, understandable and available to all parties affected by the decision;
- *transparent decision-making*: clarification to all affected parties of how assessment results were taken into account;
- *post-decision analysis*: sufficient information on the actual impacts of implementing the decision to judge whether it should be amended.

geology and relief: refer the following chapter) but is more dependent on present-day land use. Vegetation and agricultural data *can* be grouped on a landform basis, but it is not necessarily meaningful to do so. In any event, the 'land systems' discussed in the following chapter are only a component of an ecosystem. In an ecological approach the focus is on plant–soil relations in relation to agricultural practices: indeed, efforts have been made to further refine an ecological approach to agriculture in Australia and elsewhere (Chapter 9).

OTHER TECHNIQUES

Other methods such as statistical and systems analyses, surveys and interviews are not discussed here. Although widely used in environmental assessment, they are not particular to it. This is also true of GIS, which is now used extensively by environmental researchers and practitioners. Although not strictly techniques of environmental measurement *per se*, two important areas are environmental audits and environmental management systems. These are discussed below and lead to a consideration of State of the Environment Reports and environmental quality indicators.

Environmental Audits

Borrowing from accounting techniques, environmental auditing essentially covers reviews of stocks and/or assessment of performance against benchmarks or goals.

The first approach is exemplified by the National Land and Water Resources Audit, which is appraising national soil and water data and environmental, economic and social costs to the nation. This is discussed in the following chapter. The second has been employed by the Australian National Audit Office (ANAO) and others to assess the performance of a number of national environment programs. The third approach is a system which is increasingly used by business as a self-audit process to review a company's environmental performance, known as Environmental Management Systems (EMS).

The ANAO Audit

A national audit of the administrative processes of key Commonwealth environmental programs was undertaken by the Australian National Audit Office in 1997 (ANAO 1997). The programs included National Landcare, One Billion Trees, Save the Bush, Waterwatch, River Murray Corridor of Green, National Corridors of Green, Drought Landcare and Grasslands Ecology.

The methodology was based on a normative model to measure actual performance against suitable benchmarks which are consistent with better practice. Comparative information on the relative performance of different but related program elements also provided a performance measure. Techniques used included interviews, examinations of files, agreements, research reports and consultations with officials and key stakeholders. Audit criteria were developed from and in accordance with ANAO audit practice and other sources relating to public accountability standards of practice. The actual process was aimed at benchmarking administrative processes and seeking cost-efficient outcomes of the environment programs. Amongst the findings was recognition of the constraints posed by poor baseline data.

Curtis et al. (1998) used a similar range of evaluation techniques to the ANAO audit (interviews, program documentation, program performance and other indicators) to assess a major Murray–Darling Basin environmental program and the Commonwealth's DPIE Farm Forestry Program (refer Chapter 15 for some of the findings of the above two audit examples).

Environmental Management Systems (EMS)

Different techniques used in environmental auditing are well exemplified in EMS. This is a voluntary process which is increasingly used by businesses and corporations (and local governments) to track their environmental performance in relation to specified environmental goals or criteria (for example, in relation to discharge of pollutants, air and water quality, and recycling). There are various EMSs in existence, one of the better known being the ISO 14000 Series (details in Rothery 1995; Tibor and Feldman 1996; Brown n.d.) which has been adopted by the Australian Standards Office as the AS/NZS ISO 14000 Series. The Series provides guidelines, standards, techniques and specifications to help companies implement better environmental practice. Policies, goals, objectives and specific targets are set out against which progress can be measured qualitatively and quantitatively. Continuous monitoring, auditing and reviews take place and corrective action is implemented where required.

Since mid-1998, Corporations Law—under Australian Company Law—has required annual environmental reports from public and big proprietary companies. Mandatory reporting is required on environmental incidents and offences. However, this requirement has been challenged by industry lobbyists who argue that there is already a plethora of environmental laws (over 1000 pieces of environmental legislation) which regulate their activities and that voluntary reporting should suffice (Hepworth 1998).

Nevertheless, Australian company environmental reporting lags behind that of other countries, even though some major companies are recognising that better management, through EMS, offers substantial economic and social as well as environmental advantages (Deegan 1996, 1998; Russel 1998). Such benefits have also been acknowledged by more than 20 local governments around Australia which have adopted EMS as an aid to meeting increased legal requirements in environmental management and planning (Environment Institute of Australia *Newsletter* No. 3, July 1996).

But as Deegan and Russel have observed, despite the enthusiastic adoption of the technique there are a number of shortcomings to EMS (refer also Cunningham and Sinclair 1999). Standards of company reports are uneven: for example, in the UK they range in size from one sentence to 30 pages (Russel 1998). There is a tendency for some reports to be self-laudatory, lacking in relevance or substance, amounting to little more than a slick PR exercise or 'greenwash'. If adequate audit procedures are not followed by accredited or independent auditors, then there is a potential for loss of confidence in the reporting process. On the other hand, companies might also use 'commercial confidentiality' as an excuse to avoid publicising their less than desirable environmental performance (Fox *et al.* 1998). Small companies generally avoid the process altogether.

Despite the shortcomings, EMS is increasingly favoured as a voluntary and co-operative arrangement between industry and governments as a means of reducing the costs of environmental licensing and inspection, and encouraging better practice through mutual agreement. The Commonwealth government is making a determined effort to encourage annual environment reports in the drive to lift corporate environmental performance. Voluntary environmental reporting guidelines are being prepared by Environment Australia for both public and private sectors. Such moves are being given further impetus by the implementation of National Environment Protection Measures (Chapter 7) and greenhouse gas reduction initiatives (Environment Institute of Australia *Newsletter*, August 1999:15).

STATE OF THE ENVIRONMENT REPORTS

Most States and the federal government in Australia now produce State of the Environment (SoE) Reports (Table 8.3). Queensland is one of only two jurisdictions still to do so; its first SoE report was due towards the end of 1999. Although the earlier reports were produced in the 1980s, more recently they have been fostered particularly by the United Nations Agenda 21 from the 1992 Rio Environment conference and by the nationwide adoption of a *National Strategy for ESD*. Agenda 21 and ESD set an ecologically sustainable development agenda within national, State and local government policy objectives (Chapter 6). Additionally, the *Intergovernmental Agreement on the Environment* provides for a co-operative approach in environmental decision-making (Chapter 7).

The OECD (1991: refer also Parker and Hope 1992) has influenced the reporting process through the pressure–state–response model, whereby human

Table 8.3   **State of the Environment (SoE) Reports in Australia.**

WA	*State of the Environment Report December 1992*, Government of Western Australia, Perth. *State of the Environment Report 1998*, State of Environment Reference Group, Environmental Protection Authority, Perth.
Vic.	*State of the Environment Report 1988. Victoria's Inland Waters*, Office of Commissioner for Environment, Melbourne.
	1991 State of the Environment Report. Agriculture and Victoria's Environment Resource Report, Office of Commissioner for Environment, Melbourne.
NSW	*New South Wales State of the Environment 1993*, Environment Protection Authority, Chatswood. *New South Wales State of the Environment 1995*, Environment Protection Authority, Chatswood.
SA	*The State of the Environment Report for South Australia*, Environmental Protection Council of South Australia, Government of South Australia, Adelaide, 1988. *State of the Environment Report for South Australia 1993*, Department of Environment and Land Management, Adelaide. *State of the Environment Report for South Australia 1998*, Environment Protection Authority, Adelaide.
Tas.	*State of the Environment Tasmania. Volume 1. Conditions and Trends 1996; Volume 2. Recommendations, 1997*, State of the Environment Unit, Land Information Services, Department of Environment and Land Management, Hobart.
ACT	*Australian Capital Territory State of the Environment Report ACTSER '94*, Office of the Commissioner for the Environment, ACT.
Aust.	*State of the Environment in Australia* and *State of the Environment in Australia: Source Book*, Department of Arts, Heritage and Environment, Canberra, 1986.
	Australia's Environment Issues and Facts, Australian Bureau of Statistics Catalogue No. 4140.0, Commonwealth of Australia, Canberra, 1992.
	Australia State of the Environment 1996, State of the Environment Advisory Council, CSIRO Publishing, Collingwood, Vic.
Qld	Due end of 1999.

activities (pressures) may alter the environment (state) which in turn engenders societal reactions (response) (Fig. 8.4). SoE Reports are seen as vital in enabling benchmarking to occur—to track environmental change in time and space in relation to reliably defined baselines. To this end, the development of a set of national environmental indicators is seen as a key.

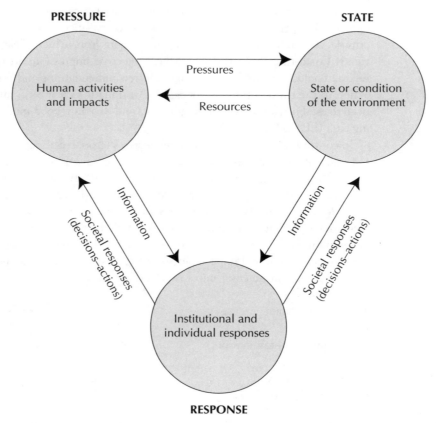

PRESSURE

STATE

Human activities and impacts

State or condition of the environment

Pressures

Resources

Information

Societal responses (decisions–actions)

Information

Societal responses (decisions–actions)

Institutional and individual responses

RESPONSE

Fig. 8.4   The OECD pressure–state–response reporting model as modified by the State of the Environment Advisory Council (1996, p. 1.7).

ENVIRONMENTAL QUALITY INDICATORS

The objectives of environmental quality indicators are to provide repeated snapshots of environmental conditions in order to identify trends, thereby providing a basis for environmental planning and management designed to maintain and improve environmental quality. The following discussion draws on Conacher (1998): see also Bossell (1999) for a discussion of theory, method and applications.

The first State SoE report was the 1988 South Australian study (Table 8.3) (Environmental Protection Council of South Australia 1988). That report included an historical perspective and then covered air, water, land, forest and wildlife resources, aquatic environment and resources, population and health, waste, noise, agriculture, energy, transport, mining and extractive industries, heritage and leisure activities. It was much more comprehensive than the Australian SoE report published two years previously.

Of interest, however, was the absence of any methodological discussion in general and, in particular, of the use of environmental quality indicators. Thus the second SoE report for the State (Community Education and Policy Development Group 1993) was somewhat unsatisfactory, since clear baseline quality criteria were not established in the first report from which clear, objective trends (as distinct from the somewhat subjective impressions in the second report) could be identified. Amongst the recommendations and priorities for action in the third SoE report for South Australia is a clear recognition of the need to establish an agreed set of indicators and to obtain good-quality supporting data (EPA SA 1998:xii).

In contrast, the much more comprehensive Victorian SoE report on *Agriculture and Victoria's Environment* contains a detailed discussion on methodology (Office of the Commissioner for the Environment 1991:Section 3). The discussion includes: the use of indicators for monitoring agricultural impacts; requirements for monitoring (trends, baselines, frequencies, spatial sampling design); a suggested monitoring program for Victoria (including the number and location of sampling sites and a pilot monitoring network), and the modelling and reporting of agricultural impacts, including the use of a GIS. The indicators of agriculture's environmental impacts are presented in some detail. They cover impacts on:

- flora and fauna;
- soil;
- chemical contamination;
- surface waters;
- groundwater, and
- additional indicators.

For each of the above impacts, both 'key indicators' and 'secondary indicators' are presented. For example, the 'soil salinity' impact has the key indicators of soil, EC sodium chloride, salt-tolerant plants (species/abundance, distribution, diversity) and watertable levels; and secondary indicators of area of salt-affected land, area of salt-prone land, area of irrigated land, and area of recharge zone vegetated and/or under natural forest. Monitoring requirements are presented for a range of indicators and consider: the minimum sampling period to establish a baseline; the minimum sampling period for trend analysis; the minimum sampling frequency for trend analysis, and whether the sampling sites should be extensive or intensive.

This excellent example does not seem to have been followed by the SoE surveys published in WA (1992 and 1998), the ACT (1994), NSW (1993, 1995) or nationally (1996) (Table 8.3). Tasmania's 1996 SoE report contains useful moves towards developing quality indicators, and there have been concerted moves towards the development of a national system of key indicators for use in SoE reporting (ANZECC 1998)—discussed further below. However, the next development in Australia was a workshop on environmental indicators at a catchment scale.

Catchment Indicators

A *National Workshop on Indicators of Catchment Health* was held in Adelaide in 1996. The workshop was sponsored by CSIRO, PISA (Primary Industries South Australia), RIRDC (Rural Industries Research and Development Corporation) and the Cooperative Research Centre (CRC) for Soil and Land Management. Two publications were associated with the Workshop. One, on *Indicators of Catchment Health* (Walker and Reuter 1996), is part of CSIRO's Multi-Divisional Program on Dryland Farming Systems for Catchment Care, and was published to assist the Landcare movement in Australia's dryland farming regions and their advisers. The second was the *Proceedings of the National Workshop*: it comprises contributed papers, outcomes and resolutions, workshop summary and a national directory of work on indicators (Anon. 1997).

The *Indicators of Catchment Health* book (Walker and Reuter 1996) is directed at landholders, catchment groups, catchment and land protection boards and rural communities. It aims to provide these individuals and groups with simple yet proven methods for measuring and monitoring the condition of the land and water resources in their catchments. It is a useful and effective practical manual, not a scientific document aimed at researchers. Following introductory sections outlining the indicator approach and means of recording the data, the main chapters describe a relatively small number of key indicators which can be measured readily by the above groups. These indicators are grouped into the following:

1   indicators of farm productivity, financial performance and product quality;
2   morphological, physical, chemical and biotic indicators of soil health;
3   physical, chemical and biological indicators of water quality, and
4   indicators of landscape integrity (the *minimum* set of indicators are considered to be bare soil, exotic weeds, tree cover and depth to water table).

Social indicators are notably absent.

In contrast, social indicators were included by the participants of the National Workshop (Anon. 1997), who also distinguished between indicators at farm and catchment scales. The indicators are listed in Table 8.4, which incorporates additions proposed after the Workshop.

Responses by delegates to a post-Workshop questionnaire indicated several reservations about the Table 8.4 list:

1   a smaller and proven list is needed, from which sub-sets can be extracted for local use;
2   the list is a start but should be treated with caution. A more rigorous evaluation is needed to develop a definitive list;

Table 8.4   **Indicators of catchment health suggested during and following the National Workshop on Indicators of Catchment Health (from Anon. 1997, Tables 1 and 2).**

	Farm Scale	*Catchment Scale*
Productivity Indicators	• % potential yield • Farm business profit/ha • Assessment of land-use capability • Grazing performance • Indicators of other farm enterprises (e.g. horticulture) • Farm inputs (chemicals, irrigation) and productivity	• % potential yield • Quality of water yielded from catchment • Assessment of land-use capability • Matching land-use practices to land capability (e.g. slope, climate, land type)
Product Quality Indicators	• Chemical residues in products • Chemical residues in soils • Records of chemical use • Achievement of industry standards for product quality	• % total production meeting market specifications for pollutants and chemical residues • Records of chemical use
Financial Performance Indicators	*Family profit* • Disposable income per family • Non-farm income *Farm productivity* • Farm income/ha/100 mm growing season rainfall • Operating surplus/land value • Operating costs/income *Resource use* • Return on capital • Land value/family • Financing cost/income	• Trends in asset value of community business • Gross value of production (GVP) for all catchment industries, including non-farm sector • Trends in asset value and GVP for sectors operating within a catchment • Viability of catchment industries • Diversity of wealth generated within a catchment • Financial equity derived from being innovative
Social Indicators	• Extent of isolation • Level of education • Level of stewardship (e.g. managerial skill) • Farm size	• Community demographic statistics • Distribution of income • Community infrastructure and its age • Expenditure on integrated catchment management including commitments from urban dwellers • Scope of leisure activities • Number of farmers and other people living in rural towns • Trends in employment statistics • Diversity of industries within catchments (e.g. tourism) • Number of businesses within catchment • Number of family breakdowns • % children returning to farm after school • Number of Landcare/farmer groups existing in catchment
Soil Health Indicators	Morphological indicators • Soil structure • Soil depth • Soil texture	• Soil fertility status • Soil erosivity • Waterlogged soils • Water table depth and EC

Table 8.4 (cont.)	**Indicators of catchment health suggested during and following the National Workshop on Indicators of Catchment Health (from Anon. 1997, Tables 1 and 2).**	

	Farm Scale	*Catchment Scale*
	• Visual observations of soil surface crusting/sealing and pugging • Soil erosion Physical indicators • Soil colour • Soil dispersion and slaking • Soil strength • Water infiltration rate • Degree of water repellence Chemical indicators • Soil pH • Total N and C • Soil organic carbon • Soil P and trace element status • Exchangeable sodium % (ESP) • Soil nutrient toxicities • Nutrient removal in farm produce Biological indicators • Presence of roots in soil profile • Earthworms • Cotton strip assay • Soil-borne root disease assessments • N2 fixing micro-organisms • Biodiversity, activity and abundance of soil organisms	• Trends in water use efficiency • % vegetation cover • Compliance between land-use practices and soil and land capability • Soil loss assessments (e.g. Universal Soil Loss equation)

Farm and/or Catchment Scale

Water Quality Indicators	*Chemical* • Electrical conductivity (EC) of surface and ground waters • Depth and EC of water table • Stream turbidity and pH • Stream N and P loads *Biological* • Signs of eutrophication (e.g. algal blooms) • Number and diversity of macro-invertebrates • Biological activity and coliform levels • Water temperature *Catchment attributes* • Prior rainfall • Water flow rate and volume • Spatial extent of riparian vegetation • Salt storage estimates within landscape (predictive tool)

Table 8.4 (cont.)	Indicators of catchment health suggested during and following the National Workshop on Indicators of Catchment Health (from Anon. 1997, Tables 1 and 2).
	Farm and/or Catchment Scale
Landscape Integrity Indicators	*Vegetation cover* • % tree cover (or % original vegetation cover for grasslands) • Number of dead trees • % surface cover • Diversity and condition of vegetative cover • Level of weed infestation • Pasture density • Spatial fragmentation of native vegetation • Biodiversity/abundance of native species *Productivity* • Trends in productivity/ha/100 mm effective rainfall • Soil resource • Measure of soil loss or erosion rate (e.g. erosion pin) • Land condition (risk assessment) *Water resource* • Depth to water table • Riparian zone dimensions

3   the social indicators need to be reassessed by social scientists;

4   the list of financial performance indicators also needs to be assessed further, for several reasons, and

5   more extensive discussion is needed on catchment indicators and on indicators for irrigated agriculture.

Other indicators and indicator approaches were also discussed and presented in papers published in the Proceedings of the Workshop. They included:

• a set of national indicators being assessed by the National Collaborative Project on Indicators for Sustainable Agriculture;

• indicators for community and council use;

• indicators used by the National Waterwatch program;

• indicators of land degradation relevant to southern NSW;

• farm economic indicators used by the FM500-FAST Project (Farm Management 500 and Sustainable Technology) in southern NSW and Victoria, and indicators for risk assessment;

• industry-based indicator kits developed by the WA Land Management Society and Charles Sturt University, and

• indicators used by a selection of community groups and State agencies operating in dryland regions of Australia.

Not surprisingly, delegates broadly supported a nationally co-ordinated effort on issues related to indicator technology, and there was 'very strong support' for establishing a National Steering Committee to guide and progress future directions on indicators of catchment health. The Committee would need to be small, representative of the interests of all stakeholders, endorsed by SCARM

(Standing Committee on Agriculture and Resource Management) and possibly work as a new subcommittee under the auspices of the Sustainable Land and Water Resources Management Committee (SLWRMC) of SCARM. The cynic might be forgiven for detecting the hand of the bureaucrat in the above. Draft terms of reference for the proposed Steering Committee are proposed in the Proceedings (Anon. 1997:16).

ANZECC's Core Environmental Indicators

More recently, an ANZECC (1998) discussion paper proposed a draft set of key indicators for reporting on the state of the environment, adapted from the OECD (1994) pressure–state–response model. One concern is to improve the effectiveness of environmental reporting and to develop indicators more appropriate to the Australian environment. It is expected that they will be relevant at both national and State levels, with core indicators able to be supplemented in each jurisdiction by additional indicators to accommodate particular management, scale or environmental issues as required. The need to monitor the core indicators consistently is crucial in both existing and new programs. However, some important variables are not measured because of data limitations or scientific uncertainty regarding the accuracy of measurement techniques. In setting out its criteria, the ANZECC report makes particular note of what is required. For example, indicators need to be: scientifically credible; useful and robust in tracking trends; readily interpretable; relevant to policy/management needs, and cost effective. They are also to be linked to and not duplicate other work, such as the Land and Water Resources Audit, the ABS environmental accounting project, and National Environmental Protection Measures (NEPMs).

Seventy-two core indicators are described under six themes: the atmosphere; biodiversity; the land; inland waters; estuaries and the sea, and human settlements. A seventh theme, natural and cultural heritage, is being developed separately. Core, key or primary indicators are those which are considered to be the most sensitive in providing a picture of change. In general they reflect ecological processes and can be subdivided into physical, chemical and biological indicators. Secondary indicators are less sensitive and provide additional information on environmental, social and economic inputs or outputs and may help to explain causal links between human activities and environmental impacts.

Many indicators overlap amongst the themes and are often complex. The report also notes that some important condition–pressure–response indicators could not be developed.

Geoindicators

Another potentially useful approach, developed in Canada by the Commission on Geological Sciences for Environmental Planning of the International Union of Geological Sciences, has been termed *Geoindicators* (Berger and Iams

1996; http://www.gcrio.org/geo/title.html). Geoindicators are employed as a means of monitoring rapid environmental changes in various ecosystems. Standard techniques used in geology, geochemistry, geomorphology and hydrology are incorporated as well as other earth-science evaluation methods such as seismicity, groundwater quality assessment and palynology. Twenty-seven indicators have been described to combine various biological, climatic and socioeconomic factors. Some of the indicators measure: coral chemistry and growth patterns; slope failure; changes in lake levels; dust storm magnitude, duration and frequency; glacier fluctuations; stream sediment storage and load, and so on. Each indicator is described in terms of its process, significance to the environment and society, causes, measurement (scale, monitoring frequency, limitations), applications and thresholds.

COMMENT

In his overview of the Australian catchment indicators workshop discussed above, Williams (1997) located indicators in the 'Monitor' part of the PRIME process of social change: Plan, Research, Implement, Monitor and Evaluate. However, Williams' later comments indicate that indicators have a broader and more important role. He noted that most indicators unfortunately are static measures, whereas from a biophysical perspective, the preferred indicator is one from which predictions can be made in time and space. This implies a move from condition to trend, which is clearly one of the objectives of State of the Environment reporting and, of course, of planning as well as research. Such a move involves a shift from symptom to system function, which requires an understanding of the processes and drivers of the agro-ecosystem as well as its properties.

Williams noted that the condition or state of a system is determined by the drivers of the system (that is, climate and management practices) which act through system processes (such as the laws of water flow) and the system properties (such as the soil's hydraulic properties). If indicators are system properties they can provide little information about the condition of the system. If prediction is required then researchers, planners and managers need to move from symptom to system function and then to system correction and repair. The language used by Williams will be readily recognised by those familiar with Chorley and Kennedy's (1971) key work, *Physical geography: a systems approach*, later elaborated in Bennett and Chorley (1978). There, very similar distinctions were drawn between system morphology, system processes ('cascading systems') and the potential for system managers to manipulate the system by identifying the key decision points (control systems). The later book goes on to deal with 'soft' systems (cognitive and decision-making systems), complex systems (physico-ecological and socioeconomic) and systems interfacing. Thus indicators are only one element in the construction of quantitative models of agro-ecosystems.

For planning purposes at least, environmental assessment needs to progress further than the search for quality indicators. The group of methods which developed under the heading of 'land evaluation' clearly overlaps with the more holistic of the ecological approaches of environmental assessment; but they were also explicitly intended to place a stronger emphasis on the 'land' component of the biophysical environment. These approaches are discussed in the following chapter.

CHAPTER 9

RURAL LAND EVALUATION

As indicated in previous chapters, comprehensive regional planning is necessary if environmental quality is to be maintained and improved. Such planning requires appropriate information and databases; on population, transportation, settlements, the economy and land use. It also requires information on the biophysical environment—on geology, soils, biota, climate, landforms and hydrology: that is, land resources. In other words there is a need to evaluate or assess the land to determine its capacity to support various land uses or biophysical changes.

Unlike the previous chapter with its broader 'environmental' focus, the emphasis of land/resource evaluation and appraisal is a utilitarian one: that is, to determine the potential for land use (refer Chapter 1). This chapter traces the increasing levels of sophistication in land evaluation techniques, anticipating the complexity of land capability and, especially, suitability techniques in Chapter 10 and integrated regional planning methods in Part 5.

Thus, land evaluation is a link between land resource surveys and land-use planning. Qualitative and quantitative data are collected on various biophysical, social and economic attributes of the land and manipulated in various ways to secure information on expected productivity, sustainability, labour requirements, gross margins and so on. Various techniques can be used in the evaluation process, including Geographic Information Systems (GIS), the Automated Land Evaluation System (ALES) and LUPIS (see Siroplan, this chapter). Data are derived from digitised climate, soils, hydrology (and other) information, from aerial and satellite imagery and from ground-based soil, vegetation and land-use surveys. Originally, land evaluation techniques were strongly focused on biophysical resources, but the need to assess social and economic factors to establish links between land condition and land use became increasingly recognised (FAO 1976, 1993b).

Several approaches are used in land evaluation, with selection of the preferred method being determined by the purpose of the exercise. The methods may be grouped under the following headings: land inventories/resource appraisals; land systems (including the South Coast study and SIROPLAN);

carrying capacity; landscape aesthetics (considered in this chapter), and land capability and land suitability evaluation (discussed in Chapter 10). There are several excellent texts dealing with the topic of land evaluation (including B. Mitchell 1979; Cooke and Doornkamp 1990; C. Mitchell 1991; Davidson 1992), and these methods are therefore considered only briefly here.

INVENTORIES/RESOURCE APPRAISALS

As the name indicates, these are simple assessments of what is there. Little analysis or evaluation is involved; essentially they are data banks. Perhaps one of the best (and earliest) examples was the very thorough Canada Land Inventory (Environment Canada 1970). As in the Canadian example, inventories may cover very large areas. Other early instances of resource appraisals included soil and geological surveys, and vegetation mapping. The main purpose of these surveys (which are still carried out in most countries) is the methodical collection of data and processing into forms (data sets) for storage or dispersal. Such data are a fundamental requirement for sound land-use planning and management processes.

In Australia, the 1920s to the 1940s was a period of extensive land and resource appraisal for soils, forests, climate, fisheries, minerals, water and land capability, with some activity continuing to the present day, although usually much more intensively. The appraisals were encouraged by the establishment of government statutory bodies (departments of forestry, agriculture, fisheries, wildlife, mines) and the national research organisation, the Commonwealth Scientific and Industrial Research Organisation (CSIRO). As a result, Australia now has a wide range of land and resource directories and inventories, though coverage is still patchy; some are listed in Table 9.1 (see also Appendix 3). Many of these data sources are subject to constant update, and have become increasingly sophisticated with regard to their collection, storage, evaluation and accessibility. Other sources of data include extensive rangeland monitoring programs which have extended over several decades using multiple methods of ground monitoring, aerial photography and satellite imagery (Joss et al. 1986; Graetz et al. 1982, 1992; Pringle 1991; Pickup and Chewings 1996). Such data have been used to describe rangeland condition and to formulate management strategies. The former New South Wales Soil Conservation Service used land resource inventories extensively for many years at both regional and property scales (Christian 1958; Condon 1968) (see Land Systems, below, and Land Capability in Chapter 10). A broadscale reconnaissance survey was used for a State-wide assessment of land degradation (NSW SCS 1989). The State was divided into grids of 5 × 10 km and an estimate of degradation made for thirteen indicators (including erosion, salinity, acidity, soil structural decline and vegetation cover). Data from aerial photographs were supplemented with field inspections and existing data.

The 1975–77 National Collaborative Soil Conservation Study devoted a major part of its report to land resource surveys and appraisals in Australia

Table 9.1   **Some national and State sources of environmental data.**

A. National

ANZLIC	Australia and New Zealand Land Information Council. Operates in the Commonwealth Department of Industry, Science and Resources. It is the Commonwealth's primary source of land information responsible for: policy, standards and co-ordination associated with the delivery of national and international land information programs; national mapping, maritime boundaries, remote sensing and geodesy; and development and implementation of ASDI (see below) (www.auslig.gov.au).
ERIN	Environmental Resources Information Network: operates through the Commonwealth Department of Environment and Heritage and provides links to Australia-wide environmental information. It maintains databases on air, biodiversity, coasts, geographic heritage, industry, mining, monitoring and protected areas, and operates environmental directories (ASDD and Green Pages) (www.erin.gov.au).
AUSLIG	Australian Surveying and Land Information Group is the Commonwealth authority on surveying, mapping and land information and provides information through products or sources such as GEODATA and NATMAP (the national resources atlas series on soils, climate, minerals, agriculture, land use, water, etc.). It is also the focal point for ASDI and works with the Commonwealth Spatial Data Committee, ANZLIC and industry groups.
ACRES	Australian Centre for Remote Sensing (Department of Industry, Science and Resources) is the national centre for the reception and processing of images and data sets generated from SPOT and LANDSAT satellites.
ASDI	Australian Spatial Data Infrastructure is a national initiative to maintain and provide better access to national geographical (spatial) information. It operates a national data directory (ASDD). Each State is to maintain a directory of key data information and how it can be accessed (see below).
ASDD	Australian Spatial Data Directory is a component of ASDI. The directory provides search interfaces and geospatial dataset information to governments and the community. It has nodes from Environment Australia, AUSLIG, BRS, ACT, SA, Vic. and WA.

B. States

Most States operate spatial databases such as: Queensland's LRIS (Land and Resource Information System) and Tasmania's LIST (Land Information System). WA's WALIS (WA Land Information System) is a consortium of over 20 agencies. It contains a Land Information Directory (LID), a computerised Internet index on land, geographical and aerial information for WA. It has links to the national land database ASDD.

(Hallsworth 1978; Department of National Development 1979). It noted that over 900 land resource appraisals (concentrating mainly on soils) had been catalogued prior to 1976, although there were many other surveys covering geology, flora and fauna and reconnaissance types. Most surveys were conducted by government agencies, mainly for agricultural purposes. After the mid-1960s, the focus shifted from land development to land management. The Collaborative Study emphasised the importance of land resource surveys to planning, land management, conservation and policy formulation, and called for improvements in data handling and agency co-ordination.

Under the National Landcare and NHT programs, volunteers use kits to obtain basic data on the health of streams, rivers and other wetlands (Waterwatch, Rivercare) and, at the individual level, farmers have been encouraged to use monitoring and data collection and evaluation as aids in property and catchment management. Through various agencies, the Australian Collaborative Land Evaluation Program (ACLEP) is gathering, reviewing and processing a wide range of land information into more standardised and accessible forms (discussed below). An inventory of Australia's vertebrates is operated by CSIRO's Division of Wildlife Ecology and includes data on taxonomy and species, and a tissue bank.

The Australian Bureau of Statistics (ABS) is an important collector of national data. But environmental data from the Bureau were limited until the establishment of an Environmental and Natural Resources Statistics Unit in 1991. Using a variety of sources, including some ABS data, *Australia's Environment: Issues and Facts* was published in 1992, presenting land, water, air and settlement data at a broad level with the purpose of providing benchmarks for debate on environmental issues (ABS 1992). An environmental data series was also initiated during the early 1990s, which saw the later publication of an environmental directory and data on environmental attitudes, the agricultural environment, and expenditure on environmental protection (ABS 1996a, b; 1997a, b). The ABS is also investigating ways of facilitating data coding from the census to harmonise with appropriate spatial (geographical) rather than statistical boundaries. To this end, a refinement of SCARM's agro-ecological regions, for example, would be particularly appropriate (ABS 1996a:8–10) and the concept could be further developed for bioregional planning (Chapter 16).

Nevertheless, despite the considerable, collective effort summarised above, much environmental data suffer from many difficulties: the data may be dispersed, unfocused, unorganised, incomplete, out of date, inaccurate or inaccessible (costly or hidden by confidentiality). The latter problem in particular was noted with concern in the OECD's assessment of Australia's environmental performance (OECD 1998). Similar concerns were raised in a data user survey by Hamilton and Attwater (1997). Increasing calls are being made for single or organised data sites such as national data banks or on-line services, and several changes have been made in recent years as the following discussion illustrates.

The IGAE

In 1992, the *Intergovernmental Agreement on the Environment* (Chapter 7) specifically addressed the environmental data problem in *Schedule 1* (Data Collection and Handling) of the Agreement. The Australian Land Information Council (ALIC—now ANZLIC) was to ensure that mechanisms were available to make data more accessible to both governmental and private sectors, and build a national data directory system (Table 9.1). ANZLIC oversees the development and maintenance of comprehensive, natural resource and environmental data sets.

An important follow-up on the need for better environmental information has been the funding, through the Natural Heritage Trust (NHT), of a National Land and Water Resources Audit (not to be confused with the Australian National Audit Office, which audits natural resource management and environment programs—ANAO 1997), and the development of the Australian Collaborative Land Evaluation Program (ACLEP).

The National Land and Water Resources Audit

Commissioned in 1996 by the then DPIE, the audit is to be an appraisal of the status of land and water degradation in Australia, including the environmental, economic and social costs to the nation. It is compiling a compatible and accessible Australia-wide information base on land, vegetation and water, to be completed by 2001, with the objective of facilitating improved decision-making relating to sustainable land management. Funding of $29.4 million was allocated from the NHT.

The first seven approved audit themes, each with its own budget, work plan, and completion dates, have been approved. Commencing in April 1998, the projects are:

1   water availability—surface and groundwater management, availability, allocation, use and efficiency of use;
2   dryland salinity;
3   vegetation cover, condition and use;
4   rangelands monitoring;
5   land-use change, productivity, diversity and sustainability of agricultural enterprises;
6   capacity of and opportunity for farmers and other natural resource managers to implement change, and
7   waterway, estuarine, catchment and landscape health.

Contracts have been let for each theme and the Audit liaises closely with ANZLIC to ensure that the data base assists the development of the *Australian Spatial Data Infrastructure* (ASDI) (Table 9.1). Websites have been established to provide updates on project activities (the Audit home page is www.nlwra.gov.au).

National Land and Water Resources (NLWR) Data Bases

The development of a national information system of compatible, consistent and accessible land, water and vegetation data is one of the six main objectives of the NLWR audit. Using data generated by the audit projects and that of other government agencies, the data are to be made available through a Web-based information system which will also provide analytical and decision support systems. An on-line Natural Resource Atlas is also proposed. Audit data are to be consistent with the distribution network component of the *Australian Spatial Data Infrastructure* (ASDI). The audit will also work closely with AUSLIG and ANZLIC (Table 9.1) to ensure that ASDI activities are closely co-ordinated. The data are to be made available through the *Australian Spatial Data Directory* (ASDD) (Table 9.1) in conjunction with State and Commonwealth information agencies. Two important NLWR data compilation projects are the nation-scale *Land-Use Mapping Project* (below) and *Australian Soil Resources System Project* (ACLEP *Newsletter* 7(3), September 1998).

The *Land-Use Mapping Project* is developing a nation-scale land-use data set for use in national statistics, resource accounting and resource modelling projects. A national land-use map is to be produced by the Bureau of Rural Resources and CSIRO using advanced very high resolution radiometer (AVHRR) data linked to ABS agricultural census data, which will produce fully attributed GIS coverage. The land-use map will be updated routinely as new data become available.

Some Other Land Assessment Programs

National Forest Inventory

The first national scientific report on Australia's forests was published by the Bureau of Rural Sciences in 1998 (National Forest Inventory 1998). With input from Commonwealth, State and Territory agencies to gather, collate and standardise the information, the report describes the broad characteristics, location and ownership of Australia's forests, their status and range of uses. It is intended that the report be updated every five years.

ACLEP

The Australian Collaborative Land Evaluation Program (ACLEP) commenced in 1996 as a national, co-ordinated effort to assess land resources. ACLEP operates through several lead agencies including CSIRO and the National Landcare Program (through AFFA), with in-kind support from State and Territory land resource agencies. A 1998 overview noted that considerable advances in information availability and innovative methods had been made over the previous six to seven years. However, the demand for land resource information far exceeded the capacity to supply it. Various land assessment programs under way around Australia are summarised below (mostly from *ACLEP Newsletter* 7(1), March 1998).

South Australia

In a joint program between Primary Industry and Environment agencies, efforts are being made to apply land resource information to planning at regional and catchment scales and to develop benchmarks for soils. A State Strategy is under way to monitor land condition. Under the terms of the *1989 Pastoral Land Management and Conservation Act*, several thousand monitoring sites on pastoral lands were expected to increase to 7000 by the year 2000 as part of a land assessment program. Using permanent photo sites and other techniques, a Land Condition Index (LCI) has been used to determine the average condition of each site.

Western Australia

A long-term rangeland survey for WA has an expected completion date of 2001. Each survey has produced a Technical Bulletin, colour land system maps (refer Land Systems, below) and individual station plans. A resource inventory and assessment of soil and vegetation condition is made and sustainable productivity is determined (example: Payne *et al.* 1998).

In the southwestern agricultural areas, soil and landscape mapping at different scales is being carried out. A soil correlator is used to ensure compatibility of survey information. Standardisation of mapping across current and historical surveys is another project, with the aim of producing seamless maps across the State at any level of the soil-landscape mapping hierarchy. An extensive drilling program, supported by remotely sensed geophysics, is under way to monitor groundwater movements and soil and water salinity. Data storage and retrieval are at an advanced stage for site and map unit data. Land resource information projects are undergoing further development, particularly in relation to catchment management. A soils interpretation manual has also been produced (Moore 1998).

Also of interest is *The Farm Monitoring Handbook* (Hunt and Gilkes 1995), which supports a kit designed for farmers and made available through a private farmer organisation, the Land Management Society. The kits include soil-testing tools as well as a penetrometer and erosion pins, step-by-step instructions and recording sheets, and are supported by advisers, workshops and government subsidies. More than 80 kits were in use by 1998. The aim is to collect standardised data on factors such as weather, soil and water conditions and productivity. Farmers may enter the information on a computer and the data are fed into a regional database. Aggregated data are then used to complement remotely sensed catchment and regional data. Farm condition and trends can be monitored and appropriate practices implemented to deal with identified problems (Anon. 1998).

Tasmania

A Land Information System Tasmania (LIST) is being developed. An integrated system, LIST covers all aspects of land information with core data being digitised. A secure, commercial Website to access land information data is to be created. A test kit for farmers to monitor soil condition has been released. Some land capability studies are complete with new ones under way to support the State Policy on the Protection of Agricultural Land. Other projects cover an assessment of salinity, monitoring of groundwater, and bench marking of Tasmania's soils.

Victoria

Detailed reports on the condition of Victoria's land resources were provided in Inland Waters and Agriculture SoE reports (Office of the Commissioner for the Environment 1989). A considerable restructuring of environmental agencies in recent years has seen the disappearance or reconfiguring of some projects, and heavy budget and staff cuts to DNRE in 1998 placed further projects in jeopardy.

The Catchment Management and Sustainable Agriculture Division within DNRE has been co-ordinating the provision of integrated information to support the management of Victoria's natural resources. This is to be achieved through the delivery of: co-ordinated land assessment (soil–land inventory, network of reference sites and adoption of national standards); co-ordinated development of essential information (soils and landforms, land system mapping, land suitability and capability, adoption rates and land use); co-ordinated monitoring and evaluation of catchment condition; strategic interpretation and analysis (such as enhanced resource assessment, economic bench marking); information packaging and dissemination (such as Websites); a decision support system for integrated management and sustainable agriculture, and linkages to national initiatives. The former Department of Conservation and Environment also produced a soil monitoring kit (McGuiness 1991).

A specialist provider, the Centre for Land Protection Research (CLPR), has a major role in survey, spatial analysis and reporting (land resource mapping, land capability and suitability assessment). CLPR is also developing an Enhanced Land Systems mapping and information framework, building on existing land systems, redefining nomenclature and reviewing a State-wide geomorphic framework. The methodology incorporates recent land and soil information, using techniques such as radiometric interpretation and digital elevation modelling. Seven major morpho-tectonic divisions have been recognised with 29 second order units (combinations of tectonics, geology and geomorphic processes such as weathering intensity, degree of dissection and fluvial and aeolian deposition). Boundaries are being refined and the enhanced geomorphic framework remains as a suitable 'umbrella' for spatially describing the nature of the land at 1:100 000 or 1:250 000 scales, as well as being an integral part of the revised Land Systems of Victoria (see below).

A Victorian Resource Atlas project is under way to present land information in both hard copy and electronic forms. A major focus will be soil and land assessment as an aid in determining agricultural suitability in conjunction with other biophysical, economic and infrastructure information. Various reference groups have been established to develop a Strategic Workplan for Land Resource Assessment 1995 – 2002.

Northern Territory

Parts of the NT were covered under land surveys during the 1950s and 1960s but they were often not completed. The Department of Lands, Planning and Environment has conducted land surveys since 1983, more recently incorporating satellite imagery. A demonstration CD-ROM on land resource assessment has been released as an example of information presentation possibilities.

Queensland

The development of better land-resource assessment methods for Queensland, with ACLEP assistance, is taking place under an Enhanced Resource Assessment (ERA) project. Initially a co-ordinating project to test new approaches such as conceptual modelling for viability in agency land resource surveys, testing has been extended to cover larger areas. Complying with ANZLIC guidelines, the Queensland Spatial Data Directory maintains a metadata directory of spatial and natural resources information, geological surveys, planning details, topographical mapping and so on, with the data being submitted by State agencies, local governments and private industry.

LAND SYSTEMS

The basis of this approach is to use data bases and other sources to provide objective assessments of land resources, initially for agriculture but more recently as frameworks for regional or national planning. The simplest, early approaches used aerial surveys with some basic ground truthing and were limited in their application. Later systems were more comprehensive.

Land systems surveys conducted by CSIRO (Christian and Stewart 1968; Stewart 1968) commenced in Australia in the late 1940s and continued to the 1970s, later refined by government agencies in some States (for example, Rowan 1990). Similar surveys were carried out by the British Ministry of Overseas Development, particularly in Africa (Land Resources Development Centre 1966 onwards), and by the French scientific organisation ORSTOM and the Netherlands International Institute for Aerial Survey and Earth Science (ITC) in many countries (Nossin 1977).

The objectives were to provide data rapidly and cost-effectively for areas where existing information was poor or non-existent. In Australia's 'outback' areas for which the Land Systems approach was designed, the only data source in most cases was aerial photography produced for military purposes at scales of 1:60 000 or 1:80 000. Survey results were intended primarily to indicate areas which might be suitable for agricultural or forestry development, and concentrated on the land resource: geomorphology, soils and vegetation (and climate where possible).

Given the nature of the basic data source, in practice initial mapping was heavily influenced by vegetative cover. Areas of similar patterns (of tone and texture) were delimited from monochrome aerial photographs. Details were obtained from ground traverses, with the routes designed to examine on the ground a representative site from each pattern identified on the aerial photographs. A hierarchy of areas of land was identified, from the smallest, the *land facet* (which could be a relatively uniform slope forming part of a catena), to a *land unit* (comprising repetitive sequences of land facets) to *land systems*, comprising land units grouped according to recurring patterns of topography, soils and vegetation. Land systems were mapped and illustrated by three-dimensional block diagrams, with descriptive material below and accompanied by text and photographs.

The first land system study in Australia was of the Katherine–Darwin region (Christian and Stewart 1953), with another early one being that for the North Kimberley region (Speck *et al.* 1960). In both cases small areas were identified as being possibly suitable for agriculture, following which research stations were established to conduct detailed field investigations and experiments. Both research stations are still operational, and in the East Kimberley the eventual outcome was the development of the Ord irrigation scheme. Debatably, these were the only positive outcomes of the series of Land Research reports. While the entire area of Papua New Guinea was eventually covered by land systems surveys, the spatial coverage in Australia was more scattered.

Weaknesses of the technique included: its reconnaissance-type of approach (which was also its strength—the ability to survey large areas of land rapidly); its static coverage (processes were not included); its geomorphic base with a strong, Davisian, denudation chronology flavour (reflecting the period and the British training of many of the scientists involved); its exclusive focus on the biophysical environment, and the crudity of the data. Soil genesis is of little importance for predicting land suitability, and denudation chronology and stratigraphy of virtually none. Nevertheless, land systems may provide a useful basis for later and more dynamically orientated work, especially when there is little background information on the environment or on ecological conditions. In Victoria, for example, geomorphic elements are being refined as part of a land systems review to provide a suitable framework for describing the spatial attributes of land—as mentioned above.

The first CSIRO land systems survey to be conducted in an area already well-developed for agriculture was in the Hunter Valley of New South Wales. This survey was also distinguished by its greater detail—made possible by the smaller area covered and the larger amount of information available (Story *et al.* 1963). But that survey also clearly exposed the limitations of focusing on the biophysical environment. In most areas, planning for future land uses does not take place in 'empty' country but in areas which are already occupied and used. Existing uses, therefore, must be incorporated in any survey designed to assist planning for future land uses.

The South Coast Study

Realisation of this self-evident truth led to the eventual abandonment by CSIRO of its limited land systems surveys (although the basic method is still used widely in various contexts), and the development of two separate but related lines of further work. One was a focus on databases, which eventually merged with the currently thriving area of research into and applications of Geographical Information Systems in land-use planning; and the other was the *South Coast Study* (also in New South Wales) (Basinski 1978). This was a very 'geographical' survey, in the sense that its approach would be familiar to any geographer brought up in the integrated (or holistic) study of regions. Not surprisingly, geographers were prominent amongst the team members—as indeed physical geographers had been in the earlier land systems surveys.

The South Coast Study, a large, multi-disciplinary project, conducted the now familiar, integrated land-systems-type survey of the biophysical environment of the 6000 km² region, but also examined present land use, population and socioeconomic aspects such as demographics, settlements, transport networks, and the economy and social structure of the region. The study explicitly set out to provide a 'rational basis for planning decisions on a wide variety of land uses', as well as investigating methods of providing and analysing biophysical and socioeconomic data for the region (Basinski 1978). This survey was a precursor of 'land suitability' surveys (Chapter 10) and regional planning

(Part 5)—and more particularly of CSIRO's Siroplan approach to land suitability evaluation for integrated, regional land-use planning.

Siroplan

Siroplan was formulated from the results of the South Coast Study. It was also a response to increasing criticism, from more informed and vocal interest groups, of planning processes which were not being accepted by the community. Agencies were also requiring more comprehensive plans but were being provided with fewer resources to undertake them. The increased demands on planners included greater public participation in the planning process, more 'transparent' plans with explicitly stated values and assumptions, plans promoting values, such as multiple use, which were not necessarily a current issue, and plans which could be seen to be consistent with State or local planning policies (Cocks, Ive et al. 1983).

There was also a realisation that data handling and analysing techniques were increasingly beyond the resources and capabilities of local authorities. In addition, local governments were being required to carry out environmental procedures as part of the local planning process, such as those which were required under the New South Wales *Environmental Planning and Assessment Act 1979*. Not only was 'best' or optimal land use sought, but this could be complicated by a range of community and political demands which needed to be recognised. As such, a 'rational' approach is not always appropriate in land use planning.

Thus, Cocks, Ive et al. (1983:335) described Siroplan as:

> a land-use planning recipe which tries to balance the demands of competing land use interests in accordance with the judgments of the clients commissioning the plan. It operates by first developing guidelines or policies which express the attitudes of interest groups on major issues, and then collecting data which allow any proposed plan (allocation of uses or controls or management regimes to locations) to be judged in terms of its effectiveness in achieving these policies. Last, it guides the clients through a progression of incrementally different trial plans towards identifying the feasible and efficient plan which best achieves their political judgments as to the relative importance of the various guiding policies, recognising the various opportunity costs associated with achieving one policy at the expense of others.

There are thirteen steps in the Siroplan methodology (Table 9.2). As stated in the description and steps of Siroplan, land uses or management regimes are allocated to land units in such a way that all interest groups are satisfied to a maximum extent. Wherever 'land use' is mentioned, 'management regime' or 'controls' may be substituted; that is, a set of permitted, conditionally permitted or proscribed land uses is allocated to land units. Reflecting the period, zoning is a more important planning tool in Siroplan than has been the case generally with the more recent planning methodologies discussed in Part 5.

Table 9.2   **Steps in the Siroplan land-use planning method (Cocks, Ive *et al.* 1983:334).**

A   Establishing Terms of Reference and Plan-making Guidelines
1   Confirm the task as being within the class of planning exercises for which the procedure is designed.
2   Identify client, study area boundaries, relevant land uses (or management regimes or controls), relevant interest groups and their demands, available land-use controls, and issues needing to be addressed.
3   Develop guidelines (policies) which suggest ways of zoning/managing various categories of land as sensible responses to the demands and issues being addressed.
4   Formulate measures of policy satisfaction which allow any zoning plan to be evaluated in terms of the extent to which it satisfies any particular policy.

B   Data Collection and Generation of Plans
5   Subdivide area into numerous mapping units which will be used:
    a   as entities against which the data will be collected and recorded
    b   as entities against which the plan will specify particular land-use controls.
6   Collect data judged necessary to allow each measure of policy satisfaction to be calculated throughout all plans.

C   Evaluation of Plans
7   Identify an initial reference plan (assignment of land uses/controls to the mapping units) judged by the client to be feasible (not unacceptable) with respect to the extent to which it achieves each policy guideline.
8   In a direction suggested by the client, search for a plan which can be judged better than the reference plan in terms of policy achievement. If successful, designate this new plan as the new reference plan.
9   Repeat step 8 until time runs out or, as judged by the client, no further attempt should be made to improve the reference plan; that is, the reference plan becomes the accepted plan.

D   Legitimation, Implementation and Updating
10   Seek interest group objections and incorporate if the client approves.
11   Allocate available resources to tasks required for plan implementation.
12   Monitor interest group demands and issues as a basis for deciding when to revise the plan.
13   Revise plan at intervals.

After such preliminary work as establishing the suitability of the project for the Siroplan process, the first step is to convert the demands of the client and interest groups into 'policies' or guidelines (Step 3 in Table 9.2). Types of land-use guidelines are (Baird 1983:33):

- exclusion
- commitment
- preference
- avoidance
- compatibility.

Commitment policies may also be termed retention policies. Exclusion and commitment policies are of the type: 'an area shall/shall not be zoned as Type A', whereas preference and avoidance policies are of the type: 'as far as possible ensure that a number of reefs/shoals zoned P are situated in the southern part of the section', or 'as far as possible avoid zoning attractive swimming areas as P or SR' (Cocks, Baird *et al.* 1983). A compatibility policy may be conditional or unconditional: 'grazing is compatible with residential farmlets'; 'grazing is compatible with rural retreats if the mapping unit contains medium density vegetation' (Davis and Ive 1985).

The area under consideration is divided into land mapping units (step 5 in Table 9.2), and there are numerous ways of doing this. Data are then obtained for each land unit to allow the suitability of the land for each use to be assessed (see Chapter 10, FAO Land Suitability Framework). Following that step, some means must be found of quantitatively measuring the extent to which policies are satisfied. For example, a policy which aims to allow a particular activity in suitable areas will be 100% satisfied if all the 'suitable' areas are 'correctly' zoned, or 50% satisfied if half the areas are zoned in that manner, and so on. An ideal plan is one which achieves 100% satisfaction of all policies. Of course, all areas will have varying degrees of suitability for different uses, not just 'suitable' or 'unsuitable'. At this stage the FAO land suitability framework may be useful, as it can provide a quantitative (if somewhat arbitrary) measure of land suitability.

The policies are weighted according to the client's view of their relative importance (which introduces a source of bias). Zoning plans are evaluated in terms of the extent to which they satisfy each policy multiplied by that policy's weighting, to produce an overall 'score'.

An initial reference plan is generated (step 7 in Table 9.2), which may be quite arbitrary, and the overall score is calculated. Small changes are then made which are considered would improve the plan, and the policy satisfaction re-calculated. These changes can be made either by re-allocating land uses to land units, or by changing the relative weights given to the policies (that is, by increasing the relative importance of those policies which the planner or client would like to see 'better achieved'). If the new plan is an improvement (satisfies policies better), it becomes the new reference plan. This process is continued until it is considered that the plan cannot be improved any further, or time and money run out.

LUPLAN was the name for the various versions of the computer package designed to implement Siroplan, and incorporates a rating and weighting pro-cedure (Cocks, Ive et al. 1983). It has since been replaced by LUPIS (Land Use Planning and Information System). Another relatively recent development is SIRO-MED, which is intended for use where land is under the control of more than one agency (Ive 1992). It operates in much the same manner as Siroplan, except that plans are generated for each of the individual interest groups, and are then combined to create a consensus plan.

As with the FAO land suitability framework, Siroplan can formulate plans for allocating a wide range of uses to different types of land; for example, forests (Booth and Saunders 1985), agriculture of various types (Baird 1983; Davis and Ive 1985) or the uses in a marine park (Cocks, Baird et al. 1983). There were some thirty serious applications of Siroplan to 1983, including one in the Rotorua Lakes district of New Zealand. By 1992 it had been used by local gov-ernments in four Australian States, and adopted by the Great Barrier Reef Marine Park Authority, the New South Wales Crown Lands Office, the National Capital Planning Authority, DCNR in Victoria, and the former Soil Conservation Service in New South Wales (Doug Cocks, pers. comm. to Clare Pullin, 1992).

The Siroplan procedure produces a zoning plan, not a management plan, although zoning plans are incorporated into management plans especially when dealing with public access. Allocation of staff and funds has not been specifically considered by Siroplan, although this could be incorporated in the procedure in a similar manner to the allocation of land uses (compile a list of demands on staff time and funds, develop policies stating which activities would best satisfy those demands, and so on). The procedure is a rationalisation of a process which may be occurring anyway in planning and possibly contains no new elements; but the structure is beneficial in clarifying to both the planners and the community the reasons for decisions. It outlines the need for public participation and states clearly where 'expert opinion' is used (weighting and re-weighting of policies and also data collection to some extent). It is perhaps too strongly 'expert' and client driven. A more technical difficulty is that the land units are not required to be delimited by the criteria used to determine the suitability of those areas for various uses; thus there may well be a spatial mismatch between land uses and land units. Nevertheless, Siroplan may be used for environmental planning and management simply by incorporating environmental planning and management goals into the Siroplan policies. However, it needs to be stressed that Siroplan is essentially a set of procedures (as listed in Table 9.2); it is not a substitute for detailed data on or knowledge concerning land suitability. For this purpose—comparing land qualities with various land use activities—other methods are required.

CARRYING CAPACITY

The above methods of evaluating land have the potential to provide increasingly detailed assessments of land for specific uses, with particular emphasis on agriculture in most cases (with the important exception of Siroplan). However, these methods do not indicate how much use, or the intensity of use, the land can support. Carrying capacity is the concept used to evaluate this: it has been defined as the 'ability of the environment to sustain human populations at given levels of technology and amenity' (Scott 1975, in Mitchell 1979:177). Although derived from agriculture, in relation to stocking rates, the concept has been applied particularly in the area of recreation and park planning. Since pressures on the land are determined by management as well as numbers, technology and the sociological factor of amenity, management should perhaps be added to Scott's definition.

According to Mitchell (1979), research in this field seeks to establish ecological and behavioural thresholds beyond which the biophysical environment deteriorates and user enjoyment declines. The questions then arise as to what facilities and arrangements are needed to increase the carrying capacity of the environment, what would be the costs of doing so and who would pay.

The role of perception is also relevant in assessing carrying capacity. Using a purely agricultural example, one of the organic farmers surveyed by Conacher and Conacher (1982) (near Boyup Brook in Western Australia) had a deliberate policy of keeping his stocking rates low. His objective was to allow the pastures to grow tall, thereby encouraging deeper and denser root development (indeed, root depth is used as an environmental quality indicator). His argument was that eventually his pasture quality was improved, erosion and soil degradation were significantly reduced, there was no need for synthetic fertilisers as the grass roots could extract nutrients from greater depths (although trace elements were applied) and his stock were healthier and more productive, all in comparison with his neighbours. Field observations supported his assessment. On the other hand his neighbours considered that his farm was under-stocked and that he was losing income.

Thus, final decisions may be arbitrary, influenced by perceptions and also value judgments and ethical or philosophical considerations.

Taking a broader view, the concept of carrying capacity is relevant at State and national scales.

To illustrate, and returning to the problem of definition: using a systems approach, the concept can be re-defined as the capacity of an ecosystem to support a particular species without damage to essential functions (which is also relevant to the ESD concept). Such a notion is predicated on biophysical factors and input/output flows of a functioning system; the systems concept of thresholds is also relevant.

This idea can be extended to human populations and measured in several ways. One model used in Australia calculates the productivity or agricultural capacity of land to support a given population at expressed standards of living. By altering certain assumptions, such as changing dietary patterns or trade flows, population numbers may be scaled up or down (Nix 1988). More recently, Newman et al. (1994) analysed eight natural resources in an attempt to calculate Australia's population carrying capacity. Other models focus on the flow of resources, energy and matter into and out of a system, noting in the case of humans that flows (such as global trade) can be some distance from the local area.

Rees (1992) used a model which employs the idea of carrying capacity, but extended to include environmental impacts of resource consumption and waste disposal or discharge beyond a population's immediate vicinity. Using the term 'ecological footprint', Rees calculated the amount of land an average person requires to provide resources and absorb wastes, noting that most developed countries run at an ecological deficit. Unlike some other models, Rees's model comprehensively measures consumption patterns. However, one fault is that it only looks at basic input/output flows and does not quantify social or livability aspects. In this regard, a more dynamic, 'extended metabolism model' is preferred in order to include these human elements (Chapter 3 in State of the Environment Advisory Council 1996). Such a model was employed by Newman and Ross (1996) to examine Western Australia's future carrying capacity.

Carrying capacity has long been a contentious matter in Australia. Nineteenth-century optimists believed in an 'Australia unlimited'—a country whose vast, empty spaces could be tamed under European agricultural practices and feed some 50–100 million people. Such notions persist: as biologist Flannery

(1998) has noted, they are mostly favoured by the 'boosters'—politicians, developers and journalists—believing in technological fixes to cater for future populations in a more politically, economically and militarily powerful nation with a population ranging from 50 to as many as 400 million people. The more restrained are supported by scientific judgments based on Australia's biophysical constraints (Gifford et al. 1975). These were set out as early as the 1920s by geographer Griffith Taylor who, with some considerable prescience, predicted a population of around 20 million by 2000 (Powell 1993b). More recent revisions of Australia's optimum population are not far short of Taylor's estimates—recognising the limits of water, soil, climate, overall productivity and ecosystem vulnerability. These are further supported by demographic studies, resource needs and quality of life issues (Cocks 1992; House of Representatives Standing Committee 1994a; Australian Academy of Science 1995, cited in Flannery 1998).

LANDSCAPE AESTHETICS

The idea of amenity in the evaluation of carrying capacity, particularly in the context of nature reserves or national parks, raises the issue of the aesthetics of landscapes—an idea embodied in McHarg's (1969) now classic text, *Design with Nature*. Again depending on the use to which the land is to be put, people's aesthetic appreciation of landscape is an important element in land evaluation. This applies not only to scenic enjoyment: aesthetics are also reflected very pragmatically in the monetary value of a farmer's land. Aesthetics are, or should be, a major consideration in devising management strategies.

Mitchell (1979) identified three general approaches in evaluating landscape aesthetics; informal approaches, descriptive methods and landscape preference studies.

Informal Approaches

In this approach, 'experts' (who are the experts in a matter such as aesthetics?) strive for consensus about the landscape attributes of an area. Measurements are rare, and choices cannot be explained or defended. According to Mitchell, the US Wild and Scenic Rivers program identified rivers using this approach. The main virtue is that the approach is quick and inexpensive.

Description of Landscapes in terms of their Different Components or in their Totality

This approach has been used by landscape architects, foresters, planners and geographers. The emphasis is on identifying and measuring critical landscape variables, determining their interrelationships and assessing their relative importance. Occasionally the relative quality of a landscape may be compared against some standards or criteria. A basic objective is to produce a map. Examples discussed by Mitchell are Litton's (1972) compositional types of landscape,

Leopold's (1969) criteria for describing the aesthetic character of a landscape, and Linton's (1968) (not to be confused with Litton) composite assessment categories. A modification of Linton's approach was used by O'Connor (1985) to assess the landscape aesthetics, in very simple terms, of the Wooroloo drainage basin in Western Australia (Fig. 9.1).

Landscape Preference Studies

The important feature of these approaches is that the researchers attempt to obtain the *community's* evaluation of landscape aesthetics rather than producing some form of 'expert' evaluation. It may be done indirectly, by inferring attitudes and perceptions from literature or art, or directly by asking people for their views.

Direct approaches have the obvious potential benefits of obtaining the assessment of large numbers of people (thereby approaching a community consensus, if it exists) and of asking people directly for their assessments of the landscape of interest. But there are a number of very real problems.

The Landscape

One difficulty is how to present the landscape for which aesthetic assessments are being sought. A common approach is to show photographs of various scenes, ask respondents to rank the scenes, and then (with varying degrees of sophistication) to identify, rank and weight the various components of the scenes. The obvious drawback of this approach is that only the visual aesthetic sense can be involved in the assessment. Moreover, the scene viewed is only a small, possibly unrepresentative square from the total landscape in which the photograph was taken. Additional difficulties for comparative purposes include the time of day, changing weather conditions and seasonality. Varying qualities of photographic skills may also influence the results: it is well established that unattractive scenes may be presented as attractive, and *vice versa*, by skilful use of the camera. On the other hand, this approach can be used with relatively large numbers of people and it is therefore possible to attempt to obtain representative samples of groups categorised in various ways (age, gender, occupation, interests, address, and so on) according to the objectives of the research.

The alternative, adopted by Cherrie (1999), is to take groups of people to the site being evaluated. For her research, using John Forrest National Park near Perth, the study area was first delimited into land units. Groups of people with different ages and interests were then taken to one or more land units for on-site assessment. One land unit was assessed by all groups for comparative purposes. This approach overcomes some of the above disadvantages of the method of using photographs (except variations in weather, time of day and season for between-group comparisons); and it has the very important benefit of being able to assess non-visual aspects of the aesthetic experience. Thus, Cherrie was able to question the respondents on their responses to sound, smell and touch, as well as the visual aspects, and also their emotional response to the

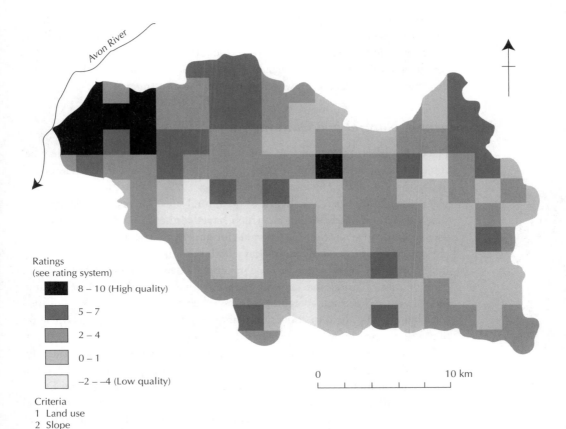

Fig. 9.1   Modified version of Linton's composite assessment categories approach to the assessment of landscape aesthetics, applied to the Wooroloo drainage basin in Western Australia. Source: O'Connor (1985, Fig. 25).

scene being assessed. This advantage far outweighs the difficulty of persuading groups of people to go into the field; but it would be impracticable for more distant areas, unless restricted to people who happen to be visiting the area at the time when the researcher is present.

The visual aesthetic sense is often thought (perhaps mainly by implication) to be the most important of the aesthetic senses. But one of the particularly interesting findings of Cherrie's work was that the four aesthetic senses of sight, sound, smell and touch were weighted approximately equally by all 13 of the disparate community groups who took part in the assessments.

Assessment of Natural and Cultural Values

Places or objects of special importance in the landscape (or in time) may be subject to evaluation in a number of ways, such as the assessments undertaken under the *Australian Heritage Commission Act 1979*. Under the Act, heritage places are natural (such as wilderness) or cultural sites (for example, Aboriginal sites), structures, areas or regions which have 'aesthetic, historic, scientific or social significance or other special value for future generations as well as for the present community'. Many places have both natural and cultural values. Heritage objects are those which provide material evidence of Australia's biophysical evolution, natural and cultural environments, or its historical or cultural life. They may be *in situ* at significant sites (such as archaeological sites or historic buildings) or held in collections (libraries, archives or herbariums) (State of the Environment Advisory Council 1996:9.5).

Various governments in Australia have passed heritage Acts which generally require the preparation of heritage listings. Items are required to meet certain criteria to be included on the lists, such as *The Register of the National Estate* (AHC 1998) (see also Chapter 6), or an assessment of forest areas (many of which are also on the National Register). An example of the latter is the regional assessment model which was developed for Western Australia's south-west forest region in 1991/92, which identified areas, provided data, assessed protection levels and developed conservation and management principles (AHC/CALM 1992).

Much of the data used in the formal assessment of the National Estate derives from extensive data sets assembled from a wide variety of sources, including government departments, industry and private organisations. As an aid to identifying important sites, objects or values for listing, protection and conservation, several other methods are also employed. They include: identification of and consultation with key stakeholders, including community and indigenous groups, through interviews and workshops; use of archival materials, maps, aerial photographs and other remotely sensed data; scrutiny of scientific papers and consultant reports; use of specialists and peer review, and field assessments.

CHAPTER 10

LAND CAPABILITY AND LAND SUITABILITY ASSESSMENT

Land capability and *land suitability* assessment differ from most methods of environmental assessment and land evaluation discussed previously, in that they are specifically conducted in relation to particular land uses. The two approaches are often confused, and in some Australian jurisdictions even used interchangeably; but they are distinct. Land capability assessment focuses on *limitations* to land uses (usually agricultural, covering a broad range of crops at moderately high levels of management), whereas the FAO land suitability framework focuses on the *suitability* of land for specified uses relating to individual crops, use or managerial practice. Levels of suitability can be established at all scales, from individual field to region; and at each scale, suitability may vary with the given or projected management practice (Schaefer and Dalrymple 1999). Land capability assessment is in fact one part of the land suitability assessment process, as will be seen. There are similarities between land suitability assessment and Siroplan; but land suitability assessment, as has been practised, seems to be much more detailed—even though it is a framework (just as Siroplan is a procedure) and not a detailed technique.

LAND CAPABILITY ASSESSMENT

Land capability assessment was devised by the United States Department of Agriculture (USDA) to provide an objective means of evaluating land primarily for agricultural uses (Klingebiel and Montgomery 1961). Unlike other methods, notably the FAO land suitability framework (below), its emphasis is on the presence or absence of *limiting factors* to agricultural land use (Table 10.1). These limiting factors refer to soils, relief and climate, with an emphasis on the first two. Soils are further evaluated in relation to texture, structure, rooting depth, drainage and stoniness, and relief in relation to slope, liability to erosion and liability to flooding. For some land uses, and depending on the objectives of the study, other properties such as load-bearing capacity and shear strength may be included. Each of these properties is then assessed, for a particular area, in terms of the limitations it poses to agriculture and the special management practices required. Where there are no or minor limitations requiring no special management methods, the land is classified as Class I. At the other end of the scale, Class VIII land poses absolute limitations to agriculture. This is the land which is often granted the status of fauna and fauna conservation reserves, and it will be readily appreciated that as a result the plant associations and fauna habitats so protected are unrepresentative of most types of land.

Table 10.1   **The United States Department of Agriculture land capability classification scheme for agriculture. From Klingebiel and Montgomery 1961. Note that each property is further subdivided into characteristics such as good, fairly good or poor. Rooting depth and slope have four such subdivisions, whilst drainage and climate have five. There are only two categories for stoniness.**

Class Physical Limitations			SOILS			RELIEF				CLIMATE
	Texture	Structure	Rooting Depth	Drainage	Stoniness	Slope	Liability to Erosion	Liability to Flooding	General Character
I None or minor										
II Some										
III Severe										
IV Very severe (i)										
V Very severe (ii)										
VI Very severe (iii)										
VII Extremely severe										
VIII Absolute										

A major problem with the land capability approach is its subjectivity: terms such as 'good', 'fairly good' and 'poor' in Table 10.1 need to be quantified to be of any real value, and this can only be done for a given area in relation to a specific land use (or uses). One of the factors which has given rise to a considerable amount of research over the years is 'liability to erosion'. The so-called Universal Soil Loss Equation (USLE) was devised by Wischmeier and Smith (1978) in an attempt to provide an objective means of estimating potential erosion by identifying and defining the main controlling factors. The equation is:

$$A = RKLSCP$$

where:

A = average annual soil loss
R = rainfall factor
K = soil erodibility factor
L = slope length factor
S = slope angle
C = cropping and management factor
P = conservation practice

Terms for the equation, and the relationships between the factors, have now been quantified for much of the United States, but for other environments it is necessary to determine values afresh, requiring a considerable amount of field plot experimental work. A difficulty is the need for accurate data on short-duration rainfall intensities; data which simply are not available in most cases. A surrogate for potential erodibility which is often used is a measure of soil aggregate stability; but this still begs the question of the nature of the land-use practices and rainfall events to which the soil is subjected.

Many countries have adopted the land capability approach to land evalua-tion, usually modifying it to suit their own environments and purposes. A sim-plified and partly quantified version from Victoria is presented in Table 10.2.

One attempt to incorporate landsurface processes in the assessment of land capability compared the process-response units of the nine unit landsurface model with the land capability of those units in part of the Western Australian wheatbelt (Conacher and Dalrymple 1977). The landsurface units are distin-guished by the pedo-geomorphic responses to particular combinations of pedo-logical and geomorphological processes. By careful delimitation of landsurface units in the field, and assessment of limitations to agricultural practices through detailed interviews with local farmers, relationships between the units of the nine unit landsurface model and land capability according to the USDA classi-fication were identified. Combinations of those units over larger areas were then grouped into process-response defined 'land systems', thereby introducing

Table 10.2   **A simplified version of the USDA land capability scheme devised for Victoria (Rowe *et al.* 1981: also see Rowan 1990).**

LAND CAPABILITY CLASSES

Criteria: one or more *physical* characteristics determine land class.

Class I

a)	Soils:	Well developed 'A' horizon.
		Water holding capacity high in terms of crumb structure.
b)	Land:	Level and well drained.

LAND MANAGEMENT
(Expected if land is being used to full CAPABILITY).

No special practices, but this would include:

i)   Fertiliser
ii)   Lime
iii) Return of manure and crop residues
iv)   Crop rotation.

Class II

a)	Soil:	Alkaline and Saline conditions slight to moderate (pH 6 to 8).
		Restrictive drainage (i.e. gleying).
		Subsurface water movements.
b)	Slope:	Gentle slopes less than 5°.

Table 10.2   **A simplified version of the USDA land capability scheme devised for Victoria**
(cont.)   **(Rowe *et al.* 1981: also see Rowan 1990).**

LAND MANAGEMENT
(over and above that of Class I)

Simple Practices

i)   Terracing
ii)   Strip cropping
iii)   Contour tillage
iv)   Rotations involving grasses and legumes.

Class III

a)   Soil: --   Low fertility (e.g. shallow depth, restrictive root zone, low water holding capacity, lack of
                       organic matter).
                       Moderate saline, acid and alkaline conditions
                       pH 5 to 9
b)   Evidence of past erosion (e.g. soil creep, rills, mild wind erosion).
c)   Slope: Moderately steep (5° to 10°).

LAND MANAGEMENT
Intensive practices: Same as Class II but more frequently, and frequent restriction in crop type.

Class IV

a)   Soils: Shallow depth.
                  Low water holding capacity.
                  Severe alkaline and saline conditions.
                  pH less than 5, greater than 9.
b)   Severe past erosion (e.g. gullying sheet wash).
c)   Slopes: steep 10°+

LAND MANAGEMENT

Suitable for occasional, limited cultivation: same as Class III but including *severe* limitation on choice of crop
limited by moisture and erosional hazards.

Class V

AREAS DELIMITED DUE TO NON-EROSIONAL FACTORS.

a)   Stony, rocky soils.
b)   Swamp and marsh.

LAND MANAGEMENT

Not suitable for cultivation: but suited to permanent vegetation.

i)   Economic value restricted to pasture and forestry.
ii)   Improvement perhaps possible (e.g. draining marshes).

Class VI

a)   Bare rock outcrops.
b)   Permanent water cover.

Table 10.2 (cont.)	**A simplified version of the USDA land capability scheme devised for Victoria (Rowe et al. 1981: also see Rowan 1990).**

1	*Weighting:*	Soil characteristics are weighted over 'Evidence of past erosion' and 'Slope'.
2	*Fertility:*	Define in terms of:
		Texture (ploughability)
		Depth (see below)
		Restrictions in root zone
		Water holding capacity
		Organic matter (content)
		Salinity Acidity Alkalinity
		Fertiliser requirements (as per farmer).
		N.B.: None of these has been quantified except pH.
3	*Shallowness of soil:*	
	Referring to depth to impermeable layer—which restricts root development of crops.	

a dynamic component to CSIRO's static land systems descriptions and enhancing the assessment of the land systems' agricultural potential (Fig. 10.1).

As discussed by Schaefer and Dalrymple (1999), by the 1980s land capability had been largely replaced by land suitability. The FAO had published a series of specific guidelines (discussed below). At the global scale, much use has been made of the FAO system to establish agro-ecological zones to map present and potential uses of world land resources. The eventual aim is to build up individual country resource bases (FAO 1993a).

THE FAO LAND SUITABILITY FRAMEWORK

The FAO Framework is based on six principles (from Davidson 1992):

1   land suitability is assessed with respect to specific land uses (the requirements of different land uses vary);

2   inputs are compared with outputs for each land unit (for example, economic costs versus returns);

3   the approach is multi-disciplinary;

4   the evaluation is made with respect to the physical, economic and social context of the area (for example, the cost of labour and the skills of the labour force);

5   land must be suitable for use on a sustained basis (with no or minimal erosion, salinisation, depletion of nutrients or other forms of land degradation: thus, land capability or even an environmental impact assessment may form part of the land suitability evaluation process), and

6   different land uses are compared on an economic basis (costs and returns are compared for each use).

Since the FAO is a 'food and agriculture' organisation, not surprisingly its land suitability framework is orientated towards food production. Circumstances which led to the FAO developing the land suitability framework were improved technology, leading to an increased capacity for land evaluation, and

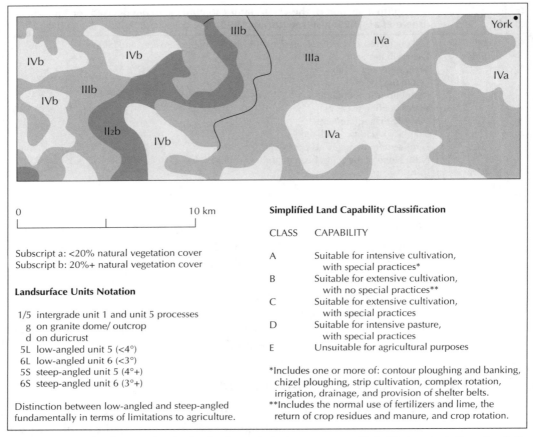

0 _____ 10 km

Subscript a: <20% natural vegetation cover
Subscript b: 20%+ natural vegetation cover

Landsurface Units Notation

1/5 intergrade unit 1 and unit 5 processes
 g on granite dome/ outcrop
 d on duricrust
5L low-angled unit 5 (<4°)
6L low-angled unit 6 (<3°)
5S steep-angled unit 5 (4°+)
6S steep-angled unit 6 (3°+)

Distinction between low-angled and steep-angled
fundamentally in terms of limitations to agriculture.

Simplified Land Capability Classification

CLASS	CAPABILITY
A	Suitable for intensive cultivation, with special practices*
B	Suitable for extensive cultivation, with no special practices**
C	Suitable for extensive cultivation, with special practices
D	Suitable for intensive pasture, with special practices
E	Unsuitable for agricultural purposes

*Includes one or more of: contour ploughing and banking, chizel ploughing, strip cultivation, complex rotation, irrigation, drainage, and provision of shelter belts.
**Includes the normal use of fertilizers and lime, the return of crop residues and manure, and crop rotation.

Fig. 10.1a   Sample of a map showing landsurface systems in an area near York, Western Australian wheatbelt. Land-systems are defined by the relationships between landsurface units (fully defined in Conacher and Dalrymple 1977) and land capability (modified from Klingebiel and Montgomery 1961), as shown in Fig. 10.1b. Refer Fig. 3.2 for location of York.

an increased awareness of, and concern for, the world's dwindling natural resources. By the late 1960s different land capability evaluation systems had been established in most countries, threatening to impede the international exchange of information. In 1970 a working group was established to develop a land evaluation framework which would be applicable to a wide range of conditions. *A Framework for Land Evaluation* was published in 1976 (FAO 1976). Detailed accounts have been published subsequently for rainfed agriculture (FAO 1983), forestry (FAO 1984), irrigated agriculture (FAO 1985) and extensive grazing (FAO 1991), and there is a useful description in Davidson (1992). The present description is modified from the simplified procedure presented in FAO (1993b): *Guidelines for land-use planning*. The procedure sets out to answer the following questions:

• for any specified kind of land use, which areas of land are best suited? and
• for any given area of land, for which kind of use is it best suited?

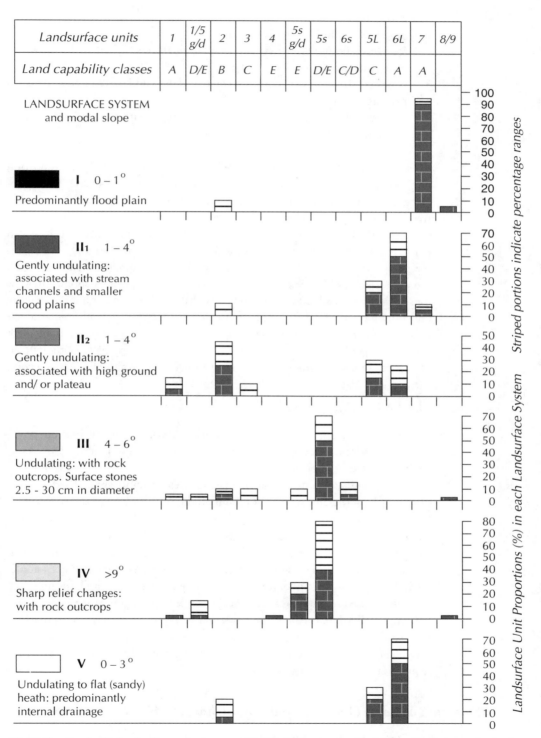

Fig 10.1b   Key for the landsurface systems mapped in Fig. 10.1a, showing the relationships between landsurface units and land capability for each landsurface system. Source: Conacher and Dalrymple 1977, Fig. 13.

The procedure:

- describes promising *land-use types*;
- for each land-use type, determines its *requirements* (for example, for water, nutrients, avoidance of erosion);
- conducts the surveys necessary to map *land units* and describe their physical properties, and
- compares the requirements of the land-use types with the land unit properties to arrive at a *land suitability classification*.

As the title states, the FAO procedure is a framework not a classification method; that is, specific classes are not defined, rather it sets out the structure of the evaluation procedure itself. The framework was developed to be applicable to any project at any scale and under any set of environmental conditions. Thus it is more flexible than the various land capability schemes to which it was in part a response. It differs from land capability schemes in other ways. For example, it uses dynamic land qualities (discussed further below) in the evaluation process instead of land characteristics. It also includes a socioeconomic analysis for suitability instead of being a purely physical classification, and it is explicitly multi-disciplinary. And its emphasis is on the *suitability* of land for particular uses, not the *limitations* as in land capability assessment: the FAO framework could be regarded as being essentially positive in its approach, although an assessment of land limitations does in fact form one step of the FAO framework.

Description of Land-use Types

A land-use type (or, in earlier FAO publications, a 'land utilisation type') describes land use in terms of its products and management practices: an example is presented in Table 10.3.

For reconnaissance surveys at the national level, highly generalised descriptions may be sufficient; such as 'sorghum production', or 'conservation forestry'. However, at district or local levels it is necessary to specify the land use in more detail, incorporating management techniques (mechanised or animal-drawn equipment), the use of fertilisers, the people undertaking the work (individual farmers or communities, agricultural agencies) and land tenure, for example. Such descriptions assist in determining the requirements of a land use; and they may also be used as the basis for extension services or for planning the necessary farm inputs. Further, the land-use types may be proposed modifications to existing uses, or even completely new uses. In these cases it is essential that the people most affected be directly involved in the suitability evaluation and subsequent land-use planning process.

Selection of Land Qualities and Land Characteristics

Land-use requirements are described by the land qualities needed for sustained production. A *land quality* is a complex attribute of land which has a direct

Table 10.3   **Example of a description of a land-use type; from FAO (1993, Table 3).**

Title	*Rice Cultivation by Smallholders*
PRODUCTION Marketing arrangements, yields	Grain for subsistence, surplus sold in local market. Straw fed to draught animals. Average yield, 2.6 t/ha. When water is not limited, wet-season yield may be 4 t/ha and dry-season yield may be 5 t/ha
MANAGEMENT UNITS Size, configuration, ownership	Family-owned plots from 0.2 to 2 ha, usually associated with as many as 4 ha of upland which may be up to 2 or 3 km distant
CULTIVATION PRACTICES AND INPUTS Labour, skill, power, varieties, seeds, agrochemicals	**Labour requirements** from 300 person-days/ha without mechanisation to 150 person-days/ha where buffaloes or tractors are used. Terraced fields need extra labour to maintain bunds and waterways **Power requirements**. Power for ploughing, harrowing and threshing may be provided by two-wheeled tractors. Alternatively, buffaloes may be used for land preparation or all work may be manual. Tractors may reduce tillage time by 60% and total time between crops by 30% **Land preparation** seeks to control weeds, create a good physical medium for rooting and reduce water seepage loss. This is achieved by ploughing or hoeing twice, followed by harrowing under flooded conditions **Recommended varieties**. Varieties are selected locally to suit specific sites and according to season. Growing period must be long enough to span the flood period and allow cultivation and harvesting under favourable conditions **Planting** rates are 20 to 40 kg/ha, seedlings are spaced from 20 × 20 to 25 × 25 cm depending on tillering capacity and length of stalks **Fertiliser**. To replace nutrients removed by a crop of 4 t/ha requires 60 kg N, 30 kg P_2O_5 **Weed control** by maintaining adequate water depth and hand weeding until the crop canopy is closed **Pests and diseases**. Chemicals used to control rice blast and stem borers. Good husbandry and resistant varieties control other fungal diseases
CROPPING CHARACTERISTICS	Rice is grown as a monoculture, one or two crops per year. Fallow land is grazed by draught buffaloes and other domestic livestock
WATER	Most crops are rainfed, with water stored in level, bunded fields. Irrigation, from tanks or by stream diversion, enables a second crop to be grown in the dry season

effect on land use. Land qualities are dynamic and comprised of interactions amongst the measurable land characteristics. Examples are presented in Table 10.4, which also indicates the *land characteristics* associated with each land quality. Thus, land characteristics are measurable attributes of land; often, the interaction of several land characteristics is required to determine a particular land quality. For example, the land quality 'availability of water' in Table 10.4 is determined by the balance between water demand and water supply. The demand is the measurable attributes of potential evaporation from the surface of the crop and the soil; and the supply of water to the crop is determined by the rainfall, infiltration, storage in the soil and the ability of the crop to extract the stored water.

Table 10.4   **Land qualities for rainfed farming. From FAO (1993, Table 4).**

Land Qualities	Land Characteristics which Measure the Quality
Availability of energy	Sunshine hours in growing season, temperature regime
Availability of water	Evaporative demand set against rainfall, soil water storage and rooting conditions
Conditions for ripening	Period of successive dry days with specified sunshine and temperature
Climatic hazards	Frequency of damaging frost, hail or winds during growing period
Sufficiency of oxygen in root zone	Soil drainage class, depth to water table
Sufficiency of nutrients	Soil nutrient levels, pH, organic matter content
Erosion hazard	Rainfall and wind erosivity set against soil cover, slope angle and length and soil permeability
Toxicity	Levels of soluble Al and Fe; pH

Only a limited number of land qualities need be selected for evaluation in most projects. Criteria for selection are:

1   the quality must have a substantial effect either on yields or on the costs of production. Some qualities, such as 'availability of water', affect most kinds of land use. Others are more specific; for example, 'conditions for ripening' is a quality which affects grain crops but not rubber, and

2   critical values of the quality must occur in the project area. If a quality is adequate everywhere, there is no need to include it. For example, most tropical crops are sensitive to frost; but since frosts do not occur in most lowland tropical areas, the land quality 'frost hazard' need not be considered in those regions.

Having selected land qualities, it is then necessary to decide which land characteristics should be used to measure them. For example, the land quality 'erosion hazard' generally requires data on rainfall frequencies and intensities, slope angle and soil properties. Since available data are often less than ideal, compromises have to be reached between land characteristics which most closely define the land quality and those which are less precise but for which data are available. Often, the choice is limited to the latter. But if there are no data on a critical land quality, then surveys need to be carried out or research initiated.

Some land suitability evaluations are carried out directly in terms of land characteristics; for example, by using rainfall instead of availability of water, or slope angle instead of erosion hazard. It may be argued that measurements of land characteristics carry an implication of land qualities, because plants do not actually require rainfall but do require water (which may be obtained from alternative sources). In practice, the FAO suggests that well-conducted surveys using either qualities or characteristics produce similar results.

Mapping Land Units and their Characteristics

Land units are the basis for the diagnosis of problems. They are defined (as in the *land systems* methodology discussed in Chapter 9) as areas of land which are relatively homogeneous in terms of climate, landforms, soils and vegetation—with an emphasis on the word 'relatively': *all* areas of land are hetero-geneous to varying degrees. The assumption is that each land unit internally presents similar problems and opportunities and will respond in similar ways to management. Appropriate land units at the national level may be termed *agroclimatic regions*; at the district level, *land systems* or *soil associations*; and at local levels, *land facets*, *soil series* or other *soil mapping units*. In the land sys-tems land evaluation methodology discussed in Chapter 9, *land units* fall between land facets and land systems in the above scale hierarchy. For land suitability evaluation, the more complex mapping units in the above hierar-chy need to be subdivided into mapping units which are no coarser than 'land units' or soil series.

The need for new surveys is determined by the requirements of the land suit-ability assessment and land-use plan and the nature and accuracy of existing information. Previously conducted soil surveys, agroclimatic studies, forest inven-tories and pasture resource inventories are major sources. For land-use planning at the national level, reconnaissance surveys at scales of about 1:250 000 may be adequate; district-level planning will need at least semi-detailed surveys at scales of around 1:50 000, and accurate work requires scales of no less than 1:10 000.

Natural resources surveys are time consuming and will delay the planning process. However, the FAO stresses that experience has shown that to proceed with land development projects in the absence of adequate resource data can lead to disasters for both production and conservation. Resource surveys and studies of land-use types can be carried out simultaneously, with frequent exchanges of information.

Setting Limiting Values for Land-use Requirements

Limiting values are the values of a land quality or land characteristic which determine the class limits of land suitability for a particular use. The standard FAO land suitability classes are presented in Table 10.5.

The first and most important decision is to separate land which is suitable for the designated use from that which is not. Important criteria on which this decision is based are 1) sustainability and 2) the ratio of benefits to costs.

1 *Sustainability*. The land should be able to support the land use on a sus-tained basis; that is, the use must not progressively degrade the land. It should be noted that many changes of land use cause an initial loss of land resources, as when forest is cleared for tea plantations or for arable farming. Habitat and wildlife are lost as well as soil and plant nutrients. However, after that initial loss, a good level of productivity must be maintained by the new system of land management. For example, if soil erosion is not controlled, the new land-use type cannot be sustained. According to the land-use type, the upper limit of the land quality 'erosion hazard' might be set in terms of slope, as follows:
    * plantation tea, high level of management: 20°
    * smallholder tea, average level of management: 15°
    * rainfed arable crops with simple soil conservation practices: 8°.

Table 10.5   **Structure of the FAO land suitability classification. From FAO (1993, Table 5).**

S	SUITABLE	This land can support the land use indefinitely and benefits justify inputs
S1	Highly suitable	Land without significant limitations. Includes the best 20–30% of suitable land as S1. This land is not perfect but is the best that can be hoped for
S2	Moderately suitable	Land which is clearly suitable but which has limitations that either reduce productivity or increase the inputs needed to sustain productivity compared with those needed on S1 land
S3	Marginally suitable	Land with limitations so severe that benefits are reduced and/or the inputs needed to sustain production are increased so that this cost is only marginally justified
N	NOT SUITABLE	Land that cannot support the land use on a sustainable basis, or land on which benefits do not justify necessary inputs
N1	Currently not suitable	Land with limitations to sustained use which cannot be overcome at a currently acceptable cost
N2	Permanently not suitable	Land with limitations to sustained use which cannot be overcome

Examples of classes in the third category
S2e Land assessed as S2 on account of limitation of erosion hazard
S2w Land assessed as S2 on account of inadequate availability of water
N2e Land assessed as N2 on account of limitation of erosion hazard

Note: there is no standard system of letter designation of limitations; first-letter reminders should be used where possible.

2   *The ratio of benefits to costs*. The use should yield benefits which justify the inputs. The user has to make a reasonable living from the land, and local experience will usually be a good indication. Alternatively, a financial analysis can be undertaken.

It is then possible to distinguish up to three classes of suitability, although this is not always necessary. Land classed as highly suitable is the best land for the specified use; moderately suitable land is clearly appropriate for the use but has limitations, while marginally suitable land falls close to (but above) the limit for suitability (Table 10.5). Unsuitable land may also be subdivided on the basis of whether the current limitations may be overcome in the future at an acceptable cost.

Construction of a table of limiting values for each land suitability class is a central operation in land suitability evaluation (Table 10.6). To construct such a table, information is needed on the performance of a land-use type over a range of sites, either from trials or from the experience of land users. Land requirements for several individual crops may be combined to assess the needs of a land-use type which includes several crops grown together or in rotation.

Matching Land Use with Land

Matching land use with land is a key step in the overall procedure. The first stage compares the requirements of each land-use type with the land qualities of each land unit. In simple terms:

Table 10.6   **Example of land requirements for a specified land-use type (bunded rice). From FAO (1993, Table 6).**

Land Qualities	Land Characteristics	Limiting Values for Land Characteristics			
		S1	S2	S3	N
Sufficiency of energy	Mean annual temperature (°C) or Elevation (m)*	>24 0–600	21–24 600–1200	18–21 1200–1800	<18 >1800
Sufficiency of water	75% probability rainfall (mm)	>1300	900–1300	500–900	<500
	Soil drainage class	Poorly drained	Imperfectly drained	Moderately well drained	Excessively drained
	Soil texture	C, ZC, ZCL, L	SC, SCL, ZL, Z	SL	S, LS
	Soil depth (cm)	>80	60–80	40–60	<40
Sufficiency of nutrients	pH of flooded soil	6–7	5–6 8–8.5	7–8 <4.5	4.5–5 >8.5
Salinity hazard	EC_e (mS cm^{-1})	<3	3–5	5–7	>7
Ease of water control	slope angle (degrees)	<1	1–2	2–6	>6
Ease of cultivation	Stones and rock outcrops (%)	Nil	1–5	5–10	>10

* Elevation is used to assess sufficiency of energy where temperature data are not available. These values apply to Sri Lanka (Dent and Ridgway 1986).

- measured values of each land quality or characteristic are measured against the class limits, and
- each land unit is then allocated to its land suitability class according to the most severe limitation (Table 10.7).

In situations where at least one limitation would cause the land unit to be unsuitable for the land use, this simple method of using the most severe limitation is valid. The FAO gives the example of maize cultivation being impossible if the soils are highly saline even though other qualities such as level land and sufficient rainfall are present.

Table 10.7   **Example of the process of qualititative land suitability classification. From FAO (1993, Fig. 8).**

Speed of Survey	Land Characteristics	Start				
			S1	S2	S3	N
Aerial photo interpretation with rapid field survey	GROUP A Slope Past erosion Surface wetness Group A only, S1	Assume:Class	S1 Y Y Y	S2	S3	N
Moderate survey speed A + B	GROUP B Surface stones and rock outcrop Unfavourable surface conditions Texture of topsoil Group A + B, cannot be better than		N	Y Y Y S2		
Slow survey speed A + B + C	GROUP C Effective soil depth Subsoil texture Group A + B + C, cannot be better than					N

However, matching can become more complex than the above simple case of comparing land-use requirements with land qualities. If the initial comparison shows that certain land units are unsuitable for a given use, the specification of the land-use type can be examined to determine whether the suitability of the land units can be improved by modifying the land-use specifications. For example, if suitability has been downgraded due to erosion hazard, a new land-use type could be designed with the addition of contour-aligned hedgerows or other soil conservation measures. As a further example, short growing seasons could be countered by using fast-growing crops. Thus by adapting the land-use requirements to match the limitations present in the land units, higher overall land suitabilities can be achieved. A further possibility is the adoption of land improvements, such as drainage or terracing. Here the characteristics of the land unit are being changed to adapt them to the land-use requirements.

Qualitative and Quantitative Land Evaluation

Some decisions require only qualitative land evaluation: the example given by the FAO (1993) is identifying the critical importance of certain areas for particular land uses such as an export crop. However, quantitative economic evaluations require estimates of crop yields, rates of tree growth or other measures of performance. Predictions of the potential performance of a land suitability class cannot be made in the absence of such data and without good knowledge of the physical characteristics of the land mapping units. Although there are quantitative models for several major crops they require good data. Even predictions based on experimental trials may be erroneous when translated to the field situation, with variations in management practices and skills. It is therefore necessary to attempt estimations of ranges of performance under the management standards likely to be encountered in the field.

Land Suitability Classification

Comparison of the requirements of land-use types with properties of land units is brought together in a hierarchical land suitability classification as follows:
- *land suitability order*: reflecting different kinds of suitability;
- *land suitability class*: reflecting degrees of suitability within orders;
- *land suitability subclass*: reflecting the kinds of limitation, or the main kinds of improvement measure required, within classes;
- *land suitability unit*: reflecting minor differences in required management within subclasses (FAO 1976:17).

Suitability is indicated separately for each land-use type, showing whether the land is suitable or not suitable, including—where appropriate—degrees of suitability: see also Table 10.5). The main reasons for lowering the classification (that is, limitations such as erosion hazard or a high water table) should be indicated. An analysis of the economic viability and possible social consequences of each land use may be made during or after the physical evaluation. In large or complex surveys involving many mapping units, land evaluation can be assisted by the use of computer packages. ALES (Automated Land Evaluation System), for example, is a computer package using the FAO land suitability framework, which brings many advantages such as increased speed of evaluation (as opposed to data collection). For example, if crop prices change, or a change is made to one or more limiting values, management or inputs, which may occur frequently, it is quicker and easier to re-evaluate the suitability of a particular land use to a land mapping unit if the process is automated. However, ALES is not a *spatial* package and therefore cannot, for example, evaluate the effects of land uses on adjacent land units on one another, or the spatial consequences of allocating particular uses to particular land units.

The main outputs of FAO's land suitability evaluation are land suitability maps, showing the suitability of each land unit for each land-use type; and descriptions of these land-use types.

Descriptions of land-use types are provided at a degree of detail appropriate to the planning scale. At a national level, only outline descriptions of major kinds of land use may be required. At district and local levels, land-use type descriptions should specify management details, inputs (such as seeds, fuel and fertilisers) and estimated production (refer Table 10.3). Such information is required later to make provision for the supply of inputs and for storage, distribution and marketing.

It is most unlikely that all the desirable information for the above steps in a land suitability evaluation will be available at an adequate level of detail. Plant tolerances to various limitations are rarely known with any accuracy. Where new land-use types are proposed for a region, trials will be required to determine whether yield predictions based on information from other regions are reliable. There may also be inadequate information on the land resources of the area being evaluated, particularly the necessary details of soil properties or climate variables.

Land-use plans can rarely be delayed until all the necessary information has been obtained, but it is also unwise to proceed if the lack of data is serious. National and international research agencies, including universities, can be approached and advised of the need for particular areas of research; and local institutions, possibly reinforced, can be asked to conduct necessary trials or surveys. Results from such research may require a minimum of three to five years, but in the life of the development of a project this is not unrealistic; and new data can progressively modify predictions and recommendations as they become available. If likely problems have been anticipated, research results are more likely to become available when they are needed.

The land suitability framework has been used quite extensively, especially in Africa (Young and Goldsmith 1977). These early applications were naturally concerned mostly with crops as the potential land uses. However, the flexible nature of the framework allows the concept to be used for a variety of projects. A later set of guidelines, *Guidelines for Land Evaluation for Forestry*, has been published and a trial application made in Australia using the LUPLAN package (Booth and Saunders 1985). In Western Australia, the land capability assessment methodology developed and widely implemented by Agriculture WA, somewhat confusingly combines the approaches of the USDA land capability assessment and FAO's land suitability framework; it is not restricted to agricultural land uses (Wells and King 1989) and has been used extensively in the preparation of regional plans (Part 5).

An alternative approach developed during the 1980s concentrated on the soil pedon and its properties rather than on soils as a component of land. The principle is that if a soil pedon in one part of the world can be classified at a family or equivalent level (according to the FAO soil system), and its management and crop yields are known, then under a similar management regime elsewhere similar yields could be expected. But as Schaefer and Dalrymple (1999) comment, detailed ground truthing is essential in all instances. Future objectives for land use must also be clearly identified and socioeconomic and ecological factors considered. Policy decisions are often flawed in their reliance on inadequate surveys.

COMMENT

It can be suggested that the FAO land suitability framework evaluates the degree of suitability of land for different purposes, and that given data on land suitability, Siroplan may be used to choose the 'preferred' land use or management regime according to a set of policies reflecting interest group requirements.

Environmental planning and management must by definition involve decisions on land use. Land-use planning is therefore a vital component of environmental planning as a whole. According to Bowman (1979), a principle of environmental management is that all land is suited to different purposes. Land evaluation is important in its own right for determining land uses that will and will not cause land degradation, as well as being a step in land-use planning.

The FAO framework produces an assessment of land suitability for different land uses, and specifies required management practices for each use on each mapping unit, but says nothing about planning, management or land allocation for the area as a whole. It may achieve a goal of environmental management through the principle that 'suitable' land use must be sustainable, but it does not take into account adverse environmental impacts caused by a land use on another part of the system (for example, a land use may cause eutrophication down-river, which may not affect the sustainability of the land use itself). It is in this context that land capability evaluation, and more particularly environmental impact assessment (the following chapter) are of particular relevance; and there is no reason why they should not be incorporated in the FAO procedure.

It should also be recognised that the multi-disciplinary approach of the FAO framework may take longer and be more costly than some other, simpler methods of land evaluation (such as land capability assessment). Thus in some contexts these other methods may be more appropriate.

CHAPTER 11

ENVIRONMENTAL IMPACT ASSESSMENT

As discussed in Chapter 7, environmental impact assessment, and the creation of environmental protection agencies, were largely seen as the solutions to the problems of inadequate methods of maintaining and improving environmental quality, and the plethora of agencies whose activities had a bearing on the environment but whose major functions were mostly related to the development or management of resources. In this respect Australia, and many other countries, followed the lead of the USA with its *National Environment Policy Act 1969* (NEPA).

Previous chapters in this Part have considered a number of other methods of assessing land, environmental quality or the state of the environment. Those methods are available to a wide range of planning and management bodies or agencies, whereas environmental impact assessment (EIA) is a more specific set of methods designed to deal with what are now recognised as being a rather restricted range of environmental problems or issues.

This chapter first outlines the purpose and nature of EIA and its methods. These topics are dealt with briefly since there are numerous texts on EIA available for those requiring further information (examples: Munn 1979; Clark *et al.* 1980; PADC 1983; Wathern 1988; Glasson *et al.* 1994; Gilpin 1995; Morris and Therivel 1995; Wood 1995; Harvey 1998; Thomas 1998). Second, EIA has now been in use for nearly 30 years, and sufficient experience has been gained to identify some of the problems with the technique. These problems are discussed in some detail, since they lead (as have other considerations earlier in this book) to the conclusion that environmental protection needs to be integrated with regional land-use planning and management. In that context, the role of EIA is more restricted than many environmental practitioners seem willing to admit. It also needs to be recognised that there is some resistance to change, on the perfectly rational grounds that it has taken many years for environmentalists to persuade governments to accept EIA and that this is not the time to change. There are also situations for which EIA as usually practised is eminently suited. But to continue with a focus on a limited method, even for good reasons, when there are better ways of doing things, seems to us to be somewhat ostrich-like and, in the long run, counter-productive. The fundamental question is: what are the best ways of maintaining and improving environmental quality while at the same time encouraging sustainable development of resources?

The following questions are posed and answered in this chapter:

- *When* is an EIA implemented?
- *Who* carries out the assessment?
- *What* is included in the EIA process?
- *How* is an EIA done?
- How is the EIA output *used*?
- What are the *advantages and disadvantages* of the method?

WHEN IS AN EIA IMPLEMENTED?

To some extent the answer to this question depends on the jurisdiction. In some cases, particularly in the UK, the specific situations for which an EIA is required are listed. This has the advantage of removing ambiguity, but the disadvantage is that every new situation has to be legislated for—and that introduces what may be unacceptable or even dangerous delays. New South Wales also specifies 'designated' developments for which an EIA is required (amongst other criteria). New Zealand's *Resource Management Act* 1991 requires an EIA to be carried out for every application for consent to develop a resource. And in Canada, according to Gilpin (1995:114), all policy decisions by the Canadian federal government are subject to EIA.

In many jurisdictions the answer to the question has been similar to that presented in the Australian Commonwealth *Environment Protection (Impact of Proposals) Act* 1974: namely that an EIA is to be carried out for any proposal (by or for a Commonwealth government agency, or in association with any other party) which may have a 'significant effect on the environment', or for 'proposals which have an environmentally controversial effect'. The difficulties with these admirable statements of principles are obvious: who decides, and in particular on what basis, what is deemed to be sufficiently 'significant' or 'controversial'? Beanlands and Duinker (1983), for example, have commented that 'significance' can be statistical, ecological, social or even with reference to the project itself. The short answer to the question is that ultimately the Minister decides, based on the advice of the appropriate agency and sometimes (notably in Canada and the USA) following public involvement in the *screening* process (discussed in Part 4). Harvey (1998:39–41) has noted that in most Australian systems the decision is either discretionary or *ad hoc*. Clearly, an approach which places considerable weight on Ministerial discretion is open to manipulation for political and other reasons. Wood (1995:97) has observed that it has been easy for Australian Commonwealth government agencies to avoid the EIA procedures due to the extent of discretion.

Other questions raised by the Australian Commonwealth Act include the definition of 'proposals' and 'environment'. The 1974 Act (currently being replaced) explicitly includes programs and policies within 'proposals'. Section 5 of the Act also includes 'the making of, or the participation in the making of, decisions and recommendations' in actions which could be subject to an EIA. However, in practice generally only individual development projects have been assessed. Of 132 directives for EIA between 1974 and 1993, none related obviously to policy, plan or program proposals (Wood 1995:284). Some writers consider the above broader aspects of 'proposals' to fall within the scope of what is termed *strategic environmental assessment* (SEA) (Buckley 1997) (discussed under 'Ecological Approaches' in Chapter 8). There is also a question as to whether 'proposals' refer to private as well as public activities. Although NEPA originally was directed at government agencies (as was also the case in Victoria), this was subsequently broadened to include private activities permitted by government (Fowler 1982).

Similarly, others refer to social, economic, technological and health impact assessments to distinguish them from narrowly defined biophysical-environmental impact assessment: although again the original US NEPA (1969) and its Canadian and Australian Commonwealth counterparts explicitly include all the above aspects under EIA under a broadly defined 'environment' (cf. Chapters 1 and 7). This is clearly to be preferred, for simplicity's sake if for no other reason. Interestingly, the Australian Inter-Governmental Agreement on the Environment (IGAE) agreed that EIA should include, where appropriate, assessment of biophysical environment, cultural, economic, social and health factors and that ecologically sustainable development (ESD) principles should also be included in the guidelines (ANZECC 1997). Thus, Victoria's *Guidelines* for EIA state that:

> For the purposes of environmental impact assessment, the meaning of environment incorporates physical, biological, cultural, economic and social factors. This definition was agreed by the Australian and New Zealand Environment and Conservation Council in 1991 (Department of Planning and Development 1995).

There is, therefore, no need to invent the term Integrated Impact Assessment (IIA—Thomas 1996:43—who nevertheless does not refer to SEA) to incorporate social and technology assessments when they are already included in EIA as originally intended and defined.

As noted above and discussed in Chapter 7, Australia's new Commonwealth *Environment Protection and Biodiversity Conservation Act* obtained Royal Assent on the 16 July 1999, but will not come into effect until well into 2000. Five pieces of legislation are replaced by the new Act. They are: the *Environment Protection (Impact of Proposals) Act 1974*; the *National Parks and Wildlife Conservation Act 1975*; the *Whale Protection Act 1980*; the *World Heritage (Properties Conservation) Act 1983*, and the *Endangered Species Protection Act 1992*. The Act has been influenced by Schedule 3 of the 1992 *Intergovernmental agreement on the environment* (www.environment.gov.au/portfolio/esd/igae.html) and ANZECC's (1997) *Basis for a national agreement on environmental impact assess-*

ment, as have developments in the other jurisdictions. There is an emphasis throughout Australia on speeding up the environmental decision-making process, providing more certainty for developers, and clarifying the roles of the States and the Commonwealth. Nevertheless, important aspects of the new Commonwealth Act are still to be negotiated with the States and there are therefore significant areas of uncertainty. However, it is clear that the Commonwealth will focus on a limited range of matters of national environmental significance and will not be involved in matters (or disputes) of only local or State significance. As far as possible, environmental assessments and approvals for developments will take place at State level. The national matters are:

- World Heritage properties;
- Ramsar wetlands;
- listed threatened species and communities;
- listed migratory species;
- protection of the environment from nuclear actions (defined broadly);
- Commonwealth marine environment;
- additional matters of national environmental significance (referring to 'prescribed actions');
- other matters involving bilateral agreements with the States or self-governing Territories, and
- proposals involving the Commonwealth.

(The Act also has extensive provisions dealing with the conservation of biodiversity but these are generally unrelated to EIA.)

Of relevance to the penultimate criterion, Raff (1995b) has raised the problem of arriving at a 'lowest common denominator' when consensus amongst States is sought in determining national environmental planning policy.

The indirect triggers for Commonwealth involvement (export approvals and funding decisions unrelated to the environment) have been removed. The Regional Forest Agreement (RFA) process (Chapter 5) is regarded as being separate and therefore activities such as wood chipping will also no longer trigger the Act. Actions in the Great Barrier Reef Marine Park, and 'actions covered by Ministerial declarations', are also excluded. There is provision for strategic assessments of policies, plans and programs, and of Commonwealth-managed fisheries. However, the former provisions are explicitly included in the current (1974) Act but have not been applied. The Environment Minister will decide whether a proposal or activity may have a significant effect on a matter of national environmental significance; and whether an activity on a Commonwealth place or for which the Commonwealth has sole jurisdiction triggers the Act. There is no reference to screening or scoping, discussed below under 'impact identification' (or to the community participation which would be associated with those activities). However, the proponent is required to invite public comment on the proposal and provide the Minister with a summary of comments received. The Minister *may* invite public comment on the draft *guidelines* for a draft public environment report (PER) or a draft EIS, and the proponent is *required* to invite comment on the draft PER or draft EIS reports themselves, within a minimum of 20 business days. There are also similar provisions for public inquiries to those in the 1974 legislation. However (and as discussed in Part 4), implementation of those provisions has been infrequent.

Fowler (1996) has commented that there was a distinct Commonwealth interest in securing co-operative arrangements with the States: the Commonwealth can handle 30 assessments a year rather than 300. Of particular concern, however, is the finding by Warnken and Buckley (1998) that of 170 EIA documents assessed by them (in relation to tourism), the best EIA and monitoring from a very poor collection were for projects falling under Commonwealth jurisdiction (discussed further at the end of this chapter). This does not bode well for a future where most EIA in Australia will be carried out under the control of the States and Territories. The discussion paper concerning the proposed Commonwealth environment protection legislation stated that the Commonwealth will be able to accredit State processes and that the proposed legislation will 'maximise reliance on State processes' (Commonwealth of Australia 1998). This concern over the future of EIA in Australia becomes magnified with the realisation that the primary emphasis of the proposed Commonwealth Act, according to the discussion paper, would be on development with environmental considerations being secondary. The environmental protection legislation will:

> recognise the need to develop a strong, growing and diversified economy and maintain and enhance international competitiveness in an environmentally sound manner (Commonwealth of Australia 1998).

This is the aim of a proposed *environment protection* Act! If implemented as indicated, this would return Australia to the situation described in Chapter 5, where environmental protection similarly took second place behind resource development. Fortunately, the new Act's objectives do not correspond with the above quote. Instead, all seven objectives are directed towards environmental protection and conservation of biodiversity, with no reference to development.

Despite the specified areas of national significance (listed above) under the new Commonwealth environmental protection legislation, it will still be possible for the Commonwealth to transfer powers to the States. This happened recently with the Hinchinbrook Canal Estate development located in a World Heritage area of the Great Barrier Reef. A commercial Deed of Agreement was signed between the Commonwealth, Queensland government and the developer which superseded the Commonwealth's *Environment Protection (Impact of Proposals) Act 1974*. Thus the Commonwealth can—and will—use its powers to exempt developers from compliance with its new environmental protection, biodiversity, or heritage laws. Such Agreements would be privately negotiated and privately enforced through reliance on voluntary compliance with environmental protection and other measures (as is now explicitly favoured by State agencies). The public will be deprived of opportunities to comment or enforce the law should infringements take place (Mould 1998; Horstman 1998).

However, as Bates (1995) has pointed out, such industry arrangements with governments, while over-riding other powers, may also attract environmental control procedures which are possibly stronger than the general control regime. Problems centre on the adequacy of the arrangements and enforcement.

In the USA, decisions on whether to proceed with an EIA can be (and often are) challenged in a court of law, in what ought to be a neutral, albeit adversarial, judicial environment. In contrast, legal challenges under the Commonwealth's environment protection Act of 1974 have been few (Munchenberg 1994).

A further difficulty is the binary nature of the decision whether to institute an EIA process in a given situation: yes or no. Some jurisdictions attempt to overcome this by having gradations of environmental impact assessment. Queensland has only one level of EIA whereas the ACT has four (Harvey 1998:51). In Western Australia, for example, evaluations of a proposal proceed from a simple Consultative Environmental Review (CER) through a Public Environmental Review (PER) to a fully fledged Environmental Review and Management Programme (ERMP) (Fig. 11.1). In Tasmania, there is a three-tiered assessment process under the land-use planning and State projects legislation (Chapter 7). The level of detail required and the extent of public involvement increase with each successive step. Relatively few proposals, especially in recent years, have been required to undergo the detailed evaluation. Exceptionally, the new Commonwealth environment protection Act has five levels of assessment. They are assessment by:

- an accredited assessment process;
- (on) preliminary documentation;
- public environment report;
- environmental impact statement, and
- inquiry.

Theoretically, then, the types of proposal which may be subject to EIA are very broad, ranging (from the ridiculous to the sublime) from a proposal to relocate a post box to the formulation of a major new government policy. In practice, neither of these examples would be subjected to EIA; the former due to its (perceived) insignificant environmental impact and the latter because governments are reluctant to subject political decision-making to independent assessment; although the arguments in favour of the latter are in fact quite strong.

WHO CARRIES OUT AN EIA?

In most jurisdictions, worldwide, an EIA is carried out by the proponent (or the proponent's consultants) of a proposal, with varying degrees of supervisory involvement by an environmental protection agency. Environment agencies are then responsible for assessing either a draft or a 'final' EIS, following which a final or supplementary EIS (respectively) is prepared. The agency then makes recommendations to the relevant Minister.

This situation has always been controversial, although it is difficult to argue convincingly the practicalities of the alternatives. The shortcomings are

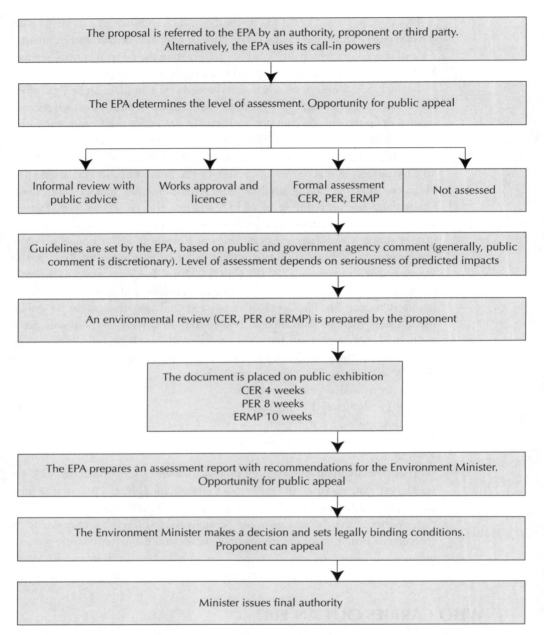

Fig. 11.1 The Western Australian environmental assessment process under the *Environmental Protection Act 1986* and the Administrative Procedures 1993 (Harvey 1998, Fig. 2.8; revised from a 1999 draft kindly provided by Sue Thomas, WA DEP, pers. comm.).

obvious: the proponent is hardly likely to put the project in an unfavourable light, or recommend that it not proceed, or that an alternative proposal should be preferred. In the woodchip case study (Chapter 2), for example, although the overall objective of the project was to enhance regional development and decentralisation in an economically depressed region, no means of achieving this other than through the woodchip industry were considered. Yet the development of tourism and recreation in particular would almost certainly achieve more substantial economic and social benefits than wood chipping, and leave the forest intact.

Similarly, a firm of environmental consultants is hardly likely to bite the hand that feeds it: the firm needs to ensure that it will not be blacklisted by clients when tendering for future contracts.

On the other hand, the direct involvement of the proponent in the EIA process is essential. Through that involvement, the originators of EIA in Australia expressed the hope that proponents would eventually come to see environmental assessment as a normal stage in the evaluation of a proposal's feasibility—at least as important as economic and technical feasibility studies—which would be carried out at the same time as the latter assessments. It was hoped that the need for environmental legislation would eventually disappear (see the second reading speech by Dr Moss Cass when introducing the Federal Environment Protection Bill in 1973).

Nevertheless, the supervisory role of the environmental agency becomes crucial. If the agency has real teeth and the full support of its Minister, then it can insist on rigorous and truthful assessments and reject environmental impact statements (EISs) (the written report which is the outcome of the EIA process) which do not meet acceptable standards. Without such political support the agency becomes ineffectual, and the Australian community (with the notable exception of New South Wales) generally has little recourse to an alternative referee such as a court of law.

From the above it will be apparent that the answer to the question, who is responsible for carrying out the EIA process, is not entirely straightforward. The discussion introduced the proponent, consultants, the environment agency, the relevant Minister and the community. In fact a number of bodies and individuals become involved at various stages of the process, as indicated in Fig. 11.2. With reference to that figure, it might be noted that there are many 'publics', often with conflicting interests and views relating to a proposal (Chapter 1); and that 'international' involvement is of particular relevance for projects with impacts extending beyond national borders.

The addition of the 'post-audit' process to the steps outlined previously for Western Australia in Fig. 11.1 is also of interest, and is in fact crucial (Bisset and Tomlinson 1990; Sadler 1990). Without that step (and action following any adverse findings) it could be argued that all the preceding steps in the process are a waste of time. This leads, then, to consideration of the next question.

WHAT IS INCLUDED IN THE EIA PROCESS?

Most of the various books on EIA listed near the start of this chapter are useful in answering this question. In general, the *minimal* requirements are that the EIA process should include the following:

	GOALS	POLICY	PROGRAMS	PROPOSED ACTION	PRELIMINARY ASSESSMENT	DECISION	DETAILED ASSESSMENT	RECOMMENDATIONS	DECISION	IMPLEMENTATION	POST-AUDIT
DECISION-MAKER											
ASSESSOR											
PROPONENT											
REVIEWER											
OTHER GOVT AGENCIES											
EXPERT ADVISERS											
PUBLIC											
INTERNATIONAL											

Fig. 11.2   Checklist for assignment of responsibilities for various steps in the environmental impact assessment process. Modified from Munn (1979, Fig. 2.2).

1   *Description of the existing environment.* This provides baseline or benchmark data for identifying future changes as a result of monitoring following the establishment of the proposed development, which can be used to assess the accuracy of the EIA predictions (Beanlands 1990). It also provides a substantial part of the basis for the predictions themselves. The section should be relatively brief and not used as an excuse to pad the EIS with largely irrelevant information.

The legislated definition of 'environment' in the various jurisdictions now becomes critical and has been discussed in some detail in Chapter 7 ('A Comment on Definition') and from a theoretical perspective in Chapter 1: refer also the early part of this chapter. Most jurisdictions now define the term broadly to include social, economic and cultural as well as biophysical aspects of the environment, and by implication refer to the environments of people (and not of some other node).

2   *Description of the proposal.* What is the full range of proposed actions and activities which may impact adversely on the environment? It is perhaps this step more than any other which requires the involvement of the proponent since no other body has the necessary information. The EIS must also include: the need for the proposal; the objectives of the proposal, and alternative means of attaining those objectives.

3   *Discussion of the predicted impacts of the proposal on the environment* (refer the next question for a discussion of how this may be done). This is the key section of the EIA process and the EIS, and should comprise the bulk of the latter. Often it does not. The section concerns the interactions of the proposal (step 2 above) on the environment (step 1).

The EIS should consider the environmental effects of all alternative means of achieving the objectives of the proposal, and should highlight unavoidable effects. The range of impacts should be discussed, considering simple, complex, direct and indirect impacts. The probability of the impacts occurring and the significance of those impacts should be evaluated (what is the answer to the 'so what?' question?). These crucial evaluations of probability and significance of impact occurrence may be subjective or objective, qualitative or quantitative, vague or precise. Too many EISs choose the first of each of these dichotomies— usually due to inadequate data, insufficient time or lack of expertise.

Table 11.1   Steps which should be included in any EIA, drawn mainly from O'Riordan and Turner (1983), Wood (1995) and Harvey (1998).

1	screening the proposal to determine whether an EIA should be conducted;
2	scoping the proposal—to determine the scope of environmental assessment, the scope of factors to be considered, the parties involved and their interests and concerns, the appropriate level of effort and analysis, and to prepare guidelines for the conduct of the EIA;
3	description of the proposal and of relevant alternatives, including doing nothing at all;
4	description of the environmental baseline, including changes *without* the proposal;
5	identification and prediction of the nature and magnitude of key environmental effects, both positive and negative, for each of the alternatives studied, over both short- and longer term periods;
6	assessment of how such effects are valued by representative sections of society;
7	testing of impact indicators as well as the methods used to determine their scales of magnitude and relative weights;
8	prediction of the magnitude of these impact indicators and of the *total* impact of the proposed project and relevant alternatives;
9	proposal of mitigation measures including identification of critical thresholds;
10	EIS presentation, public consultation and participation, including feedback;
11	recommendations for project rejection, or acceptance plus advice on how to reduce or remove the most serious of the impacts as measured and as socially valued, or adoption of the most suitable alternative scheme;
12	decision-making;
13	post-decision monitoring and auditing both during project construction and after project completion to ensure that environmental effects are minimised and to compare actual with predicted impacts;
14	amelioration if thresholds are exceeded or unacceptable impacts are identified, and
15	longer period reviews (5 years, 10 years, 20–25 years), requiring comprehensive assessment and adaptive management. A period of 25 years is a common limit for licence periods—'in perpetuity' arrangements should be avoided.

In contrast to the above minimal requirements for the EIA process, Table 11.1 (page 231) presents a more useful and comprehensive list of 14 steps which should be included in any EIA process.

HOW IS AN EIA DONE?

Much of the EIA methodology derives from or has developed from the Leopold matrix (Fig. 11.3), and it provides a convenient launching platform to discuss not so much the details of EIA methods *per se* but the difficulties involved. The full matrix contains 100 'project actions' across the top, grouped into 10 categories: modification of regime; land transformation and construction; resource extraction; processing; land alteration; resource renewal; changes in traffic; waste emplacement and treatment; chemical treatment, and 'accidents, others'. Down the left-hand-side of the matrix are listed 88 'environmental characteristics and conditions', grouped under: physical and chemical characteristics; biological condition; cultural factors (land use, recreation, aesthetics and human interest, cultural status and man-made facilities and structures); ecological relationships, and 'others'. The square at the intersection of each 'project action' and 'environmental characteristic' is split by a diagonal. A number between 1 and 5 is placed in each triangle created by the diagonals, with the number in the top left triangle indicating the 'magnitude' of the impact of the action on the environmental characteristic, and the number in the lower right triangle indicating the 'importance' of the impact. A plus or minus sign before the number indicates beneficial or adverse impacts, respectively. 'Magnitude' indicates

Environmental Characteristics and Conditions	Project Actions	A. Modification of Regime a–m							B. Land Transformation and Construction a–g
		a	b	c	d	e	f.....		
A. Physical and Chemical Characteristics									
1. Earth									
a. Mineral Resources			2 / 1						
b. Construction Materials			6 / 4		4 / 7		3 / 2		
c. Soils					8 / 6				
d. Landform				6 / 1					
....									
f.									

Fig. 11.3 Extract from the Leopold environmental impact matrix (Leopold *et al.* 1971).

the intensity of the impact; for example, the magnitude of the immediate impact of clear felling on the forest would rate 5 out of 5, since the forest is totally removed. 'Importance' is a measure of significance; is the impact merely of local significance or is it of regional importance? Despite a magnitude of 5 for a clear-felled coupe of (say) 100 ha, its importance may rate as only 1 or perhaps 2 out of 5, since the impacted area is small in relation to that of the total forest.

The inclusion of 'cultural factors' in the Leopold matrix is of interest: contrary to the views of environmental agency personnel in some Australian jurisdictions, it demonstrates that 'environment' has never been restricted to the biophysical environment in the country which initiated the EIA process: the USA. The Leopold matrix, nevertheless, is dominated by the physical environment—perhaps reflecting the geomorphological orientation of the principal author. In drawing attention to this, Munn (1979) listed the following 'Areas of Human Concern' or impact categories which he considered should be included in an EIA (Table 11.2). With this and perhaps other modifications, the major benefit of the Leopold matrix is its provision of a checklist of factors which need to be considered in undertaking an EIA. Not every factor will be important in any particular situation, of course; the value lies in providing an *aide memoir*.

The quite extensive range of problems associated with the actual use of the Leopold and other matrices and checklists has given rise to numerous modifications and other and (perhaps) more sophisticated methods. These problems and responses to them are now considered, including:

- the determination of boundary conditions in space and time;
- impact identification;
- the level of resolution;
- impact measurement;
- impact evaluation;
- the spatial problem;
- the problem of indirect effects;
- the problem of cumulative effects;
- the problem of dynamics, and
- the types of project actions for which EIA is appropriate.

Boundary Conditions

One of the requirements of EIA is the determination of boundary conditions in space and time. Where, spatially, does impact assessment cease? With reference to the woodchip industry case study (Chapter 2), for example, should the EIA have considered only the woodchip licence area, or should it also have considered the environmental impacts of: wood chipping and clear felling on surrounding forest habitats and other land uses; increased power generation and therefore coal mining in the Collie basin; the new railway line constructed to take the woodchips from the chip mill near Manjimup to Bunbury, and the new inner harbour constructed at Bunbury to cater for (then) the woodchip berth and only one other user?

Table 11.2   **'Areas of Human Concern' or Impact Categories which should be included in EIA. Source: Munn (1979, Table 3.3).**

1	Economic and occupational status	Displacement of population; relocation of population in response to employment opportunities; services and distribution patterns; property values
2	Social pattern or life style	Resettlement; rural depopulation; change in population density; food; housing; material goods; nomadic; settled; pastoral; agricultural; rural; urban
3	Social amenities and relationships	Family lifestyles; schools; transportation; community feelings; participation vs alienation; local and national pride vs regret; stability; disruptions; language; hospitals; clubs; recreation; neighbourliness
4	Psychological features	Involvement; expectations; stress; frustrations; commitment; challenges; work satisfaction; national or community pride; freedom of choice; stability and continuity; self-expression; company or solitude; mobility
5	Physical amenities (intellectual, cultural, aesthetic and sensual)	National parks; wildlife; art galleries and museums; concert halls; historic and archaeological monuments; beauty of landscape; wilderness; quiet; clean air and water
6	Health	Changes in health; medical services; medical standards
7	Personal security	Freedom from molestation; freedom from natural disasters
8	Religion and traditional belief	Symbols; taboos; values
9	Technology	Security; hazards; safety measures; benefits; emission of wastes; congestion; density
10	Cultural	Leisure, fashion and clothing changes; new values; heritage; traditional and religious rites
11	Political	Authority; level and degree of involvement; priorities; structure of decision-making; responsibility and responsiveness; resource allocation; local and minority interests; defence needs; contributing or limiting factors; tolerances
12	Legal	Restructuring of administrative management; changes in taxes; public policy
13	Aesthetic	Visual physical changes; moral conduct; sentimental values
14	Statutory laws and Acts	Air and water quality standards; safety standards; national building Acts; noise abatement by-laws

Today the answer to these and similar questions would almost certainly be affirmative. Raff (1997), for example, has identified and substantiated ten basic principles of quality in EIA which can be distilled from hundreds of cases involving this area, and which he consequently terms the *common law* of NEPA-style EIA (Table 11.3). The first principle is that an assessment cannot be restricted to site-specific environmental effects. Principle 3 is that 'greater plans' must be assessed in addition to single phases in their execution (which leads to programmatic or strategic EIA); and the fourth principle is the 'rule' against segmentation of the project. In other words, the EIA must not solely assess individual aspects of the overall proposal which on their own will yield negligible impacts; the sum or total effect of the proposal on the environment must be identified and evaluated (refer Step 8 in the list of 14 steps in the EIA process, in Table 11.1).

Over how long a period should the EIA be restricted? In the case of the woodchip industry, the EIA considered a period of 25 years, despite the fact

Table 11.3 **Raff's (1997) ten principles of quality in EIA: the *common law* of NEPA-style EIA (slightly modified).**

1	An assessment cannot be restricted to site-specific environmental effects.
2	The assessment must contain a statement of alternative courses of action and their environmental significance even if it is beyond the power of the proponent to implement them.
3	Greater plans must be assessed in addition to single phases in their execution.
4	The project must not be segmented.
5	Time frames must be viable: the range of alternatives that must be discussed in an EIS is limited to realistic alternatives that are reasonably available within the time the decision-maker intends to act.
6	Alternative courses of action must include the option of doing nothing.
7	The assessment must engage in a real inquiry and not merely dispose of alternatives in favour of a decision which has already been arrived at.
8	Environmental effects are not to be disregarded merely because they are difficult to identify or quantify.
9	The EIA must take a 'hard look' at the environmental consequences of the project and what that entails. While the EIS is not required to be perfect: it must not be superficial, subjective or non-informative; it should be comprehensive in its treatment of subject matter and objective in its approach.
10	The findings of the EIA must be presented in clear language and the methods used to arrive at them must be explained.

that the relevant Minister was reported as stating that the industry would continue 'in perpetuity'. Impacts on the biophysical environment, the tourist industry and the local community of clear felling over a 25-year period will be considerably less than those over a much longer period of time, when a much larger total area and proportion of the forest under consideration, will have been clear felled and regenerated. Raff (1997) refers to the 'rules' against 'science fiction' and 'crystal ball inquiries' in discussing his fifth principle. This states that the time frame should be realistic and 'reasonably' available within the time the decision-maker intends to act (Table 11.3).

Impact Identification

This is usually done by the use of checklists (such as the Areas of Human Concern in Table 11.2) or matrices (such as the Leopold matrix). Gilpin (1995:43) has commented that of all methods, checklists have tended to survive as a guide to potential impacts. He provides very useful sample checklists for a wide range of topics, including: air quality; ecology; economics; environmental health; hazard and risk; noise; social impact assessment; urban impact; water quality; community centre; expressway; hydro-electric power; marina; nuclear power plant; open cut mining; sewerage systems, and electricity generation.

Two of the methods used to assist in impact identification are screening and scoping. *Screening* is an integral part of the environmental assessment process (step 1 in Table 11.1) and is used to determine whether (and if so, at what level) an issue should proceed to the stage of requiring an EIA or a public inquiry. Public participation in screening is widespread in Canada but the public is rarely involved in Australia. There do not appear to be formal screening criteria in Australia, as a result of which the system is discretionary or *ad hoc*. New South Wales and the ACT are exceptions. New South Wales has a list of 'designated developments' and the ACT a list of 'defined decisions' which must be subjected to an EIA, although these mechanisms are not true screening processes. The ACT has Preliminary Assessments, which Harvey (1998) has noted are often comprehensive so as to avoid a full-scale EIA.

Following that stage, *scoping* (step 2 in Table 11.1) is then used to assist in determining what should be included in an EIA. Scoping refers to surveys of community and expert opinion to determine what matters of value to a community should be included in an EIA and to prepare publicly available guidelines for the preparation of the EIA. These give the proponent some certainty in terms of what is expected of the EIA investigations, which is important since the proponent is accountable for the standard and scope of the EIS. A major difference with screening is that the public and the proponent are often involved. The US CEQ introduced scoping in 1978 and it is part of the NEPA regulations. In Canada, scoping was first used in 1985 in relation to the second nuclear reactor at Point Lepreau, New Brunswick; it is now a common practice at the federal level and appears to be implemented to varying degrees in all Provinces. Review panels hold scoping workshops to learn the nature and importance of issues and concerns about a proposal before issuing guidelines for the preparation of an EIS. There are four aspects to scoping according to the Canadian guide. These are to determine:

- the scope of environmental assessment;
- the scope of factors to be considered;
- the parties involved and their interests and concerns, and
- the appropriate level of effort and analysis.

Screening and scoping are, or can be, important means of incorporating public participation in EIA and environmental decision-making. These are discussed further in Part 4.

Impact Resolution

At what level of detail should the EIA be conducted? It is very difficult to answer this question without knowing the magnitude and importance of the impacts—a *Catch 22* situation (Heller 1962). With reference to the woodchip case study again, is it necessary to evaluate every minor (first order fingertip) tributary and swamp and determine the effects of clear felling and subsequent regeneration on discharge, water temperatures, nutrient flows, dissolved and suspended load, and the effects of any changes to these and other properties on aquatic flora and fauna? An affirmative answer would require considerable time, effort and expense. A very good indication of the kind of research required to provide such data is provided in Moldan and Cerny (1994).

The answer, therefore, hinges on what the likely impacts are and what are deemed to be their ecological and social significance. The question of significance will raise questions of (for example) the presence of rare and/or endangered species of indigenous flora and fauna, or whether the waters are likely to be required for human use and if so, what the appropriate water quality standards are for the anticipated use. EIA assessors (consultants) draw upon their experience and knowledge of research results obtained in other, comparable situations. But the answers also hinge on the community's assessment of what is of value in the impacted area, as identified in steps 2 and 6 of the 14 steps in the EIA process (Table 11.1). There is no short, easy answer to the question, except to note that increasingly, scoping is being undertaken as a normal part of the EIA process, for the above reasons. In Australia, scoping is available in about half the jurisdictions. Some, such as South Australia, also provide the opportunity for public input into the EIA guidelines document, but it is not required.

Impact Measurement

Preferably, impact measurement includes measurements of the existing environment—incorporating recognition that it is dynamic and not static—and modelling to predict the magnitude of impacts on it. Impact predictions should be objective and clearly defined and should include assessments of the probability of the impact's occurrence and the statistical significance of those occurrences. These ideals unfortunately require a level of knowledge of environmental (including the cultural environment) characteristics and processes which is often unavailable or incomplete.

The Leopold and similar matrices and checklists are particularly deficient in their subjectivity, although techniques are available which can reduce some of the guesswork. Specifically, with reference to Fig. 11.3 for example, who decides which number from 1 to 5 should go in each triangle, and on what basis is that decision made? No single person could possibly have sufficient knowledge on all aspects of every project action and its impact on all the environmental characteristics to even hazard a guess as to what the likely impact will be. Even a committee of scientists would find it difficult to do so—and a collective decision is still subjective.

The Delphi approach is one method designed to overcome this problem. Individuals from a group of experts familiar with the particular environmental characteristic under question first independently graph the levels at which that characteristic becomes a problem. Munn (1979) provides two examples from Dee et al. (1973); one referring to the level of dissolved oxygen in water, and the other to stock carrying-capacity of a farm paddock (Fig. 11.4). With the first, as the dissolved oxygen content decreases, so does the quality of the water. In the second case, however, very low *and* high stocking rates are both considered to be harmful to pasture quality, as reflected in the different-shaped graph. The experts are then asked to assign a number from 0 to 1 to the actual measurements (of dissolved oxygen and stocking rates in these examples). In this way wholly unlike measures of environmental impact can be standardised (Fig. 11.4). After the first round, the experts are shown the responses of all the members in the group and then asked to repeat the exercise. Differences between them are thereby narrowed and consensus is usually achieved. The procedure is described in Table 11.4. Nevertheless, carrying out such an exercise on the

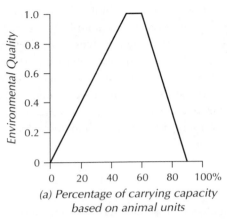

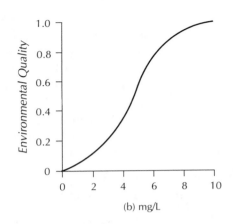

(a) Percentage of carrying capacity
based on animal units

(b) mg/L

Fig. 11.4   Examples of value functions converting parameter estimates from diverse units to comparable meas-urements on an 'environmental quality scale'. a) converts carrying capacity, where the maximum adverse effects come from both zero and 100% stocking rates in a paddock, with the optimum capacity being around 50–60%, in the opinion of experts. b) converts water quality, measured by dissolved oxygen content. The shapes of the two functions are quite different but their conversion to a common scale of 0 to 1 means that in terms of environmen-tal quality, the totally different parameters can be compared directly. The method is outlined in the text and in Table 11.4. From Munn (1979, Fig. 4.4).

huge range of all possible impacts in the existing Leopold matrix—or any other version—would be a massive exercise. Moreover, the process is still essentially a subjective means of obtaining a consensus view amongst a group of 'experts'.

Modelling environmental impacts is an alternative, and Glasson *et al.* (1994) have presented various types of models for EIA, including mass balance models, statistical models, physical, image or architectural models, field and laboratory experimental models, and analogue models.

Table 11.4	Procedure for determining value functions as in Fig. 11.4; from Munn (1979:50).
Step 1:	Obtain information on the relationship between the parameter and the quality of the environment.
Step 2:	Order the parameter scale (abscissa) so that the lowest value is zero.
Step 3:	Divide the quality scale (ordinate) into equal intervals between 0 and 1, and determine the appropriate value of the parameter for each interval. Continue the process until a curve can be drawn.
Step 4:	Ask several different specialists to repeat steps 1 to 3 independently. Average the curves to obtain a group curve. (For parameters based solely on value judgments, use a representative cross-section of the population.)
Step 5:	Show curves to all participants, and ask for a review if there are large variations. Modify the group curve as appropriate.
Step 6:	Repeat steps 1–5 with a separate group of specialists, to test for reproducibility.
Step 7:	Repeat steps 1–6 for all the selected parameters.

With reference to the salinity problem, for example, on the basis of existing data it is possible to model the effects of replacing native vegetation with crops on transpiration rates and hence on changes to the depth to groundwater.

Empirical modelling based on statistical relationships is very common in the natural sciences. There have also been significant advances in physically based mathematical modelling, for example, in the substantial MEDALUS project (Mediterranean Desertification and Land Use) (Brandt and Thornes 1996). In several chapters of that book, researchers discuss the modelling of: slope cate-nas, looking at the interactions amongst hydrology, ecology and land degrada-tion; the effects of climate and land-use change on basin hydrology and soil erosion; and the effects of desertification on short-term water resource trends. These are all discussed with reference to Mediterranean Europe and are built on a considerable body of empirical work. Earlier, Bennett and Chorley (1978) examined a wide range of systems models in their text on environmental sys-tems, including the interaction of 'human' with 'biophysical' systems. Grant *et al.* (1997) have discussed different techniques for ecosystem modelling for plan-ning purposes—understanding system dynamics and altering variables and assumptions to determine impacts and changes.

With specific reference to EIA, de Broissia (1986) presented a useful discus-sion of mathematical modelling in relation to EIA in Canada. Models discussed relate to air dispersion, hydrology and hydrodynamics, water quality (including groundwater), erosion and sedimentation, oil slick and liquefied natural gas spills, risk and pathways analysis and biotic models. More recently, Panizza *et al.* (1996) briefly discussed modelling the effects of land-use changes on overland flow, channel flow, runoff and coastal cliff erosion; as well as GIS and environ-mental and geomorphic indicators. Simulation modelling in EIA has been con-sidered by Bisset (1990). Based on work by Holling (1978) and his associates, the key component is a workshop in which participants (scientists, decision-makers and computer modelling experts) have to reach consensus on the important features and relationships which characterise the system likely to be affected by the proposed development. This would seem to be similar to the Delphi technique discussed above. Nevertheless, similar empirical and mathe-matical modelling of the potential impacts of project actions on the socioeco-nomic and cultural aspects of the environment has not been developed to the same extent as for impacts on the biophysical environment.

In either situation there is some doubt about the efficacy of modelling when it comes to making hard decisions (for example, on a particular farm, referring again to the salinity case study). Mathematical modelling is at best an indica-tive and not a prescriptive tool. In general, therefore, less formal predictive models are used, and in practice there is a heavy reliance on expert opinion (Thomas 1996:146).

Returning to the problem of subjectivity, 'total impacts' as required in step 8 in Table 11.1 cannot be arrived at validly by summing the columns (or rows) of an impact matrix (as in Fig. 11.3). If columns or rows are summed, then the implication is that all actions and environmental characteristics are of equal importance, which is extremely unlikely. It also does not take into account the possibility that a particular impact (or group of impacts) may be totally unac-ceptable, to the extent that regardless of the significance of all the others, the project should still not proceed. The response to this shortcoming is to carry out a weighting exercise; but this topic introduces the next heading.

Impact Evaluation

The difficulty here concerns the issue of value judgments. The environmental field in general and EIA in particular are heavily politicised and very strongly influenced by groups of people having widely disparate views on what is important, what society's objectives should be, on quality of life issues and so on. It is therefore particularly important that EIA and decisions based on its findings should not be the preserve of any particular interest group—or at least that bias should be accounted for. Public participation is discussed in Part 4, but the point raised here concerns the means whereby broader views than those of the EIA team can be incorporated in EIA.

As discussed by Canter (1983), numerous weighting-scaling and weighting-ranking checklists have been developed for EIAs. An early weighting-scaling checklist was developed by the Batelle Institute in the USA for water resources projects, as reported by Munn (1979). The method contained four categories, subdivided into 17 components and 78 environmental factors. The four categories of environmental impacts are: ecology; physical/chemical environmental pollution; aesthetics, and human interest/social. Each of the categories, components and factors was assigned an importance weight by an inter-disciplinary team using a ranked pair-wise comparison technique, resulting in parameter importance units (PIUs) (Table 11.5). PIUs can then be multiplied by scaling numbers (as discussed above using the Delphi technique) to derive an overall value, or environmental impact unit (EIU), for each environmental factor (or component or category). A further development of this method is multi-attribute utility theory, discussed by Bisset (1990).

The more precise are the measured attributes (size, density, shape, roughness, numbers) of the 'objects' to which weighting scales are applied, then the more precise will be the final EIU. Nevertheless, there will always be an element of subjectivity. The important point here, particularly for consultants defending their work in a court of law, is that the procedure followed can be described precisely, explained and justified. More subjective methods which involve neither scientific measurement nor group consensus but which rely wholly or largely on the opinion of the consultant, are much more difficult to defend.

A related, important point is that the community can be involved directly in evaluating the importance or significance of an environmental impact; it is not necessarily reliant on 'expert' (or, worse, political) opinion.

The Spatial Problem

A particularly important problem is that project actions and their environmental effects do not occur uniformly over an area or region. The woodchip case

Table 11.5   **The Batelle environmental classification for water resource development projects. Bracketed numbers are relative weights (PIUs) for the 78 'environmental factors'; from Munn (1979, Table 4.4).**

ECOLOGY
Terrestrial Species & Populations
– Browsers and grazers (14)
– Crops (14)
– Natural vegetation (14)
– Pest species (14)
– Upland game birds (14)
Aquatic Species and Populations
– Commercial fisheries (14)
– Natural vegetation (14)
Pest species (14)
– Sport fish (14)
– Waterfowl (14)
Terrestrial Habitats & Communities
– Food web index (12)
– Land use (12)
– Rare & endangered species (12)
– Species diversity (14)
Aquatic Habitats & Communities
– Food web index (12)
– Rare & endangered species (12)
– River characteristics (12)
– Species diversity (14)
Ecosystems

AESTHETICS
Land
– Geologic surface material (6)
– Relief & topographic character (16)
– Width & alignment (10)
Air
– Odour & visual (3)
– Sounds (2)
Water
– Appearance of water (10)
– Land & water interface (16)

– Odour & floating material (6)
– Water surface area (10)
– Wooded & geologic shoreline (10)
Biota
– Animals—domestic (5)
– Animals—wild (5)
– Diversity of vegetation types (9)
– Variety within vegetation types (5)
Man-made objects
– Man-made objects (10)
Composition
– Composite effect (15)
– Unique composition (15)

PHYSICAL/CHEMICAL
Water Quality
– Basin hydrologic loss (20)
– Biochemical oxygen demand (25)
– Dissolved oxygen (31)
– Faecal coliforms (18)
– Inorganic carbon (22)
– Inorganic nitrogen (25)
– Inorganic phosphate (28)
– Pesticides (16)
– pH (18)
– Streamflow variation (28)
– Temperature (28)
– Total dissolved solids (25)
– Toxic substances (14)
– Turbidity (20)
Air Quality
– Carbon monoxide (5)
– Hydrocarbons (5)
– Nitrogen oxides (10)

– Particulate matter (12)
– Photochemical oxidants (5)
– Sulphur oxides (10)
– Other (5)
Land Pollution
– Land use (14)
– Soil erosion (14)
Noise Pollution
– Noise (4)

HUMAN INTEREST/SOCIAL
Education/Scientific
– Archaeological (13)
– Ecological (13)
– Geological (11)
– Hydrological (11)
Historical
– Architecture and styles (11)
– Events (11)
– Persons (11)
– Religions and cultures (11)
– 'Western Frontier' (11)
Cultures
– Indians (14)
– Other ethnic groups (7)
– Religious groups (7)
Mood/Atmosphere
– Awe/inspiration (11)
– Isolation/solitude (11)
– Mystery (4)
– 'Oneness' with nature (11)
Life Patterns
– Employment opportunities (13)
– Housing (13
– Social interactions (11)

study (Chapter 2) provides a particularly good example of this fact. Clear felling (with its associated snig tracks, log landings and logging roads) is by far the most environmentally important project action. But activities are not spread evenly. Instead, each year clear felling (and subsequent regeneration) takes place in several coupes (areas of up to 900 ha depending on the forest type), with large areas of uncut forest remaining. Only over many years is the entire forest (less reserves) cut over, and then in a mosaic form. In such a situation it is impossible to complete a Leopold-type matrix: how can one number be placed in one box to indicate the environmental effect of clear felling, relevant to the entire forest? And yet, this is precisely the method that was used in the EIA of this project (Forests Department, n.d.).

Probably the most obvious solution to this problem is the overlay technique popularised by McHarg (1969) in his book *Design with Nature*. The potential effects of project actions on environmental attributes (soil, vegetation, topography and so on) were mapped on transparencies which could then be overlaid physically. The darkest areas are those with the greatest impacts. This is still very subjective, and physical limitations restrict the number of overlays to about six. With the advent of GIS, not only are the number of overlays theoretically limitless, but numerical values can be used with composite values calculated for each pixel (smallest identifiable spatial unit).

Indirect Effects

Environmental impacts are rarely a simple matter of a single cause having a single effect: rather there is normally a more complex chain of multiple causes and effects. Network analysis recognises that effects are often much more complex than the simple cause/effect relationships implied by matrices or checklists: a relatively simple example is presented in Munn (1979), reproduced with some modifications in Fig. 11.5.

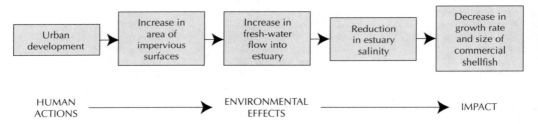

Fig. 11.5   Example of a flow chart used for impact identification; from Munn 1979, Fig. 4.1.

With reference to the secondary salinity case study for a further example (Chapter 3): clearing of indigenous vegetation leads to several hydrological changes, with a cascading effect of groundwater tables rising closer to the surface; soluble salts accumulate in surface soils rendering the land agriculturally unproductive (with some farms becoming unviable and the farmers going out of business, contributing to rural depopulation and the decline of small country towns); the increased salts are washed into streams, rivers and dams; water impounded in some dams becomes unfit for human consumption or irrigation; new dams have to be constructed with their own network of environmental impacts—and so on. Matrices cannot deal with this kind of situation, one which in an increasingly complicated world is more common than not.

Constructing first a conceptual and then more realistic networks is the appropriate means of dealing with this, but it is easier said than done. Canter (1983) has a useful discussion of network analysis. A fruitful approach is first to recognise the need for networks to be identified, and second to attempt to construct at least a conceptual, systems-type model of the situation under investigation (Bennett and Chorley 1978: refer also to the discussion on systems diagrams by Bisset (1990)). Unfortunately, the resultant 'spaghetti diagrams' often can be more confusing than helpful; but then that is the nature of the real world. Bisset (1990:57) noted that systems diagrams are usually restricted to ecological impacts: 'Attempts to incorporate socioeconomic impacts are still fraught with conceptual and practical problems.'

Cumulative Effects

This problem refers to the fact that only very rarely does the environment change simply in response to one project action. Instead, environments are subjected to numerous human actions and natural processes, all impinging on one another in space and over time. Changes can be subtle and occur gradually over long periods of time, with their importance being overlooked until a threshold is reached.

Often, cumulative effects are discussed as referring to a situation where individual actions may each have environmentally insignificant impacts, but where numerous such actions may collectively have serious adverse effects. A simple example is urban air pollution. The effects of each individual vehicle on air quality are negligible, but their combined effects may have serious consequences for human health, as well as on ecosystems and some building materials. The loss of urban wetlands is another, slightly different example. Here it is often difficult to attribute their loss (or degradation) to any one action: rather, it is a result of a combination of actions in a metropolitan region: clearing of natural vegetation; draining; building roads and houses; infilling for rubbish disposal or playing fields, and so on.

A related example is provided by the woodchip case study. Adverse effects on the forest ecosystem by clear felling for the woodchip industry are compounded by other activities and processes: mining and quarrying; road building; expansion of towns; recreation and tourism, and the spread of forest diseases. These do not occur uniformly over space or time. Canada's CEARC has provided a useful typology of cumulative environmental impacts which takes these and other aspects into account. It is presented in Table 11.6.

In another CEARC publication, Peterson et al. (1987) discussed the following four pathways of cumulative effects:

1   persistent additions from one process, without interactive relationships;
2   persistent additions from one process, interactive because of biomagnification;
3   compounding effects involving two or more processes, additive through non-interactive multiple impacts, and
4   compounding effects involving two or more processes, interactive through synergistic relationships.

Peterson et al. then provided a conceptual framework for cumulative effects assessment, subdivided into the ecosystem component, the research component and the management component. Importantly, this was followed by a discussion

Table 11.6   A typology of cumulative environmental effects (from Sonntag *et al.* 1987).

Type	Main Characteristics	Examples
1 Time crowding	Frequent and repetitive impacts on a single environmental medium	Wastes sequentially discharged into lakes, rivers or airsheds
2 Space crowding	High density of impacts on a single environmental medium	Habitat fragmentation in forests, estuaries
3 Compounding effects	Synergistic effects arising from multiple sources on a single environmental medium	Gaseous emission into the atmosphere
4 Time lags	Long delays in experiencing impacts	Carcinogenic effects
5 Extended boundaries	Impacts resulting some distance from source	Major dams; gaseous emissions into the atmosphere
6 Triggers and thresholds	Disruptions to ecological processes that fundamentally change system behaviour	The greenhouse effect; effect of rising levels of CO_2 on global climate
7 Indirect effects	Secondary impacts resulting from a primary activity	New road developments opening frontier areas
8 Patchiness effects	Fragmentation of ecosystems	Forest harvesting; port and marina development on coastal wetlands

of alternative institutional arrangements for responding to perceived cumulative effects where, amongst other alternatives, they considered the creation of separate agencies and links with environmental planning.

The above considerations again raise the question of who should be responsible for carrying out the EIA, discussed earlier in this section. Why should the forestry industry, for example, be concerned with, or be required to consider, the cumulative impacts of its actions with those of other activities in which it has no interest and for which it is not responsible?

The difficulty with cumulative impacts goes beyond the particular EIA methods being used: there is no clear methodological answer. It is in fact a fundamental problem which raises serious doubts concerning the efficacy and relevance of EIA itself, and the types of projects for which it is appropriate, discussed below. It also demonstrates very well the need for and importance of applying the precautionary principle in environmental planning (defined below, under 'Decision-Making'). We (and others) consider that the solution lies in incorporating environmental protection/enhancement in the much broader land-use planning process—a major point to which we return in Part 5. Gilpin (1995:32) observed that the role of EIA would be facilitated by strengthening regional environmental planning, especially in its capacity to handle cumulative and secondary environmental effects, and through the establishment of regional databases. Peterson *et al.* (1987) took the point further, commenting in their Executive Summary that:

> The terms of reference for this review posed the following question: can cumulative effects management be improved by contributions from regional environmental planning and area assessments? The question was answered

in the affirmative, but the review focused more attention on the reverse question: can regional environmental planning and area assessments be improved by deliberate incorporation of cumulative effects assessment criteria? The answer to this question was also in the affirmative.

The Problem of Dynamics

Simply stated, this difficulty refers to the fact that environments are constantly changing. Biophysical (natural) environments are driven by solar energy, weather and climate, none of which is constant over time. Ecosystems do not have simple equilibrium states. The human aspects of environments are similarly dynamic. Thus the problem faced by the EIA consultant when initially identifying and describing the baseline environment with which the proposed project will interact, is first to recognise that the environment is dynamic and then to describe its temporal variations accurately. The baseline (or benchmark) description should not be that of a static entity.

This is, of course, very difficult if not impossible to do, if for no reason other than the unavailability of much of the necessary data. And it makes it doubly difficult to predict with any pretence at accuracy the effects of proposed project actions on an environment whose attributes and properties are changing continuously.

A contemporary example is that of the so-called enhanced greenhouse effect. There is no disagreement that CO_2 concentrations in the atmosphere are increasing. There is also no doubt that climates have been warming since the last glacial period peaked some 16 000 years ago and have continued to warm in historic times. Valley glaciers in both hemispheres have retreated during the twentieth century, with abandoned terminal moraines identified and dated historically at, for example, Edith Clavell glacier in the Canadian Rockies and the Franz Joseph glacier in New Zealand's South Island (Plate 11.1). What is difficult if not impossible to identify and quantify is the significance of the human impact on global warming relative to natural processes. For example, according

Plate 11.1   Retreating snout of Franz Joseph Glacier, South Island, New Zealand: a) in 1967, photographed by David Hopley, and b) in 1983, photographed by Arthur Conacher from the same position. The waterfall on the right-hand-side of the photos provides a reference point.

to Gray (1998), atmospheric CO_2 measured at Mauna Loa increased at a *constant* rate of 3.1 gigatonnes each year from 1971 to 1996; yet annual CO_2 emissions from the combustion of fossil fuels *increased* by 54%, from 4.23 to 6.51 gigatonnes per year, over the same period.

With reference to socioeconomic–cultural aspects of 'the' environment, the effects of the late 1997 meltdown of the Asian economies had massive repercussions worldwide. Yet it seemed to have taken everyone (including the 'expert' economists) by surprise (is this chaos theory at work?). Additionally, social systems are heavily dependent on historical factors.

Recognition of the inherent uncertainty and unpredictability of future environmental states and conditions led to the idea of 'adaptive management' (Iles 1996). Introduced by C.S. Holling and his colleagues, 'adaptive management' is an approach to environmental policy which treats policy measures as experiments to be learned from. It needs a new kind of management authority with an explicit mandate to experiment, or a new framework which will allow a number of pilot, small-scale policies to be tried before proceeding to wider implementation. Management policies need to be constantly revised in the light of feedback, and management structures and processes need to be supple and dynamically changing. Iles (1996) provides a detailed example from the Florida Everglades. He comments that although adaptive management trials in the USA are being undermined by lawsuits, paradoxically, lawsuits can be agents of creativity.

Types of Project Actions for which EIA is Appropriate

The following overlapping typology of environmental problems was listed in the Introduction to Part 2:
- sectoral or regional problems;
- project-specific problems;
- numerous, dispersed, discrete actions that have an impact on the environment; and
- environmental issues.

The salinity problem in Western Australia's wheatbelt is a good example of the first type of problem (Chapter 3). It is a regional problem relating to the agricultural sector, but it is not amenable to EIA-type evaluations. There is no particular 'project' which can be identified for which an EIA assessment is appropriate—and in any event it is too late for that: the problems have already developed as a result of a range of actions taken by many farmers over a long period of time. Processes are now well understood but it is still difficult to provide concrete advice to any individual farmer. Solutions require concerted actions at State, local government, community and individual levels, and the actions will vary from place to place. As O'Riordan (1985) has pointed out, changes to the agricultural sector as a whole are required.

The woodchip industry (Chapter 2) illustrates the second type of problem. Here there was a clearly identified proposal for a new project, for which predictions of environmental impacts were required and for which EIA is eminently appropriate. The third type of problem listed above is cumulative impacts, discussed in the previous section. The relevance of EIA in such situations is dubious; and there are clear similarities between these kinds of situations and the regional, sectoral environmental problems listed first. Finally, environmental issues, discussed in Chapter 1, are characterised and defined by politicisation, publicity and conflict. The woodchip industry case study is also an illustration of an environmental issue: but once the industry had been approved, EIA was no longer relevant as a means of dealing with the ongoing conflict. Certainly, EIA was not designed with conflict resolution in mind (prevention, perhaps). Instead, there is increasing use of mediation, especially (in Australia) in New South Wales (Naughton 1995). There, the *Courts Legislation (Mediation and Evaluation) Amendment Act 1994* has inserted provisions for consensual, court-referred mediation and neutral evaluation into the six Acts governing the various courts of the State, including the Land and Environment Court, if the parties agree.

Thus, in short, only one of the four types of environmental problems categorised above is really amenable to EIA-type approaches—namely, project-specific problems. Although these are important, in the overall context of the world's major environmental problems and issues (including the effects of acid rain, stratospheric ozone depletion, increasing greenhouse gas emissions, marine pollution, land degradation and desertification, crime, poverty, drugs and armed conflicts) it is evident that alternative methods of maintaining and improving environmental quality are required.

HOW IS THE EIA OUTPUT USED?

The Environmental Impact Statement (EIS)—the output of the EIA process— can be used for public inquiries, as a planning tool, as a basis for decision-making and as a reference point for monitoring and auditing project implementation.

Public Inquiries

In those jurisdictions where public inquiries are held in relation to environmental problems, the availability of a competently produced EIS is invaluable. It provides the necessary information and recommendations from which discussion and disputation can proceed and, it is hoped, a basis for reaching agreements between conflicting interests. In such cases there should be a final EIS following the public hearings. That final EIS should record the findings of the hearing—recording the more important arguments, the responses to them, and the recommendations of the hearing Commissioners. It is stressed that public inquiries in this context refer to a procedure *prior* to the approval of a proposal. Public inquiries are discussed further in Chapter 14.

Planning

An EIS can also provide a valuable input into the planning process, and its place in that process is considered further in Part 5. It can also be used in land suitability evaluation (discussed in Chapter 10). However, all the various short-comings discussed above (and summarised below) should be taken into account: EIA is an incomplete and imperfect planning tool. Nevertheless, the public can respond to the EIS, as most planning legislation makes provision for this. Thus the EIS is the public document which demonstrates the need for the project, identifies a series of planning options, evaluates them in the light of their environmental effects, and arrives at tentative preferred options. An example is given in Fig. 14.1.

Decision-making

For the EIS to have maximum benefit in decision-making, and as agreed by all commentators (but not always implemented), the EIS must have been prepared and considered *before* final or critical decisions are taken. Once decision-makers have assessed a project and determined what should be done, it is too late to undertake an EIA. Not only does human nature make a change of mind difficult, but political decision-makers are busy people who, once an issue has been dealt with, have other matters to turn to. They will not reconsider a mat-ter unless there are very good reasons for so doing.

In Australia, the (former) CEPA, which became part of the Environmental Protection Group within Environment Australia in 1996, was required to pro-duce an 'assessment report' for the Environment Minister separately from the proponent's EIS. In WA, too, the EPA produces a separate report and recom-mendations for the Minister. These documents are available to the public. Fed-erally, the Minister was required to make recommendations within 42 days, and these usually coincided with those in the assessment report. Wood (1995) noted that by the end of 1991, only three or four proposals which had been through the full Commonwealth EIA or PER procedure had been refused. On the other hand, virtually every proposal had been modified to reduce impacts. Some unsatisfactory ones may drop out, or the conditions imposed are so oner-ous that the proponent withdraws. The main aim is to mitigate impacts. 'A refusal rate of 2% of an average of eight EISs/PERs annually over 19 years is a remarkable testimony to the negotiating skills of the EPA staff' (Wood 1995:193). More recently, most review and approval processes have been speeded up.

Presented with a range of options in an EIS, decision-makers may decide to apply the 'polluter pays' principle, the 'user pays' principle, the 'precautionary principle', or the 'inter-generational equity' principle—with the latter two principles recognised explicitly in the new Commonwealth environmental leg-islation, amongst five principles of ecologically sustainable development. These principles are reasonably self-explanatory except perhaps the 'precautionary principle'. This is well defined in both Australia's ESD strategy and by the IGAE (ANZECC 1997), from Principle 15 of the Rio Declaration:

> where there are threats of serious or irreversible damage, lack of full scientific certainty shall not be used as a reason for postponing cost-effective measures to prevent environmental degradation.

Decision-makers may select various options including an unqualified approval of the project, or a qualified approval with various management methods required to be implemented, or the project may be rejected (and possibly alternatives suggested). The minimal use is to see the EIS as simply improving the design of a pre-selected project. Thus the terminology used in Western Australia—an Environmental Management and Review Program (ERMP) rather than an EIA—has given rise to concern for many years. It implies that the project will be approved and that the purpose of the EIA is merely to identify management methods whereby adverse environmental effects may be 'minimised' (an over-used and rarely defined term). There is a clear danger of running foul of Raff's (1997) seventh common law principle of EIA: namely that the assessment must engage in a real inquiry and not merely dispose of alternatives in favour of a decision which has already been made (Table 11.3). What is to be assessed, he states, should be a *proposed* action, not a *fait accompli*.

However, Wood (1995:70) noted that six of the seven EIA jurisdictions he reviewed did not require either the EIS or an EIS review document to be a central determinant of decision-making. He further commented that Western Australia was unique amongst all Australian States and Territories in that all conditions set by the Minister for the Environment are legally binding.

Project proponents fear the delays which EIA may introduce in the project approval process, with some justification. But they still tend to see EIA as an *additional* requirement. Instead, and again as emphasised by all researchers (and indeed by the governments which introduced the EIA requirements), EIAs should be carried out simultaneously with economic analyses, engineering designs and other feasibility studies, and most emphatically not as an afterthought or add-on requirement.

Monitoring and Auditing

It also needs to be emphasised that the EIS is not the end of the environmental assessment process. There should be post-implementation monitoring and auditing of the project and its actual as distinct from predicted environmental effects, modification of the project where such post-audit procedures indicate the need (Fig. 11.6), and use of the post-audit results to improve the EIA process.

Harvey (1998:78–9) usefully distinguished between and elaborated on monitoring and auditing, both of which he considered to be major weaknesses in Australian EIA processes.

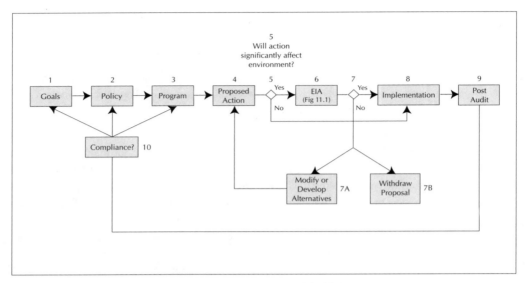

Fig. 11.6   Flow chart—the place of EIA in decision-making, modified from Munn 1979, Fig. 2.1.

Monitoring is of two types: implementation or impact monitoring. Implementation monitoring involves assessing whether a project or action has been implemented in accordance with the conditions of the approval, with respect to environmental impact mitigation measures. Impact monitoring, on the other hand, usually refers to assessing the environmental impacts resulting from the project or action. The emphasis is on collecting and analysing environmental data.

Harvey (1998) commented that there is some provision for monitoring in some Australian jurisdictions, but it is largely discretionary and has not been an important part of the EIA process. Malone (1997) also observed that EIA in Australia does not incorporate monitoring programs in a systematic fashion and commented that it is the weakest link in EIA procedures. He examined the need for monitoring, identified hurdles, discussed what is required and who should do it, and looked at legal and other measures which already exist to monitor environmental conditions. Provisions for monitoring compliance with the Act are included in the *Commonwealth Environment Protection and Biodiversity Conservation Act 1999*, but there are no provisions for impact monitoring as defined above.

Auditing also has more than one definition. Impact auditing compares the predicted impacts from the EIA with the actual impacts and is particularly useful for improving the quality of EIA prediction methods. Environmental management auditing refers to assessing the performance of corporate programs and structures for environmental management, and assessing corporate risks and liabilities (environmental management systems—EMS—are discussed in Chapter 8). The new Commonwealth environment protection Act includes provisions for the first two of the above types of audits. The first refers to a situation where the holder of an environmental authority may have contravened a condition of that authority; the second applies where there is the possibility that environmental impacts may be greater than those predicted.

With reference to the first definition of auditing in the previous paragraph, one of the major difficulties in Australia is the lack of verifiable predictions in EISs (Buckley 1989). This lack clearly renders impossible a measured comparison of actual with predicted impacts. This and other difficulties are discussed further in the following section.

EVALUATION OF EIA

Overall, EIA as a methodology for maintaining and improving environmental quality is very unsatisfactory. The technique is limited spatially and temporally; it is subjective; cumulative impacts pose almost insuperable difficulties; the method is *ad hoc*—responding to project proposals in a piecemeal fashion, rather than proactively setting policy frameworks in which developers and planners can operate; effective public participation (essential to resolve human conflicts) has been patchy at best—discussed further in Part 4; and (a fundamental problem) there has often been a political failure of will to act effectively.

In Australia, Gilpin (1995:21–2) quoted Smyth, a previous Director of the New South Wales Department of Environment and Planning, with regard to deficiencies of EISs to 1987:

- they fail to address real environmental issues relevant to the local community;
- much of the information is descriptive and superficial;
- they have a narrow focus on engineering aspects of proposals, rather than a broader, environmental planning approach;
- information is misleading, absent, contradictory and biased and there is a lack of objective findings;
- they use complex technical jargon which is incomprehensible to the community;
- they fail to justify adequately the proposals and to address alternatives, and
- they offer only a justification for a previous decision.

Smyth argued that more than 50% of all EISs failed to address the issues relevant to decision-makers, and that more than 75% failed to communicate adequately with the public.

Gilpin (1995) considered these criticisms to be somewhat harsh. But Wood's more recent, detailed review identified similar shortcomings. He found that EISs in Australia were often of indifferent quality, limited in their effectiveness, difficult for the public to understand, neglected cumulative impacts, did not treat alternatives properly and often neglected the no-action alternative. The documents were essentially proponent statements (Wood 1995:176). With reference to Commonwealth EIAs, the main weaknesses were found to lie in their coverage, screening, decision-making, monitoring, public participation, system monitoring and SEA (Wood 1995:291).

More recently, Warnken and Buckley (1998) examined the scientific quality of EIAs conducted for tourism-related developments (the main growth industry in Australia). They obtained all documents relating to 170 of the 175 tourism developments subjected to EIA in Australia from 1979 to 1993. In their analysis they considered only the 115 EIAs from 1987, as the earlier ones were piece-meal and the results were of even poorer scientific quality.

Only 17 individual impact predictions were testable by monitoring, and of those, only seven were accurate. Warnken and Buckley (1998) considered that while the value of the EISs as public information tools was not in doubt, as exercises in applied science they fell a very long way short of the most basic standards. In summary, the main shortcomings were:

- the scientific quality of the EIAs was very poor;
- methods were inadequately specified;
- sampling was inadequately replicated in space or time;
- significant parameters were often ignored;
- impact predictions were rarely quantitative or testable and 'frequently' inaccurate, and
- monitoring programs were generally inadequate for detecting likely impacts.

The following recommendations to improve the quality of EIA are modified from Warnken and Buckley (1998:7):

1 the EIS should be subject to independent, scientific peer review;
2 scoping should be refined, and
3 consultants should show that they have appropriate qualifications and be required to state that the level of funding for the study is appropriate for the likely impacts and their significance.

More generally, Wood (1995:298) identified the following areas in which improvements needed to be made in EIA practice: coverage; integration into decision-making; impact monitoring and enforcement; public participation; system monitoring; SEA, and quality of reports. Harvey (1998:210) considered the weaknesses of Australian EIA systems as being: the lack of monitoring and auditing; restricted appeal rights; inadequate consideration of cumulative and trans-boundary effects, and lack of quality control in relation to EIA consultants and EIA documentation.

Wood (1995) noted that EIA needs to be placed in the context of the ESD goals and the requirements of Agenda 21. He suggested this could be achieved by incorporating strategic environmental assessment (SEA) into EIA to achieve goals such as 'no net environmental deterioration' or, preferably, 'net environmental gain'. Harvey (1998) asked whether Australia should adopt formal SEA or focus on better planning. He commented that the EU requires SEA to accompany structural fund applications (for broadly based projects related to strategic or regional planning), and that the World Bank requires SEA for regional and sectoral development activities. As noted previously, some provision for strate-

gic assessment is made in the Commonwealth's 1999 new environmental protection Act. In fact there are many examples of Australian approaches to strategic planning which incorporate SEA principles but are not so identified.
Examples include the Great Barrier Reef and the Murray–Darling Basin.

As discussed by Conacher (1988, 1994a), different groups have differing perceptions of the roles of EIA. Agencies see it as an administrative requirement;
conducting an EIA fulfils required procedures and guidelines set down in legislation. Developers see EIA as a necessary hurdle which must be overcome in
order to obtain approval and licensing for their project. Environmental consultants, who actually do the work, are expected to practise good science in what
is essentially a politically motivated and economics-biased system. And the
community, not only conservationists and environmentalists, hopes that EIA
will protect environmental quality. The end result of all this is what Beanlands
and Duinker (1983) called a 'shotgun approach' to EIA. There is a tendency to
cover everything in a rather unsystematic way, because the object of the exercise is seen by those most closely involved as a bureaucratic/political requirement, not as a genuine attempt to improve the quality of the environment. But
because the practitioners are often closely scrutinised by members of the community, they also need to 'cover their backs'.

A consequence is the considerable length of some EISs. According to Harvey (1998), the expectation in the USA is that EISs should not exceed 150
pages. In Australia, the EIS for the Multifunctionpolis in South Australia was
more than 400 pages long, and that for the third runway of Sydney's airport 720
pages. Who has the time to read these massive documents? Further, the costs of
the reports have to be subsidised if they are to be made available to the public
at a reasonable price—in South Australia, for example, the practice is to keep
costs to below $30.

The main problems with EIA have been summarised by Conacher (1994a)
as follows:

- Many EIAs lack a central, conceptual or analytical theme or recognisable
  investigative design to guide the collection and interpretation of data.
- Predictions, if made, are commonly vague and of questionable value to project decision-making. Indeed, some scientists argue that ecological perturbations cannot be predicted (Marshall *et al.* 1985; Munro *et al.* 1986).
- Similarly, significance levels and probabilities are rarely indicated except
  in the most superficial, subjective manner.
- EIA is often applied as a reactive and discrete activity, only loosely related
  to the broader process of environmental decision-making. In particular,
  the co-ordination of assessment procedures with subsequent regulatory
  responsibilities (issuing permits and licences, for example) is often poor.
- EIA is almost exclusively applied to project-specific developments and not
  to the other, more complex and more common categories of environmental problems.
- Project-orientated impact studies are not necessarily protecting the environment (Couch 1985).

- As usually practised, EIA cannot handle the cumulative effects of piece-meal development.
- EIA is inappropriate for resolving environmental issues.
- The intensity of environmental disputes has continued to grow (in Canada—and in Australia?) despite the incorporation of EIA in planning processes.
- EIA is also inappropriate for many, major environmental problems such as those associated with acid precipitation, ozone depletion, toxic waste disposal, water contamination, and large-scale agribusinesses with extensive consumption of pesticides, fertilisers, and surface and groundwater resources. Putting it another way,

> It is increasingly recognised that traditional mechanisms for management of natural resources and the resolution of conflict over their use and development are inadequate in terms of recognition of the complexities of ecological integrity; in terms of co-ordination of the resource management functions of various agencies; and in terms of providing opportunity for the consideration and assimilation of the views of various interests within society (Grinlinton 1992).

The above shortcomings of EIA are also clear arguments for moving towards the more comprehensive, integrated, regional planning approach. Part 5 considers the way in which environmental protection has been or is being incorporated in the regional planning process in Australia; but before moving to that topic, Part 4 first looks at the crucial and pivotal question of public participation.

Part 4

Public Participation in Environmental Decision-making

As discussed previously, by definition environmental issues involve conflict amongst groups of people with different viewpoints. Full, meaningful participation by the community in the entire process of environmental protection is essential in order to resolve such conflicts or to prevent them from arising in the first place.

Additionally, there has been increasing alienation between governments and the governed with regard to the relationships between people and their environments since the 1960s and 1970s. Initially, this was associated with publications such as Silent Spring (Carson, 1962), Ralph Nader's (1965) Unsafe at any Speed and the subsequent development of 'Nader's Raiders', and the arguments over zero population growth in relation to the Club of Rome report (Meadows et al. 1972) concerning an increasingly unsustainable use of the world's resources. Zero Population Growth Inc. was founded in 1968 to promote an end to US population growth through lowered birth rates as soon as possible and, secondarily, to encourage the same goal for world population (Ehrlich et al. 1977:744). Following the moon landings and the view of earth from space, there developed an acute realisation that this is the only home we have, providing further impetus to the formation and rapid growth of international, activist environmental movements such as Friends of the Earth and Greenpeace.

In Australia, the Northern Territory's Ranger uranium inquiry, the battles over mineral sands mining on Queensland's Fraser Island, the dispute over the ACT's communications tower on Black Mountain, Tasmania's Lake Pedder and the Gordon-below-Franklin dam environmental issues, and wood chipping in New South Wales, Victoria, Tasmania and Western Australia were all confrontations with national implications. The disputes involved national conservation groups such as the Australian Conservation Foundation and the Wilderness Society, federal environment legislation and/or the Australian High Court (Davis 1972; Doyle and Kellow 1995; Bates 1995). Environment interest groups and the Greens political party also took up the causes of minorities.

Distrust between governments and the governed was a major factor underlying the growth of the public participation movement, which the late Derrick Sewell (pers. comm.) identified as resting on two fundamental considerations:

1   ethical considerations—those involved in the problem and its outcomes have the right to be consulted and involved in decision-making, and

2   pragmatic considerations—support for policies and programs depends on those who will have to pay for them.

Many other writers have discussed the need, desirability and rationale for public participation in environmental planning and decision-making.

McAllister (1980), in his book Evaluation in Environmental Planning: Assessing Environmental, Social, Economic and Political Trade-offs considered that the classical arguments emphasising the importance of active citizen participation in a democracy remain fundamental. To these first two arguments (following) he added a further four on the basis of his extensive experience.

1   Through citizen participation, the tendency for people to misuse power, given the opportunity, will be checked.

2   Citizens will become more informed about public affairs and thereby make wiser choices at the ballot boxes.

3   Planning should serve the interests of the public, which is difficult to do when those interests are not known. Plans have been abandoned due to lack of public support. (Similarly, Self (1998) has commented that planning is tied to the conception of the individual as a citizen, and not just as a consumer of services.)

4   Citizens are demanding a stronger voice in decision-making. They want to have the opportunity to participate actively in discussing major issues and developing programs for addressing them. Simply selecting decision-makers every four or so years is not enough.

5   The process of citizen participation, if properly organised, can help promote conflict resolution.

6   The community represents a huge pool of valuable knowledge that should be tapped in the problem-solving process. General education levels have risen steadily over the decades, and citizens include all of our experts, only a small fraction of whom officially assume this role in each public planning action.

However, McAllister also commented that exclusive use of citizen judgment has the serious weakness of lacking scientific rigour in estimating certain types of impacts. Therefore there is a need to design evaluation processes which include both citizens and experts.

More recently, the widely experienced Alan Gilpin (1995:63 and 161), made similar observations in the Australian context. He stated that democracy, which he defined as majority rule with respect for minority rights, is increasingly seen as a continuous and dynamic process. Governments carry ultimate responsibility but only with the most careful scrutiny. Constant vigilance is necessary because power corrupts. The idea that governments were elected to govern and should be free to do so without public 'interference' has tended to wither in democratic countries. The days have gone when satisfactory development decisions could be made behind closed ministerial and cabinet doors; when it was thought that the only rights people had was to come out every few years at election time to pass judgment on the quality of government.

Thus, communities increasingly expect that national, State and local authorities will respond to their concerns and that they (the communities) will have the opportunity to participate at policy and planning stages. Growing recognition of the desirability of public participation in decision-making led to an apparent increase in public consultation in Australia. Governments here and in other countries responded in three main ways:

1 they moved to make more effective use of existing institutions and legislation;
2 they introduced new legislation and new policies and introduced new ways of involving the community, such as task force workshops, and
3 they increased the opportunities for public participation by (for example) providing funds to attend hearings, or setting up special environmental tribunals and courts to hear disputes.

As has so often been the case, the USA provided the lead. A key event in the citizen participation movement was the requirement for 'maximum feasible participation' in the US Economic Opportunity Act 1964—the poverty program. Since 1964, model cities, transportation, urban renewal, education, health, environmental and most other programs in the USA have required public participation in their planning and management (McAllister 1980).

But there have been large variations between and within countries, and in Australia many would argue that the responses have been little more than tokenism. Politicians and bureaucratic executives generally have been unwilling to relinquish any real power, and perhaps feel threatened by an informed and vocal community.

This Part comprises three relatively brief chapters which consider: the allowance for and effectiveness of public participation in decision-making in environmental and resource legislation; public participation methods, and the political influence in downgrading public participation in Australia in the 1990s.

CHAPTER 12

PUBLIC PARTICIPATION IN AUSTRALIA

This chapter draws on Canadian examples by way of comparison with Australia. The two countries are similar in a number of respects (size, peripheral populations, reliance on primary industries, Commonwealth countries, federal systems of government, marginalised minorities), yet Canada appears in some respects to place more emphasis than Australia on maintaining and improving the quality of its environment.

THE ALLOWANCE FOR PUBLIC PARTICIPATION IN DECISION-MAKING IN RESOURCE AND ENVIRONMENTAL LEGISLATION

The relevant environmental and resources legislation and agencies were considered in Part 2. In general, there is little scope for public participation in Australian natural resource legislation except in provisions for objections (in mining legislation) and appeals (in planning legislation). Objections in mining legislation normally arise in the context of competing mineral claims at the exploration stage, and are heard in Mining Wardens' courts. Wardens' recommendations are forwarded to the Minister who makes the final decision. It is rare for objections to an exploration permit to be raised on environmental grounds, because the legislation generally makes scant reference to this aspect. It is even more rare for a Mining Warden to adjudicate on such disputes. First, it is very difficult for an environmental objector to obtain a hearing, and second, Mining Wardens do not have the skills, qualifications or experience to adjudicate on environmental matters even if the legislation makes it possible.

Nevertheless, as discussed in Chapter 5, in recent decades changes to mining Acts have imposed some significant restrictions on exploration activities. In Queensland, for example, where a lease is subject to a hearing of objections, the Warden in deciding whether to recommend to the Minister the granting of a lease, must take into account whether there are likely to be any adverse environmental effects caused by the operation (Bates 1995:216).

Planning legislation in all Australian jurisdictions makes provision for planning and development proposals to be made public and for public objections and appeals. They are heard through boards or tribunals which judge the merits of a case. Points of law are generally directed to courts of law. But in general such objections and appeals are finally determined by the planning Minister, who was often the original decision-maker. Members of the community understandably have little confidence in such a process. On other occasions, objections and appeals are raised against decisions made by local governments, often by a developer whose proposal has been rejected by a local Shire (or County) Council. These are difficult situations. Local governments arguably are grass roots democratic institutions, much closer to community views on an issue than the Minister. In Western Australia at least, there is a perception that in recent years the Minister has supported developers far more frequently than the democratically elected local government body.

Environmental protection legislation appears to favour more effective public participation in decision-making than most previous Acts relating to the use of natural resources—insofar as EIA is essentially a public process. A good example is the provision for public inquiries in response to public comments in relation to major projects or those likely to have a substantial environmental impact in the Federal *Environment Protection (Impact of Proposals) Act 1974* and again in the new 1999 Commonwealth *Environment Protection and Biodiversity Conservation Act*. But powers under the new Commonwealth legislation still appear to be limited in respect of public rights. Introducing the primary legislation in 1974, the (then) Minister for Environment, Dr Moss Cass, said that it would:

> ...not give individual citizens the power to stop bad projects or to set conditions for moderate ones...(but)...it will enable the public to argue a case publicly, to have the case published, and to force governments to justify their decisions (cited in Bates 1992:94).

These restrictions still apply in the new Act, whilst ministerial decisions, and the reasons for them, are to be made public through postings on the Internet.

Under the 1974 Act, EISs and PERs were open to public comment, although exceptions can be made on the grounds of public interest, national security or commercial confidentiality. The Minister could seek a public inquiry; but only five were ever initiated under the Federal Act, in relation to: uranium mining in the Northern Territory (the Ranger project) in 1975; the export of mineral sands from Fraser Island in Queensland (also 1975); broadcasting facilities in the Ulladulla Hinterland NSW (1989); resource use at Shoalwater Bay in Queensland (1993), and the East Coast Armaments complex in Victoria (1995)—Harvey (1998:170).

The 1999 Commonwealth environment protection legislation retains the pr vision for the public inquiries. As commented on in the previous chapter, draft guidelines of the preparation of an EIS or a PER *may* be referred for public comment, but the draft reports themselves must be made public and the proponent is required to summarise any comments made. The environment Minister will have the power to monitor compliance of a proponent with an environmental management strategy, enforce conditions and take action if the proponent defaults. Such conditions will be publicly available but with potentially extremely broad exceptions (as before) based on issues of commercial confidentiality, national interest, national security and foreign policy. It is almost as though the new Act has been designed to prevent the public from getting in the way of new developments—in marked contrast to the central role of public participation in the US NEPA process.

The biodiversity component of the new Act requires the Minister to make conservation plans public and to consider all comments. Three months are given for this process, unlike the 20 working days for comment on draft PERs and EISs.

In the States and Territories, environmental protection and planning legislation generally allows for public participation at various stages of an EIA process. There are varying degrees of public participation in administration, policy formulation, information, decision-making, monitoring, reporting and enforcement (Taberner et al. 1996). As Wood (1995:225) commented, 'EIA is not EIA without consultation and public participation'. However, many Acts use terminology such as 'may' rather than 'shall', which more often than not leaves public participation to the discretion either of the agency or the Minister, thereby weakening the intent and influence of public participation. This discretion may also apply to publication of decisions, reports and outcomes of inquiries.

In Canada, on the other hand, public involvement in EIA at the scoping stage, in the preparation and review of an EIS, in hearings and in commenting on panel reports is well established. The *Canadian Environmental Assessment Act 1995* widened and improved the scope for public participation, and placed greater emphasis on the effectiveness of public reviews. The environment Minister's powers were enhanced (Gilpin 1995:114). The public participation funding program is particularly significant, having been formally established in 1990 after several years of practice. With annual budgets exceeding Can$2 million, the program employs researchers to prepare and argue a case. The *Canadian Environmental Assessment Act 1995* also provides for setting up a 'public registry' of all documentation relating to every project subject to environmental assessment. There is a computerised index which is accessible electronically. Most documents have been available to members of the public at no cost (Wood 1995:237). In general, EIA procedures in Canada's Provinces are similar to its federal procedures (Thomas 1996:114). And in New Zealand, the degree of public participation in resource management planning and decision-making is a significant component of the *Resource Management Act 1991*. Hearings and mediation of disputes are a feature of procedures, and public submissions are sought as part of any inquiry (Gilpin 1995:142).

Although provision for public participation in Australia is often permissive rather than mandatory (as indicated above), public participation is more pervasive than the requirements specify; but it relies on written rather than oral presentations. There is consultation and participation *after* a draft EIS has been produced and proponents are usually required to respond to these comments in the final EIS (or in a supplement). There is only very limited public intervention funding (Wood 1995:238): Victoria's EIA guidelines state that funds may be made available to help community groups prepare submissions (Thomas 1996:18). Unlike Canada, there is no central collection of EIA reports for reference purposes (Wood 1995:250). However, Queensland, South Australia and Tasmania require information held by relevant authorities to be kept. The jurisdictions vary in their requirements, but the type of information includes copies of licences, works approvals (including their conditions), details of environmental reports, results of monitoring programs, details of environmental management programs and environmental protection orders, clean-up orders, details of pollution incidents and civil and criminal enforcement action. The registers must be available for inspection by the public. There is also a non-statutory register in Victoria (Taberner *et al.* 1996).

New South Wales was distinctive in that an express objective of its *Environmental Planning and Assessment Act 1979* was to increase opportunities for public involvement and participation in environmental planning and assessment—recently diminished (discussed in Chapter 14)—hence the use of the past tense in this paragraph. Written notice of development applications was to be provided to neighbours, people detrimentally affected and to local newspapers. Applications could be inspected and any person could make a submission and inspect the EIS. There were detailed provisions for public inquiries (Fowler 1982). Further, the Land and Environment Court recognised that reasonable opportunities for public participation in the plan-making and development processes are crucial to the integrity of the planning system. It has declared invalid decisions which fail to satisfy the public notification requirements in the Act (Bradbury 1995). In Victoria, the 1995 *Guidelines* state that when an Environmental Effects Statement (EES, equivalent to an EIS) is required, the Minister for Planning 'usually' establishes a Consultative Committee to guide the preparation of the EIS. Community and environmental groups are represented on the committee. A public scoping phase is carried out under the auspices of the committee and the EIS is exhibited for two months. The Minister for Planning will 'usually' appoint a public inquiry panel for proposals which are the subject of an EIS to provide advice as part of the assessment process (Department of Planning and Development 1995).

A HIERARCHY OF PUBLIC PARTICIPATION EFFECTIVENESS IN DECISION-MAKING

The previous section discussed the allowance for public participation in decision-making on environment-related matters in Australia. Here we consider the *effectiveness* of the various modes of that participation. The underlying assumption is that if environmental issues hinge on human conflicts, then the people directly involved must be included in the environmental management process. But that involvement must be an effective one, not merely window dressing or a one-way transfer of information from an agency to the community. Other terms for this undesirable *modus operandi* include 'top-down', autocratic and 'command-driven'. In other words, effective public participation does not mean 'getting the public on side'; rather, it means full, co-operative involvement in the actual process of making decisions—and it must be seen to be such. As will be argued, few public participation methods achieve that objective.

For the purpose of the discussion, the various avenues for public participation in environmental decision-making may be grouped into the following hierarchy of increasing effectiveness (from Conacher 1988):

- submissions;
- objections and appeals;
- public hearings and inquiries;
- setting objectives and agendas, and
- legal rights and mediation.

Submissions

Any member of the public can make a submission to any government agency or politician on any matter and 'the public' is often invited—even importuned—to do so. It is often in the interests of an agency (or sometimes a consultant) to be able to demonstrate that a large number of submissions was received on a particular matter—although one good submission potentially may be more useful than a hundred bad ones. The question nevertheless becomes, how effective are those submissions?

Harvey (1998) has commented that formal public involvement in environmental matters in Australia usually takes place through submissions on EIA documents. The Western Australian *Environmental Protection Act* requires the EPA to consider and respond to submissions: it cannot ignore them. But in practice it is rare for a person making a submission to receive anything other than a formal, standardised acknowledgment. In New South Wales, under the 1979 *Environmental Planning and Assessment Act*, anyone can make a submission relating to, or inspect, an EIS. More generally, a fairly common outcome is for a report to be produced which lists the various comments received on draft EISs and the proponent's response to them. This is actually a *requirement* in Canada but not in NSW, although it is often done in practice according to Harvey. Queensland has no requirement for public review of their Impact Assessment Statement (IAS—equivalent to EIS) under the *State Development and Public Works Organisation Act*, but in practice public review is usual and encouraged. This is also the case for EISs under the *Local Government (Planning and Environment) Act* (now superseded by the *Integrated Planning Act 1997* (Chapter 19). As a consequence some assessments have been termed 'environmental studies' to avoid public input (Harvey 1998:53). South Australia requires public meetings for the exhibition of EISs and PERs, but there is no requirement for comments to be recorded or for issues raised to be taken into account. Each level of assessment is subject to public review in Tasmania, but again there is no formal requirement for proponents to respond, unlike Victoria where they are required to do so (Harvey 1998:64–8). Extracts from submissions may even be included in the final report's Appendices. But it is: a) usually difficult to ascertain the connection between the various submissions and the report's recommendations (unless a submission happens to coincide with the agency's—or report writer's—views); and b) even more difficult to identify whether the recommendations have actually had any bearing on government regulations or legislation.

To be effective, the contents of submissions should be summarised in the ensuing report and the agency's (and, in the interests of transparent decision-making, the proponent's) responses to those contents given. If this is not done, then submissions become (almost) totally ineffectual means of achieving public participation in decision-making; rather, they are mere public gestures to make both the agency and the submitter feel good. But one might well ask—why bother to go to the time, expense and frustration of such exercises if they are to be little more than apparently reluctant public relations exercises?

The other point about submissions to re-emphasise (which in fact applies to most methods of public participation) is that, as discussed by James (1991) (Chapter 13), they are unlikely to be representative of community views. However, public participation has various objectives, one of which may be to tap expertise within the community. In that case the lack of representativeness is irrelevant, and certainly better than no response at all. The challenge then lies in recognising and balancing the value of particular contributions.

Objections and Appeals

As discussed by Gilpin (1995) and Harvey (1998), some 10% of all federal EISs in the USA were subject to lawsuits in the first 13 years of NEPA, peaking at 189 in 1974. They were often brought by citizen groups based on the claim that the EIS was inadequately prepared. There were accusations that litigation was often used as a stalling tactic. It has also been claimed (in the Victorian Parliament, during debate on the *Environmental Effects Act 1978*) that the US requirements had produced a lot of unnecessary work in relation to some relatively insignificant projects and insufficient investigation for some major ones. Furthermore, the fate of some important projects was settled in law rather than by elected decision-makers (Raff 1995b:253). But in a comparison of US and German EIA procedures for similar projects, Kennedy (1986—cited in Harvey 1998) found that the US system actually facilitated development. This is probably because of greater certainty in the US system. With an established history of environmental litigation, US lawsuits proceed against a background of precedents and court cases. In contrast, (West) German environmental groups caused significant delays to development projects subject to EIA requirements.

Litigation has also been criticised for the cost to the litigants, and of course on the grounds that lawyers and judges have little expertise in environmental matters and therefore decisions are made on the basis of what may well be environmentally irrelevant legalistic arguments.

With the exception of NSW, which has completed more EIS documents than all the other States combined (Harvey 1998:185), Australian EIA has generally explicitly avoided US-style judicial review in public law courts and the availability of appeal rights. The key difficulty is to obtain 'standing'. In contrast, various steps in the environmental assessment process in the Canadian federal system are open to varying degrees of public participation and challenge in the courts. There had been more than 50 court cases by mid-1994 (Wood 1995:82).

NSW is unique in Australia in its generous standing and appeal rights for EIA-related development applications. Under the *Environmental Planning and Assessment Act 1979*, anyone dissatisfied with a decision can appeal within 28 days to the Land and Environment Court (Fowler 1982:56) (but, as discussed in Chapter 14, these rights were being eroded in the 1990s). According to Taberner *et al.* (1996), it is possible to bring both civil and criminal proceedings in relation to environmental protection and planning law in New South Wales, and civil proceedings in Tasmania, South Australia and Queensland; but neither in Victoria and Western Australia. There have been legal challenges outside New South Wales, but often they are not directly linked to EIA legislation or the adequacy of environmental investigations. In the ACT, any person who has made a submission in relation to EIA documents and applications, may appeal to the Land and Planning Appeals Board established under the *Land (Planning and Environment) Act*. There are no provisions for third party appeals in the Northern Territory. However, if the proposal was initially lodged under the Planning Act, citizens may appeal to the Planning Appeals Tribunal with regard to the conditions. Similar arrangements apply to mining permits. Queenslanders have no scope for appeal under the *State Development and Public Works Organisation Act*, but appeal rights are present under Part 7 of the *Local Government (Planning and Environment) Act* (now the *Integrated Planning Act 1997*). Both applicant and objectors may appeal to the Planning and Environment Court. South Australians have no appeal rights under the *Development Act*. There are first and third party appeal rights to a Resource Management and Planning Appeal Tribunal under Tasmania's *Environmental Management and Pollution Control Act*, but they are available only to those who made representations during the public exhibition period. Victorians generally have no rights of appeal. However, limited appeals to the Victorian Civil and Administrative Tribunal are possible: if the proposal is co-ordinated with the EIA process and an inquiry is held (which happens fairly frequently), that becomes an alternative to the appeal process.

Provisions for public involvement and appeal under Western Australia's environment protection Act are considered generous, and include third party rights. Most EPA assessments do not warrant extensive public review, but appeals under Section 100 of the Act against the level of assessment and against the advice in the EPA's assessment report can be made to the Minister for the Environment within two weeks of the completion of an assessment report. About half of all appeals to the Minister relate to the assessment level (Carew-Hopkins 1997). However, once the appeals process is cleared, only proponents may appeal against conditions set in the final decision. In the past, statutory public appeals provisions under the Act were regarded as an indicator of the health of WA's EIA system, and a steady increase in the number of appeals was evident during the 1990s. But Bache (1998) has observed that although a large number of appeals can be interpreted as indicating an accessible system of objection, it may also indicate community disenchantment with the EIA process.

Finally, under the 1974 Commonwealth legislation, the public can review the project-specific guidelines prepared by Environment Australia but there is no provision for appeal against the level of determination or the final decisions (Wood 1995:126; Harvey 1998:82–6;169). With reference to the new Commonwealth environmental legislation, there are provisions for standing for individuals or organisations with environmental interests in relation to judicial review of administrative decisions made under the Act, or failure to make a decision, or the decision-making process.

The obvious problem with objections and appeals is that, by their very nature, they occur after the event: the decision has already been made. Consequently, the process inevitably becomes adversarial rather than participatory. Therefore appeals are hardly an effective means of involving the community in making the decision. Furthermore, once a decision has been made, then it is human nature to defend that decision rather than to admit (in response to an objection) that the decision-maker has made a mistake. It is rare for a government agency or a government to make such an admission. There is also the difficulty of dealing with 'group' rather than 'individual' dynamics, as well as the constraints of the legislation under which agencies operate.

Other weaknesses of tribunal or appeals systems relate to: standing (there is no guarantee that an appeal will be heard); findings may not be open to public scrutiny; decisions are final (unless points of law are raised), and all the facts may not be presented.

Thus, potentially, objections and appeals can be effective only if they are made to a higher authority than the one which made the decision; as occurs in the formality and hierarchy of courts of law, or in appeals to State planning Ministers over local government decisions. Even though an appeal or objection may be made to a different State government Minister from the one who made the initial decision, any change of mind on a significant issue will involve the entire Cabinet or, effectively, the State government; and this is very difficult to achieve. In relation to some major environmental disputes, Australian conservationists have relied on the federal government to override State government decisions, but with limited success (Munchenberg 1994; Bates 1995). The 1999 Federal *Environment Protection and Biodiversity Conservation Act* further removes the Commonwealth from such avenues of appeal.

To reiterate, appeals or objections are not effective means of involving the community in decision-making. The time for involvement is *before* a decision has been made and positions have hardened—although that second line of defence does need to be available.

Public Hearings and Inquiries

The first problem with this approach is that public hearings and inquiries are rare in Australia in the context of environmental protection legislation. The five public hearings conducted at the federal level, within the provisions of the 1974 federal environment Act, in 24 years from 1974, compare unfavourably with more than 40 completed panel reviews under the provisions of Canada's environmental legislation at the federal level in 16 years from 1977 to mid-1993, and another 13 actively in progress at the latter date (Wood 1995:60). Revealingly, one of the conclusions reached by the Study Group on Environmental Assessment Hearing Procedures in Canada (1988:41) was that:

> A public hearing is not a privilege granted to the population, but a service requested of the public by the government to help the government make a better decision and to favour a harmonious relationship between economic development and environmental protection.

As might be expected, public hearings are widespread throughout the US planning systems. Perhaps more surprisingly, public inquiries with full public participation are quite common in Britain—in relation to town planning, airports, motorways and energy installations (Gilpin 1995:64). Similarly, government projects (such as road, air or rail) in Germany are subject to special procedures, generally including inquiries involving display of plans and public participation (Raff 1995b:254).

There are provisions for public inquiries in the Australian States. In Victoria, inquiries are often held (at the discretion of the Minister) at the end of the public exhibition stage of an environmental effects statement (EES). New South Wales also undertakes inquiries and hearings as a normal part of the environmental planning and EIA process. New South Wales employs full-time commissioners, and Victoria has a panel system of more than a hundred qualified and experienced people. From the 1980s to 1992 there were 250 public inquiries in New South Wales into environmental planning and heritage matters, and a procedural handbook was published (Gilpin 1995:64). There has also been an increasing emphasis on 'planning focus meetings' and 'community focus meetings' to identify key issues to be considered in an environmental assessment (Harvey 1998). However, in 1985, amendments to the environment legislation gave the New South Wales Minister a clearer option to exclude public participation in the preparation of a State Environmental Protection Policy (SEPP); and public involvement in both planning and assessment were decreased overall. As Justice Stein commented in 1998 (discussed in Chapter 14), this exclusion process (which also took place in the other States) continued into the 1990s, increasing the likelihood of disputes reaching the courts.

Harvey (1998:184) has stated that in practice, no inquiries have been called by any of the States and Territories, which appears to contradict Gilpin's information, above. Harvey comments (pers. comm. 1999) that it is a question of definition, and that his statement refers to 'higher level' inquiries (above an EIS) as provided for in the 1974 and 1999 Commonwealth legislation. This is supported by Fowler, who noted as early as 1982 that inquiries in New South Wales conducted by the Commission were initially judicial, and then became round table discussions with invited participants. The public was relegated to the role of observers. Such inquiries are not full, open, public hearings in the manner of the procedures employed in the USA or Canada, or as provided for in the Commonwealth legislation.

On a more positive note, the Australian Senate (the Federal Parliament's upper house) has conducted numerous public inquiries into an extensive range of environmental matters, including wood chipping, water resources, hazardous chemicals, coastal management, mining, land degradation and land use policy (Table 15.1), and the authors contributed to several of those listed. The following comments are made on the basis of that experience, as well as from visits to Canada and with reference to some of the Canadian literature.

The effectiveness of public hearings and inquiries depends largely on the way in which they are structured. Australian hearings tend to be formal, with the person or persons making the submission on one side of a long table, facing the hearing Commissioners or Senators. The proceedings are recorded *verbatim* and published as *Hansard* reports. One of the Senators (usually from the political party forming the government, although the composition of the hearing Senators is usually bi-partisan) chairs the meeting. Following an opening statement from the Chair explaining the purpose and structure of the proceedings, the person(s) appearing at the hearing is invited to make an opening statement. The Senators then take turns to ask questions. It is rare for Senators to interrupt, so it becomes very much a formal, one-on-one situation. Proceedings are generally open to the public although the person appearing before the inquiry can ask for evidence to be submitted *in camera*.

These proceedings are generally fair but in our view suffer from a number of disadvantages. It can be intimidating for someone lacking experience to appear before such an inquiry. It can also be very tiring being subjected to a sometimes hostile, sequential cross-examination from five or more Senators in the presence of public onlookers or interested parties (also sometimes hostile). Lawyers are not present, although it is debateable whether an inexperienced person would benefit from having legal representation. More useful, from an environmental point of view, would be the provision of expert assistance to laypersons. The Canadian *Environmental Assessment Act 1995* includes a participant funding program to help individuals and organisations to involve themselves in public reviews of projects (Wood 1995:64). Funds can be made available to disadvantaged communities to hire experts and, less frequently, lawyers, to assist them in presenting their case. In Australia, however, there is only some very limited public intervention funding (Wood 1995:238).

Another point is that although hearings in Australia are generally open to the public, it is not possible for witnesses to question one another. Yet some of the most beneficial exchanges take place informally, during tea breaks, or outside the meeting room. A less formal hearing, as conducted in Canada, makes such interaction possible and assists conflicting groups to better understand each other's point of view—providing that the Chair is firm. The Canadians' experience has been that once that improved understanding occurs, many of the disagreements (not all) are resolved, which greatly simplifies the job of the independent Canadian Panels (who are appointed by the government and whose personnel change from one issue to another, with their skills and expertise reflecting the nature of the issue under consideration) (Study Group on Environmental Assessment Hearing Procedures 1988).

Wood (1995:192) has stated that in the early years of Canada's Environmental Assessment Review Process (EARP), panel recommendations were accepted by initiating departments. Conacher (1988) noted that in only one instance from 34 hearings in 1977–86 were a Hearing Panel's recommendations not accepted by the Canadian Federal government. However, according to Wood (1995), as EARP was applied subsequently on a less voluntary basis, initiating departments tended to deviate from panel recommendations. Decisions were made so late in the design/construction process that the 'no go' option was virtually precluded. Wood noted that the Chairman of FEARO found it necessary to state in 1993 that decisions should be made only *after* environmental factors have been assessed. Nevertheless, Wood also reported that there have been many successes involving panel reviews. In contrast, Australian Senate inquiries rarely have much influence. The committees are appointed by the Senate, not the government; and their personnel are politicians, not (usually) experts in the area under inquiry. In the woodchip case study, for example, when the second Senate committee inquiry in Western Australia determined that the Forests Department was not complying with its own undertakings, it became evident that the Senate was powerless to do anything about it (Conacher 1983). On the other hand, recommendations from the agricultural and hazardous chemicals inquiries probably influenced the framing of subsequent legislation and the development of codes of practice and national/State policies.

A further point is that Canadian Panels will conduct hearings in isolated locations, rather than expecting poor and dispossessed minorities to travel often considerable distances to a city, as is usually the case in Australia. A notable exception has been the outback locations of some of the Native Title Tribunal hearings under the skilled guidance of Justice Robert French.

Canada has conducted a considerable amount of research into means of improving the public hearing process in order to make it more effective (for example, Study Group on Environmental Assessment Hearing Procedures 1988). (It should be noted that the Canadian Environment Assessment Research Council—CEARC—was a highly effective organisation which was much admired. It produced some 150 research reports on various aspects relating to environmental assessment, but regrettably was disbanded in 1992 largely as a result of public expenditure cuts—Wood 1995:61.) There is no equivalent body responsible for carrying out research with regard to the Australian federal environment legislation, as noted previously; and we are unaware of any federally initiated research which would improve the quality and effectiveness of Senate inquiries. In this context, the loss of universities' previous neutral objectivity in their research as a result of their increasing reliance on external sources of funds and, closely related to that, the large costs (and sometimes complete unavailability) of data, are having increasingly serious, adverse consequences for the quality of environmental management in Australia.

The conduct of public inquiries at the State level is more varied, depending very much on the nature of the inquiry and who is conducting it. As commented previously, Harvey (1998) observed that no public inquiries have been called in any State or Territory. The 'public' hearings tend rather to be inquiries held with selected, invited members of the community. They also tend to be overtly political. Thus a 'public' meeting held in 1992 to inquire into the Western Australian *Environmental Protection Act 1986* comprised an invited audience, chaired by a prominent businessman. Not surprisingly the inquiry produced recommendations watering down the power and roles of the EPA chairman and director in line with the (then Labor) Government's, business community's and mining lobby's well-known views, and implemented with alacrity by the incoming (1993) Coalition Government (Conacher 1994b). An earlier example was the process by which the Western Australian State Conservation Strategy was developed. Directed by a prominent environmentalist under the auspices of the Environmental Protection Authority, and involving extensive community consultation through a series of well-attended workshops, the outcome was a report containing views unpalatable to the then governing Labor Party. The Minister disclaimed the results in an attachment to the front of the report and a duly sanitised version was produced a few years later. Meanwhile, the producer of the original, more useful strategy 'retired' to the south coast. A final point with regard to inquiries, and relevant to the theme of this book, concerns the general reluctance of State and federal governments to act on inquiry recommendations for natural resource management and land-use reforms (Chapter 15).

Potentially, then, public hearings and inquiries can be an effective means of obtaining meaningful public participation in environmental decision-making; but it depends on how they are done and, as before, there is always the question of the extent to which recommendations from such inquiries are implemented. The answers depend on the politicians of the day.

Setting Objectives Or Agendas

Effective public involvement in setting objectives or agendas is probably the most meaningful form of public participation in environmental decision-making. This is because the input is obtained early in the process, long before any decision has been made, and therefore the contributions can influence the actual nature and content of what is to be considered by the ultimate decision-makers.

Additionally, the input can be obtained (if desired) in a way that is truly representative of the community as a whole, by conducting carefully structured surveys as well as the more usual workshops.

As discussed in Chapter 11, the processes of screening and scoping are effective means of obtaining this public input. Screening and scoping, now in regular use in Canada, are both valuable and necessary means of involving the community in decision-making. The 1974 Commonwealth environment protection Act had permissive powers for screening and scoping but, unlike in Canada, those processes did not appear to involve the public. Under the 1999 Act, only the lowest level of environmental assessment—the 'Assessment on Preliminary Information'—requires the proponent to obtain public comment on information included in the referral and to summarise for the Minister any comments received. The higher levels of 'Public Environment Reports' and 'Environmental Impact Statements' state only that the Minister *may* invite anyone to comment...and take account of the comments received. The terms 'screening' and 'scoping' do not appear to be used.

More generally, *screening* (using questionnaires, workshops and public hearings, for example) can be used to determine the level of community concern over an issue, and hence the level of detail an investigation should go into (such as a preliminary, summary inquiry versus a full-scale EIA process). With regard to regional environmental planning, screening can be used to assist in deciding what the issues and objectives are. *Scoping* (again soliciting community input) is then used to identify the nature and importance of specific issues and concerns, which assists in determining what should be included in the investigation (Everitt and Colnett 1987). Unfortunately, as discussed previously, the public is rarely involved in the screening process in Australia, whilst scoping is available in only about half the jurisdictions. Even then, community involvement is discretionary.

Legal Rights

Arguably, the right to have a dispute resolved by a neutral and impartial judge (or jury) in a properly constituted court of law is a fundamental democratic right. This has always been the case in the USA, and the principle was applied fully in NEPA. However, as noted by Harvey (1998:6), in the early stages of EIS in the USA the emphasis was on procedures—what would stand up in court—and not on content, or what would be useful to decision-makers.

In contrast, in Australia the public generally does not have the 'right' or 'standing' to take a government agency to court in an attempt to force it to adhere to the provisions of its enabling legislation. Indeed, as discussed earlier in this chapter, the Federal *Environment Protection (Impact of Proposals) Act 1974* was written deliberately so as to prevent the public from having access to the courts over environmental issues. However, access to administrative tribunals, boards or specially constituted courts, applies under certain circumstances (refer Objections, Appeals and Inquiries). Such cases are judged on merit and generally centre on administrative matters. Only points of law, should they arise, are referred to courts of law. On the other hand, the same government agencies can enforce their requirements on members of the public in a court of law. This is particularly true of most environmental legislation, in cases of wilful damage, negligence or non-compliance (Chapter 7).

Canada, too—as is still true of most countries—did not provide its citizens with the right of access to the courts over environmental matters. However, the situation in Canada changed after the late 1980s. For example, if the Minister decides that a public review is not required, that decision can be challenged legally (Wood 1995:125). Indeed various steps in the Canadian environmental assessment process are open to varying degrees of public participation and challenge in the courts.

Amongst the Provinces, Ontario has established a Bill of Environmental Rights. In developing the proposed Bill in 1992, the Task Force, in its terms of reference, was governed by the following policy objectives and principles:

1   the public's right to a healthy environment;
2   the enforcement of this right through improved access to the courts and/or tribunals, including an enhanced right to sue polluters;
3   increased public participation in environmental decision-making by government;
4   increased government responsibility and accountability for the environment, and
5   greater protection for employees who 'blow the whistle' on polluting employers (Jeffrey 1992).

The arguments against extending to citizens the right to have environmental disputes resolved in a court of law are instructive. They usually compress into two areas. First, legal solutions to environmental issues are costly and time consuming. It is difficult to disagree, except that the argument ignores the important principles involved. For instance, international agreements over several decades have stated that a healthy environment is a basic human right, although draft principles have yet to be legally codified (Simpson and Jackson 1997). In the USA it is often said that public access to the courts has very rarely stopped a development proposal. Instead, conservationists have been able to use the courts to impose costly (to the developer) delays until the courts have been satisfied with the level of environmental protection to be undertaken. This is not necessarily a bad thing.

The second line of argument has been presented previously—namely that lawyers are not qualified to determine the 'best' (in a scientific sense) outcome of an environmental issue. This is also true—and the use of 'expert' witnesses is not without its difficulties either (Bell and Williams 1998). But this argument ignores the fact that most environmental issues revolve around *conflict* between groups of people, not necessarily scientific accuracy, and that in a sense a court of law is a major form of formally structured mediation between conflicting parties. Moreover, the 'best' outcome is often one that leaves most people satisfied, not necessarily that with the optimum scientific result. And again, of course, there is the question of the entitlement of citizens to have access to such a neutral forum.

Mediation is another technique for involving the community: hundreds of major public disputes have been resolved in this way in the USA (Gilpin 1995). It is a potentially valuable mechanism for resolving contentious environmental matters, and Cocks *et al.* (1996) have proposed a prototype mediation support system in the context of resolving forestry issues in Australia. In Canada, the Minister may decide that mediation will take place instead of a Panel Review (Wood 1995). The Australian Resource Assessment Commission (RAC) outlined the possible role of mediation in its public inquiry procedures before the Commission was disbanded after only four years; but it is understood the procedure is still alive federally and in general is now encouraged. In 1991, the NSW Land and Environment Court introduced the option of mediation of disputes in several classes of its jurisdiction (Gilpin 1995; see also Naughton 1995).

The cost of legal challenges is a major problem. Only the wealthy and the poor (the latter through access to Legal Aid—and that has declined in recent years, as has funding to Environmental Defenders' Offices) have access to the courts. The great bulk of the population, the middle-income earners, cannot afford lawyers and are ineligible to obtain Legal Aid. This is a problem which must be resolved—perhaps by the legal profession, rather than relying on government. Environmental Defenders' Offices (EDO) perform a valuable role in this context. An example is the comprehensive and helpful guide published by the WA office (EDO 1997), covering various statutory public rights in relation to environmental law and the avenues available for public participation. A special issue of the *Environmental and Planning Law Journal*, on ten years of EDOs, also offered some insights into public interest environmental litigation in Australia. Matters discussed included the role of EDOs, difficulties in securing standing, costs and reform, and the large discrepancies amongst the States (*Environmental and Planning Law Journal*, 13(3), 1996).

Access to the courts to resolve conflicts is an effective means of public participation in that all persons appearing before the court do so theoretically on an equal footing. However, for the main objective of involving the public meaningfully in environmental decision-making it is really an adjunct, or last resort, not the prime method.

We return therefore to the earlier proposition that good public participation is implemented most effectively by involvement early in decision-making processes. The following chapter outlines various strategies for achieving this.

CHAPTER 13

METHODS OF PUBLIC PARTICIPATION

There are various objectives for involving the community in environmental decision-making and no single technique can attain all of them. Indeed some can thwart others. The objectives may be to:

1   increase public awareness of environmental problems and issues;
2   mutually educate the governors and the governed;
3   reduce suspicion of the decision-making processes;
4   seek greater transparency and accountability;
5   tap expertise in the community;
6   reduce conflicts between different interests and seek consensus;
7   ensure that a plan or proposal will be accepted by the community, or
8   redefine the goals of government.

Numerous methods can be used to achieve these objectives, as listed by Munn (1979) and presented in Table 13.1.

Related to the above are the following eight steps required of agencies and decision-makers to improve public participation in environmental decision-making (developed from Mitchell 1990 and especially James 1991). The basis of the process is that all parties should feel satisfied that decision-making and outcomes have been achieved fairly. Effective participation ensures that people have an 'investment' in seeing that decisions are implemented.

1   Provide timely, comprehensive, comprehensible and accessible information.
2   Provide opportunities for participation at different stages, importantly including the initial stages of plan or policy formulation.
3   Provide a range of participation methods and determine which best suit given situations, including: direct contact with agency staff in an informal atmosphere; public meetings; public surveys; meetings with organisations; consultative committees; advisory groups; workshops, and submissions (refer Table 13.1). Avoid 'top-down' approaches. Ensure that options are explored adequately.
4   Take into account the characteristics, needs and experience of a range of 'publics', with the differing views being valued or validated. Avoid or

Table 13.1 Methods of public participation (from Munn 1979:Table 4.2).

	Objectives					
Techniques	***Inform Educate***	***Identify Problems/ Values***	***Get Ideas/ Solve Problems***	***Feedback***	***Evaluate***	***Resolve Conflict/ Consensus***
Public hearings		x		x		
Public meetings	x	x		x		
Informal small group meetings	x	x	x	x	x	x
General public information meetings	x					
Presentations to community organisations	x	x		x		
Information co-ordination seminars	x			x		
Operating field offices		x	x	x	x	
Local planning visits		x		x	x	
Class action litigation	x		x	x		x
Information brochures and pamphlets	x					
Field trips and site visits	x	x				
Public displays	x		x	x		
Model demonstration projects	x			x	x	x
Material for mass media	x					
Press releases inviting comments	x			x		
Letter requests for comments			x	x		
Workshops		x	x	x	x	x
Advisory committees		x	x	x	x	
Task forces		x	x		x	
Employment of community residents		x	x			x
Community interest advocates			x		x	x
Ombudsman or representative		x	x	x	x	x
EIS review by public	x			x	x	

diminish conflict: use independent and skilled facilitators where possible; find areas of common interest, and separate people from problems (Fisher *et al.* 1991).

5 Explain environmental and planning procedures in clear, simple terms, and provide advice or even financial support (to attend hearings and/or to hire consultants or lawyers, particularly for disadvantaged groups: this may include the need for lawyers, consultants and agency representatives to travel to distant groups rather than the groups being required to attend hearings).

6 Make an honest commitment to act on the public's views and suggestions where possible, given the context and constraints. Demonstrating that decisions taken are appropriate and responsible, sensitive to community concerns, needs and values, is part of accountability.

7 Include regular feedback, reporting on progress fully and openly and demonstrating that the public participation program is not merely trivialising a required process.

8 The post–decision-making period (over the longer term) needs to remain open and 'consultative'. Mechanisms for dealing with future changes or concerns, as they arise, need to be clear and mutually acceptable to decision-makers and the community.

Mitchell (1990) raised a further interesting point. Good outcomes depend on participants wanting things to happen. Enthusiasm and co-operation will often make a faulty system work well, whereas a well-designed system will falter if participants are unable to work with one another.

Nonetheless, a few years later, Mitchell (1997) appeared to have developed some reservations about aspects of public participation. In addressing the Second National Workshop on Integrated Management in Australia, he found the idea of sitting all interested parties around the table did not work: not everyone has the time or the inclination. In Canada, the Province of Ontario has realised that not everyone can be satisfied and it now considers only selected, interested parties. However, as already discussed, there are many other means of obtaining effective community involvement.

THE JAMES' STUDY OF PUBLIC PARTICIPATION PROCEDURES

All the above eight steps are important and various refinements can be made. Further elaboration is made with reference to a particularly interesting study by James (1991). She investigated public participation procedures in relation to environmental management planning in New Zealand. Her aims were to find out a) who participated in management planning, and b) how effective public consultation was from the viewpoint of those who made submissions. The matters examined were the Tararua Forest Park management plan and the Tongariro National Park management plan review. Submissions came from throughout New Zealand, and James contacted 421 of those people who had provided complete names and addresses, achieving a 53% response.

James observed that one of the major problems in any consultation exercise is that those who respond are not usually representative of the community. This point is well illustrated by her study; participants tended to be male, middle class and well educated (Table 13.2). Further, 91% of individuals making submissions belonged to an outdoor recreation and/or conservation organisation. According to James, other studies have shown that such people differ significantly from unaffiliated outdoor recreationists in their perceptions of the value of outdoor recreation and the environment. She noted that it is important to seek views from other sections of the community, and that those analysing submissions need to be aware of the backgrounds of people making submissions, because it cannot be assumed that their views cover the range of opinion and interest.

James commented that it is usual for notification of the preparation or review of a management plan to be made in the Public Notices section of major newspapers, and that this is a highly selective means of informing the community. Further, for many groups representation is restricted by lack of resources—of time, personnel, finance and expertise. Large voluntary associations are more likely to have the resources to mobilise participation amongst their members, whereas the interests of low-status and low-income groups tend to be ignored in public consultation. Methods of participation also have a major impact on the

Table 13.2	Characteristics of those who made submissions (in relation to two New Zealand environmental management plans) compared with the general population (per cent). Source: from James (1991).	
	Participants	*General Population*
Sex		
Male	78	49
Female	22	51
Ethnicity		
Maori	<1	12
Non-Maori	99	88
Education		
University qualification	48	3
Trade/vocational	25	16
Occupation		
Professional/technical	57	15
Managerial/administrative	21	5
Production workers	2	31
Income		
$60 000 and over	19	)
$40 000 – 59 000	24	)

ability of groups to participate. Groups which have status, are familiar with the workings of the system, and are well organised and articulate, inevitably gain advantages. (However, our experience has also shown that the wealthy are disproportionately averse to door-step interviews.)

The type of opportunity offered is also important. Some procedures offer only a limited, reactive role for participants. For example, if people are asked to make submissions on proposals formulated by the agency, or a questionnaire seeks people's views on pre-defined issues, then it is difficult for participants to develop alternative ideas. Similarly, displays or exhibits, information sheets and news items are useful ways of disseminating information, but they are essentially public relations exercises which facilitate the flow of information to the public. Public meetings, too, are generally more successful at disseminating information from the agency to the public than conveying the range and detail of community perspectives back to the planner. Participants are 'prompted' or 'directed' through the process rather than being allowed to contribute spontaneously—although the latter would require skilled guidance from the Chair of the meeting in order to avoid chaos and loss of focus.

In contrast, methods used for scoping and screening (discussed in Chapter 11 in the context of EIA), such as workshops and advisory groups, require a more sustained input from participants. They are particularly important where issues are complex and there may be conflicting interests. To be successful they require from the agency a commitment of staff, time and resources, as well as skilled and objective guidance, and from the community appropriate expertise and a commitment to being involved.

These methods do not necessarily widen the basis of community participation and so may not solve the problem of representation, although scoping methods can deliberately seek broad representation.

James (1991) commented correctly that it is wrong to assume lack of interest on the part of those who do not participate or who have not previously made their views known. It is much more likely that there are other reasons—including lack of awareness of the issue, shyness, feelings of inadequacy and so on—which prevent people from participating. But if the environmental planner is interested in planning and managing for the community as a whole, and not only a particular segment or interest group, then the effort has to be made to obtain the input of those who do not volunteer their views. Properly designed random surveys of communities are the only means of achieving this objective; and then careful attention must be paid to the design and wording of the questionnaires and the training of the interviewers (Dixon and Leach n.d.).

It was found by James (1991) that, overall, participants did not feel that they had received sufficient feedback on progress being made with the plans. This is a common occurrence, and 12 years previously, Sewell and Philips (1979) also found this to be the area where most public participation exercises fail. Without follow-up, which shows the community that their legitimate concerns have been resolved in a logical and fair way, it is inevitable that people will become cynical about the sincerity or value of consultation. Poor feedback, probably more than any other factor, influences participants to judge their experience negatively. Participants do not necessarily expect their views to change the plan, but they do expect to be told what decisions are made and, particularly, why. Perhaps even more important is the question of whether the concerns of the community have in fact been heard and acted upon in a democratic manner.

SOME REASONS FOR THE LACK OF PUBLIC PARTICIPATION

If EIAs and EMSs are able to formalise and build in an audit or review process in decision-making and planning tiers, then it should not be too difficult to achieve public participation in environmental decision-making more generally. The question can then be asked—why doesn't it happen? Possible answers include:

- political and administrative decision-makers fear loss of control or power;
- lack of bargaining or negotiating skills;
- fear of additional delays in reaching decisions;
- cost of review processes;
- difficulty of access to key information. *Freedom of Information* legislation (FOI—some cynically say Freedom *from* Information) can be very costly and difficult to access and is therefore used with reluctance;
- consensus within a community is rarely achievable and therefore a form of 'guided democracy' is preferred by the lead agency;
- the process becomes too bureaucratised and people lose interest;

- people become cynical over being 'milked' for information, or making submissions which are disregarded;
- loss of political support if decisions go against the prevailing powers;
- fear of providing ammunition to the 'opposition' (political or environmental groups), or creating unfair commercial advantage;
- democracy is a myth—it is seen as inefficient and time wasting in the prevailing economic rationalist context: hence the 'fast tracking' and other planning/development processes taking place around Australia.

If time and cost are key reasons for the lack of public participation follow-up, then a review of present planning and budgetary procedures is clearly needed.

In the authors' experience with community groups, they have found that those which are more successful tend to exhibit the following characteristics. They have:

- sound leadership, which includes organisational skills, a flair for conducting meetings effectively and fairly, and the ability to harness and encourage people's enthusiasm and skills;
- adequate resources to carry out tasks, or the ability to identify sources and obtain funds, relevant information and other necessary resources;
- a good relationship amongst the players, including moral and practical support from local authorities;
- confidence in setting attainable goals;
- stability and certainty;
- regular meetings, reviews and feedback, and
- flexibility in seeking alternative solutions and strategies.

It follows that groups which lack all or some of the above characteristics are likely to fail. In relation to several of the above points, it needs to be emphasised that environmental conflicts rarely occur between two monolithic groups of people and that the dynamics of human behaviour need to be taken into consideration. Environmentalists or 'greens' are very rarely united—as very well documented in *Warriors of the Rainbow*, the entertaining account of the early years of Greenpeace (Hunter 1979). Egos get in the way. Similarly, there is often conflict within (and amongst) government agencies and large business corporations, partly for structural (organisational) reasons as well as those pertaining to human nature. But perhaps some tension is healthy?—nothing can move forward or be shaped in a vacuum. And besides, creative solutions often arise from conflict.

Some of the above comments have implied that public participation is not working as well as it can or should. The following chapter looks at this question more closely, focusing in particular on political factors.

CHAPTER 14

IS PUBLIC PARTICIPATION BEING DISMANTLED?

One person's balance is another's bias

Since the mid-1960s in North America and more recently in Australia, and despite the problems outlined in the previous chapter, the partial adoption of public participation has had some achievements. It has:

1 greatly improved public involvement;
2 influenced policy—though this is very difficult to quantify;
3 stopped or delayed environmentally harmful projects, or at least forced reviews;
4 concentrated on a few phases of the planning process—operational aspects rather than normative or strategic planning (but that is changing), and
5 tended to refine rather than innovate, particularly with regard to methods of public participation.

There continues to be a division of opinion over the extent to which public participation should be required rather than permitted. Should it be legislated for? There is a danger in Australia that an emphasis on a discretionary rather than a required approach for public participation is leading to a serious diminution in its use and effectiveness.

The first part of this chapter considers the question of whether political opinion leads or follows community views in relation to environmental matters. This is followed by consideration of recent (mostly in the 1990s) changes which have taken place around Australia, partly in response to the economic rationalist ideology, which are seriously reducing the opportunities for and effectiveness of public participation in the democratic process.

POLITICAL AND COMMUNITY ATTITUDES TO THE ENVIRONMENT

It would probably be fair to say that environment agency personnel are more inclined than governments to involve the community in environmental decision-making. But the effectiveness of an environmental agency in this and other respects ultimately is determined by the level of political support it receives from its Minister in particular and Cabinet more generally.

This point raises the interesting question concerning whether governments lead or follow in relation to environmental matters. Surveys repeatedly show that the public values 'the environment' highly, particularly over the longer term (crime, health, education and jobs dominate short-term considerations—ABS 1997a). In 1992, 75% of Australians aged 18 and over were concerned about at least one environmental problem. This percentage dropped to 68% in 1996 and increased again to 71% in 1998 (EIA Newsletter February 1999:12). Ranked concerns in 1996 were: air pollution; ocean pollution; freshwater pollution; destruction of trees and ecosystems; rubbish disposal; the ozone layer, and extinctions of species. The proportion of people expressing concern increased with their level of education. Thus, 61% of people with no formal qualifications were concerned about environmental problems, as compared with 75% of those with vocational qualifications, 83% of full-time tertiary students, 85% with a degree and 91% of those with postgraduate qualifications. Twenty-three per cent of respondents thought that the environment had improved over the previous 10 years, whereas 44% thought it had become worse. Issues more important than the environment were crime, health, education and unemployment; but 89% stated that environmental protection was equal to or greater in importance than economic growth (ABS 1998)—and, of course, crime and some aspects of health are also environmental problems.

Yet right-wing governments, by their nature, tend to give higher priority to development than environmental protection (although relatively left-wing governments are emphatically not blameless in this regard). For example, within months of the return of the Liberal/National Coalition to government in Western Australia in February 1993, following 10 years in opposition, Conacher (1994b) summarised the following 'news stories' (and the story for Victoria and Queensland over a similar period was not dissimilar, as later chapters show).

- In March 1993, the new Government cancelled an arrangement under which 7500 km² of the Rudall River National Park, in the east Pilbara region, was excluded from mineral exploration, and ended a moratorium on new exploration permits over another 65 000 km² in and near the park. This is Australia's largest terrestrial national park. About 300 Martu (Aboriginal) people live in the park.

- The Government also threatened to review the previous Labor Government's decision not to allow coal mining in the Mt Lesueur National Park south of Geraldton. However, it should be acknowledged that the Labor Government had controversially permitted mineral exploration in five national parks, including parts of the Rudall River.

- The new Premier proposed to introduce legislation which would protect mineral leases from the impact of the Australian High Court's Mabo land rights ruling. That ruling ended the legal myth that Australia was uninhabited when first settled by Europeans in the eighteenth and nineteenth centuries, and opened the door to land rights claims by Aboriginal people who could prove uninterrupted occupation of the land in question.

- In April, the Minister for Mines stated that the Government would open the way for oil exploration in the northeastern section of the Ningaloo marine park south of Exmouth. The park includes 260 km of nearshore coral reefs with more than 217 species of reef-building coral, 501 species of fish and 600 species of molluscs. Humpback whales, dugongs, three species of sea turtles and the whale shark also inhabit the area.
- The Minister for Planning indicated that coastal 'developments'—marinas, hotels, golf courses, shopping centres and residential areas—should not be fettered by restrictive environmental considerations. Fourteen projects were listed as likely to be implemented if existing controls were eased.
- Flooding of the Fitzroy River in the Kimberley region triggered a promise by the Premier to construct three dams, despite the demonstrable economic failure and adverse environmental consequences of the Ord dam and irrigation projects and the total failure of the private Camballin scheme on the Fitzroy.
- It appeared that the ban on duck shooting would be lifted, even though the Australian Conservation Foundation claimed that public opinion is 2:1 against duck shooting.
- Also in April, it appeared that pastoral leases might be granted perpetual tenure (later mooted indirectly by the Federal conservative Government in relation to the Wik legislation in 1998). Some 38% of Western Australia's 2.5×10^6 km^2 is held in as few as 500 pastoral leases, with a significant proportion owned by mining companies and international corporations. For comparison, only about 6.3% (16×10^6 ha) of the State is held by the conservation estate. It is not irrelevant that the Deputy Premier is the leader of the rural-based National Party.

Conacher (1994b) commented that during the same period as this stream of pro-development announcements, there was no formal government response to the comprehensive Western Australian *State of the Environment Report* published in December 1992 (Government of WA 1992). That report contained positive findings concerning the attempts being made to retain ecologically important areas in: 60 national parks; 100 nature and conservation reserves; and six marine parks, but it also highlighted: the deteriorating quality of Perth's air; soil degradation; the endangered state of much of the State's flora and fauna, and the destruction of wetlands and estuaries.

The question, then, is how these government attitudes might be modified by the community—particularly if governments are out of step with community opinion. Public participation in decision-making is crucial; but if governments ignore the community, will not permit it to become involved, or fail to provide people with the necessary information, there is a real danger of governments creating an increasingly disaffected electorate. Indeed this has already happened.

The remarkable rise of the extremist One Nation political party in 1998, although unrelated to environmental matters, was widely interpreted as indicating citizens' dissatisfaction with the major political parties' unwillingness to listen to their concerns, however misplaced some of them may have been. Ironically, the near collapse of One Nation a year later was due to essentially the same reason, with its own members (including Queensland MPs) resigning primarily because of undemocratic processes within the party organisation.

THE POLITICAL INFLUENCE

Since the mid-1980s there has been a marked slowing down of the public participation movement. There is increasing political and some bureaucratic distaste for it on the grounds that it is costly, time consuming and inexpert. It is also seen as a threat to the so-called 'Westminster system' of Australia's government, whereby governments are elected by (or given a 'mandate' from) the community to make decisions on the community's behalf, based on knowledge often unavailable to the community: but see the discussion in the Introduction to this Part.

For example, in Western Australia, the 1996 *Planning Legislation Amendment Act* set out new provisions for the role of the Environmental Protection Authority (EPA) in relation to the environmental implications of schemes or amendments made under the Act (Fig. 14.1). These must be referred to the EPA for assessment and the EPA decides whether to proceed. The worthwhile aim is to integrate environmental protection into planning schemes. In the event of disputes arising from EPA decisions, any disagreements between the Ministers for Environment and Planning are to be resolved at Cabinet level.

However, the Conservation Council of Western Australia (a peak council which represents conservation groups in the State) strongly opposed the Amendment Act because they feared it could exclude the public from the environmental assessment of land-use planning proposals. After the first six months of the Act's operation, the co-ordinator of the Council wrote to *The West Australian* (31/3/97) stating that:

- if the EPA decides not to assess a proposal for rezoning, the public cannot appeal;
- once rezoning has been completed, neither the public nor the EPA can raise objections at the development phase unless it involves serious pollution issues;
- the EPA was assessing only 5% of the rezoning applications it receives, and
- the EPA no longer lists the rezoning proposals it decides not to assess, on the grounds that the public has no right to appeal (another example of *Catch 22*).

In support of these concerns, only 16 proposals were assessed from 344 proposals referred to the EPA by the date of its 1996–97 *Annual Report* following passage of the amended Act in 1996. However, among the unassessed proposals

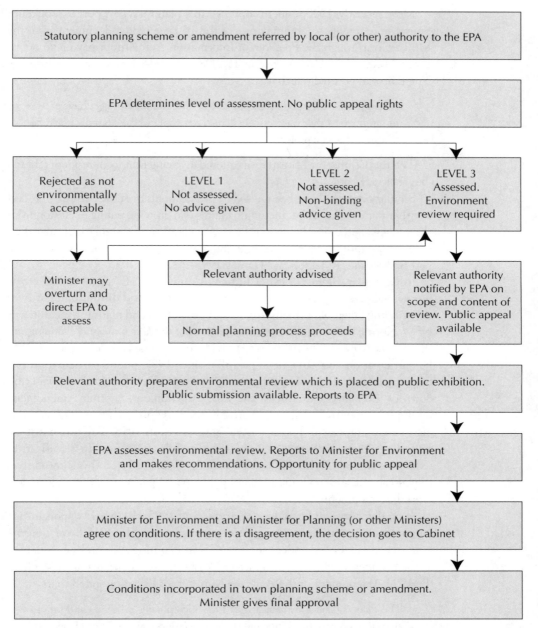

Fig. 14.1 The EIA process for statutory planning schemes and amendments in Western Australia under the *Planning Legislation Amendment Act 1996*. Compiled from EDO (1997, Ch. 5) and a DEP draft kindly provided by Sue Thomas, 1999.

were trivial planning matters, and the EPA recommended modifications to the process so that referrals for assessment would include only those with a potential for environmental impact *and* that third party appeals to the Minister be made possible in the event of the EPA's refusal to assess a proposal.

More recently, 1000 residents opposed to a plan for a new port in Cockburn Sound were unable to obtain copies of the 16-page tender document from the Minister or through the Freedom of Information Act without paying an outrageous $1000 (*The West Australian* 18/2/98, 19/2/98—including an Editorial). The Editorial pointed out that

> ...businesses with a potential commercial interest in the project have access to information that is effectively denied to people whose lifestyles could be threatened by the project...

The then Transport Minister was quoted as saying that local residents should 'buy a bloody copy'.

Similar concerns have been raised by professionals over the planning process in other parts of Australia, including claims of a growing public exclusion from the process, autocratic behaviour by governments, and systematic dismantling of the important gains in environmental and planning laws and policies of the past few decades (Gleeson and Hanley 1998).

In particular, Stein (1998), a Justice of the Court of Appeal in the NSW Supreme Court, expressed his deep level of concern and disillusionment over the winding down of both environmental protection and planning measures in NSW. Noting that one of the main objectives of the *Environmental Planning and Assessment Act 1979* (NSW) was to provide for increased opportunities for public participation and ESD, he argued that since 1988, public participation has been diminished and ESD paid mere lip service. The first significant watering down of the Act occurred with the 1985 Amendment, another Amendment resulted in the virtual demise of Part 4 of the Act in December 1997, and the plan-making process in Part 3 was being reviewed in 1998. The thrust of that amending Act, which was to commence in July 1998, was to significantly reduce community rights in decision-making. Stein believed this showed that the bureaucrats and politicians regard public participation as a nuisance which slowed development.

A similar story can be told for Victoria (Raff 1995a, b), where opportunities for public objections have a long history, going back to the 1915 *Local Government Act* and extended under the State's Town Planning Acts and subsequently in the *Planning and Environment Act 1987*. But in the second reading speech to the 1993 Amendment of the Act, the Minister for Planning stated that:

> The *Planning and Environment Act 1987* incorporates processes and provisions that prevent flexibility and efficient decision-making. The balance has been tipped too far in favour of objectors who are able to hold up developments even though they may not be directly affected by the development. That type of system may have been fine for the 1980s but does not suit the community needs of the 1990s and beyond into the 21st century (quoted in Raff 1995b:75).

According to Raff (1995a, b), amendments to the *Planning and Environment* and *Environmental Effects* Acts under the Kennett Government of the 1990s have reduced opportunities for public participation and objection. The general effect has been a dilution or neutralising of the Acts' powers as development and planning proposals are increasingly removed from public scrutiny, judicial review or merits appeal (refer also to Chapter 19).

Returning to New South Wales, by 1988 the environmental planning instruments (SEPPs, REPs and LEPs) had been stripped of their effectiveness by notions such as 'flexible planning', non-prescriptive planning, performance-based assessment and discretionary decision-making (notions which, incidentally, are also fully embraced in Queensland's new *Integrated Planning Act 1997*). SEPPs are no longer required to be exhibited or receive public input, they are sometimes prepared without such inputs and have even been made without consulting affected agencies or their Ministers. Stein (1998) cited eight examples of Governments passing project-specific legislation to overcome the results of decisions in the Land and Environment Court and the Court of Appeal. A series of Acts of Parliament have been passed to create new decision-makers at the expense of local government and with diminished opportunity for public participation—such as the Darling Harbour Authority and the Olympic Co-ordinating Committee (these have their parallels in the East Perth and Subiaco Re-Development Authorities in Perth, and in other States, particularly Victoria and Queensland, in attempts to by-pass primary environmental protection legislation, albeit with conditions attached).

Stein also commented that governments have become almost obsessed with the corporatisation and privatisation of public utilities at the expense of the general public's right to participate and principles of accountability. There is also a sense of never saying 'no' to a proposal, however irrational or environmentally damaging it may be. Controversial proposals are massaged, coerced or 'mediated through' to a 'yes'. The changes have all been part of a move to 'privatise' planning decisions, premised on the argument that there is a need to introduce competition into planning. The inescapable result will be a diminution in public participation and an increase in disputes reaching the courts (Stein 1998).

In the same volume, Self (1998) observed that the future of planning depends on the belief that it is possible and desirable to shape our future through *collective action*. Without such a belief, planning becomes at best a band-aid operation to modify market forces which are beyond the planner's control, or at worst an ally in the stabilisation and maximisation of real estate values.

The above and other changes coincided with the worldwide spread of the economic 'rationalist' ideology (more appropriately termed 'economic fundamentalism'). To some, opponents of such values are seen as disaffected neo-Marxists; others see the changes as a serious erosion of basic human rights. One can only hope that the ensuing environmental damage will not have reached disastrous and irreversible levels before sanity prevails once again in an environment where people, equity and justice matter. It should be noted that it has made little difference which of the two major political parties (Liberal and Labor) has been in government in the various Australian jurisdictions.

COMMENT

Stein's (1998) answer to the problems of 'winding back' the effectiveness of environmental planning and public participation in New South Wales was to take the politics out of planning. State and local governments need to respect each other and the value of public participation in making planning decisions. Truly State or regional issues should be the province of State government, and truly local issues should be resolved at the local level.

He suggested further that the Legislature should agree not to pass legislation which circumvents planning and environment laws without it lying on the table of both houses of parliament for a reasonable length of time and being subject to public exhibition, to allow fully informed public discussion. This would counter the present situation of rushed Bills, gagged debates, *in-camera* decision-making and inadequate time for Members and the public to comprehend the complexities of legislation to be voted upon. It would allow parliament to take action where proposed legislation has unintended or undesirable effects, tempering the tendency of rule by bureaucrats, and the by-passing of parliament, and 'restore some balance to our fragile democratic system and uphold the values which we prize so highly'.

But if politicians won't take any notice…?

Part 5

Integrated, Regional, Environmental and Land-use Planning and Management in Australia

There has been a shower of paper emanating from the bureaucracy which has new-found vigour, purpose and latitude. It remains to be seen how much of this new material, breathlessly released, will result in changes on the ground. An interesting observation concerning this flurry of publicly disseminated documents is the blurring of the conventional distinction between discussion papers, green papers and white papers, as well as confusion as to whether papers are issued by the bureaucracy as opposed to those of ministerial direction (Choy and McDonald 1993:388: commenting on the period following the election of a Labor Government in Queensland in 1989).

Previous chapters have considered a wide range of various methods used in environmental management. All the methods discussed have a role to play and all are important. But they need to be nested within a broader policy framework. Consideration of the case studies in Part 1 and the numerous deficiencies of the most widely used environmental protection technique (EIA) in Chapter 11 has also demonstrated the same need. Many commentators have indicated that incorporation of environmental protection within a regional planning policy framework, in particular, is essential. Environmental problems and issues have spatial components and therefore involve land-use (or resource allocation) planning and management. Environmental planning and management also need to be more proactive—that is, not merely reacting on a case-by-case basis to development proposals or to specific environmental problems, but setting the context within which such proposals may proceed. Additionally, project proponents normally are not in a position to consider the huge range of difficulties created by cumulative impacts and there is therefore a need for direct involvement of an over-arching agency or body.

   Recognition of these facts led most Australian jurisdictions to develop various means of integrating environmental protection with regional planning. This Part considers some of these developments and evaluates their effectiveness. But it must be recognised that the entire area is extremely complex and subject to rapid and considerable change; to provide a coverage of all the jurisdictions means that the treatment is unable to do full justice to the work that is being done. Our coverage focuses on non-metropolitan developments.

   A major challenge facing environmental and natural resource management, concerns the best way of integrating the various complexities of the biophysical environment with appropriate social, economic and cultural frameworks. This was also a primary aspiration of Agenda 21, Australia's ESD strategies and the Inter-governmental Agreement on the Environment (IGAE) (Chapters 6 and 7).

However, Australia faces a number of constraints and uncertainties in achieving effective integration in environmental planning and management, or in framing proactive, comprehensive and flexible policies. They include:

- the biophysical environment—the size of the country, the wide range of climatic zones and bioregions, and the often harsh and unpredictable elements (climatic extremes, poor soils and water deficits);
- political/economic pressures—changing governments placing environmental matters lower on their agendas; the pace (or effectiveness) of structural and legislative reform, and the effects of distorting policies (pro-development, driven in part by macro-economic and global imperatives);
- problems of scale—difficulties of dealing with multilevel governments; different policies, conflicting agendas, inadequate (or inappropriately targeted) resourcing; and differing management units, from farm paddock to continent;
- social/cultural factors—identifying the real barriers to change; changing attitudes to the environment; the cost/price squeeze influencing environmental priorities; recognition of land rights; effects of gutting rural communities; potential for social disturbances (war, revolt, migration); impacts of technological change, and uncertainties regarding demographic changes.

The legislative responses to land degradation in Australia have taken place at many levels and over a period of time. Previous chapters have discussed some of these responses.

Australia is signatory to various environment conventions, incorporating desertification, climate change, biodiversity and wetlands (Chapter 6). It has adopted internationally recognised ESD principles in its environmental, national land degradation and natural resource management policies and legislation. Also influential in achieving behavioural and procedural change has been a range of codes, standards and regulations designed to encourage self-regulation and better management in land use. Economic incentives can be found in tax relief for landcare works; or rewards in national landcare achievements through government, community or industry recognition of excellence. National government support in addressing land degradation problems and seeking better management has extended from government inquiries to funding across a range of research sectors, notably CSIRO, LWRRDC, RIRDC, CRCs and universities, again focusing on land, water, vegetation and sustainable management. Of interest here have been projects directed to a 'cleaner', alternative or innovative agriculture (RIRDC 1996, 1997), 'sustainable production systems' (LWRRDC 1998), and a co-operative program, the National Land and Water Resources Audit. But the most influential development has been a major injection of national funds into the National Landcare Program and the Natural Heritage Trust (NHT—discussed in Chapter 15).

State responses to land degradation are also linked to many of the national programs, and the States and Territories have representation on special ministerial councils such as ANZECC and ARMCANZ. All States and Territories are or have been revising and restructuring their related natural resource legislation and agencies. Goals and strategies have been broadened and strengthened to include integrated management and sustainable development principles and to harmonise with national goals and strategies. Key State legislation dealing with land degradation and environmental management (some discussed in more detail in the following chapter) broadly covers: soil conservation; water quality; pollution; vegetation cover (clearing controls, revegetation); conservation (flora and fauna biodiversity); water extraction, use, drainage and runoff; exotic pests and diseases, and land-use controls. Devolutionary and reform processes now confer greater responsibility on regional and local authorities with respect to environmental management, while public agencies are expected to show a 'duty of care' towards the environment and public, and avoid, minimise or compensate for any damage (although the law is not always consistent here). This shift is exemplified by Agriculture WA's refocus from production to environmental matters: the agency now allocates almost half its budget to resource protection and sustainable rural development (Department of Environmental Protection 1997, No. 8:9).

During the late 1980s the development of a remarkable rural-based phenomenon occurred—Landcare. This is discussed together with soil conservation, the development of Land Conservation Districts, and the move to integrated catchment planning (ICM), in Chapter 15. These were all manifestations of the emergence of regional planning and environmental management in non-metropolitan Australia. As Landcare matured, more refined structures and policies emerged in more integrated catchment management and regional planning models. Chapter 16 discusses the moves towards legislative and structural reforms in environmental and natural resource management and, especially, regional, non-metropolitan planning around Australia.

Subsequent chapters look specifically at the integration of environmental management with regional, non-metropolitan planning in the various jurisdictions. Western Australia is treated in some detail in Chapter 17. It provides a base, with case studies, for comparison with the other jurisdictions: the Northern Territory and South Australia in Chapter 18, and Queensland, New South Wales and Victoria in Chapter 19. Tasmania is given special treatment, together with New Zealand, in Chapter 20, as both are considered to provide something of a model for other jurisdictions to consider.

Two reports provide helpful overviews of legislative and administrative arrangements for planning systems and sustainable regional planning around Australia. One comes from the National Office of Local Government (NOLG) and the other from Greening Australia.

In 1998, NOLG (which is located in the Commonwealth Department of Transport and Regional Sciences), reported on a comparison of planning systems around Australia (Collie Planning and Development 1998). Its prime objective was to explore prospects for an intergovernmental agreement on development assessment. Legislative and administrative structures (covering primary legislation, planning strategies and policies), development control documents, development approval processes, the appeals system and building controls are tabulated in detail for each jurisdiction. Environmental matters are not emphasised, except within the context of regional and environmental planning policies and plans. The report's usefulness lies mainly in its documentary nature.

The *Greening Australia* report (Dore and Woodhill 1999) is more germane to Part 5 of this book. It provides useful background on the history of and contemporary arguments for sustainable regional development in Australia, incorporating natural resource management and environmental protection issues. Various jurisdictions are covered separately, followed by case studies. The latter half of the report contains conference proceedings and commentaries on the strengths and weaknesses of regional arrangements. Recommendations are made for enhancing sustainable regional development through various strategies, including improved co-ordination, leadership and resourcing.

Part 5 concludes with an overview of the pros and cons of regional environmental planning and management, followed by the overall Conclusions to the book.

CHAPTER 15

LANDCARE, LAND CONSERVATION DISTRICTS AND CATCHMENT PLANNING

Most environmental protection work has a planning component. This chapter considers some of the legislative and community responses and policies in relation to land degradation for Australia as a whole, before moving on to the main topic of the means of incorporating environmental protection in regional planning and management in subsequent chapters.

Early concerns over land degradation in Australia are reflected in the various soil conservation Acts and programs which were introduced in the 1930s and 1940s, undoubtedly influenced by the US dust bowl experience of the 1930s (Bradsen 1988). By the 1970s, an increased understanding of land degradation problems and their management, a number of inquiries and reports, and a greater community awareness, contributed to the emergence of a range of federal and State legislation and policies to deal with rural and urban land-use concerns (Table 15.1 and Fig. 15.1).

The Commonwealth Government was especially active over the 1970s, with a number of major national inquiries or reports dealing with land use, rural policy and land tenure (Appendix D in Senate Standing Committee on Science, Technology and the Environment 1984). Amongst the issues covered, the reports called for more co-ordinated land use, natural resource management and environmental management policies. Various national and State structures were proposed, with some recommending the establishment of regional commissions or councils to guide development, land-use planning and management. The importance of environment in the planning process was acknowledged and the need for improved land information was a common complaint.

The findings of the Commonwealth Government's 1978 national collaborative study on soil erosion in Australia had a major impact, raising government, agency and public awareness of the nature and extent of Australia's growing land degradation problems. Of particular significance was the estimate that more than half of Australia's agricultural and pastoral lands required treatment for erosion or vegetation degradation. Improved management was all that was required for just under half of this land, but structural works were needed on the remainder at an estimated repair cost of $675 million (Department of Environment,

Table 15.1   **Selected examples of land degradation and environmental issues reported by government committees of inquiry or agencies in the 1970s and 1980s.**

State of the environment (Commonwealth)
Department of Arts, Heritage and Environment 1986a
    and b
(refer also Table 8.3)

Soil erosion, conservation, land degradation
Department of Environment, Housing and Community
    Development 1978
House of Representatives Standing Committee 1989
NSWSCS 1989

Hazardous chemicals
House of Representatives Standing Committee 1982
Senate Standing Committee on Science and the
    Environment 1982

Agricultural/veterinary chemicals
State Pollution Control Commission 1980
ASTEC 1989
Senate Select Committee 1990 (inquiry set up in 1988)

Greenhouse effect (agriculture)
Lottkowitz 1989

Effects of drought, drought policies
DPIE/ASCC 1988

Rangelands condition
LCC 1989

Water supply, quality
Senate Standing Committee on Natural Resources 1978
Steering Committee/DRE 1983
Office of the Commissioner for the Environment 1989

Salinity, waterlogging
Public Works Department 1979
MDB Ministerial Council 1987

Irrigation effects
Joint Committee 1979

Land-use inquiries
Commission of Inquiry 1973
Department of Environment and Conservation 1974
Senate Standing Committee on Science, Technology
    and Environment 1984

Land-use pressures
House of Representatives 1976

Forests, use, loss of
Senate Standing Committee on Science and the
    Environment 1976, 1978
DCFL 1988

National estate
Commonwealth of Australia 1974

Coastal zone
House of Representatives Standing Committee 1980

Flooding of Lake Pedder
Lake Pedder Committee of Inquiry 1974

Wildlife conservation
House of Representatives Select Committee 1972

Housing and Community Development 1978; Woods 1983). By the late 1980s, this figure had increased to more than $2 billion—excluding other types of land degradation as well as offsite costs (House of Representatives Standing Committee 1989). By that time it had become evident that land degradation problems were not confined to the various processes or effects of soil erosion, but that they also included loss of soil quality, increasing water and soil salinisation, acidification and waterlogging, loss of prime agricultural land to other land uses, and problems associated with the use of synthetic agricultural chemicals (Conacher and Conacher 1995).

Thus, the timing of the release of the *National Conservation Strategy* (Department of Home Affairs and Environment 1984), which in turn was a response to the *World Conservation Strategy* (IUCN/WWF 1980), could not have been better. The National Strategy was quickly followed by equivalents in the various Australian States and Territories (Appendix 3). In 1988, the *National Soil Conservation Strategy* (ASCC 1988) listed a set of land management goals, and funding was made available under the *National Soil Conservation Program*

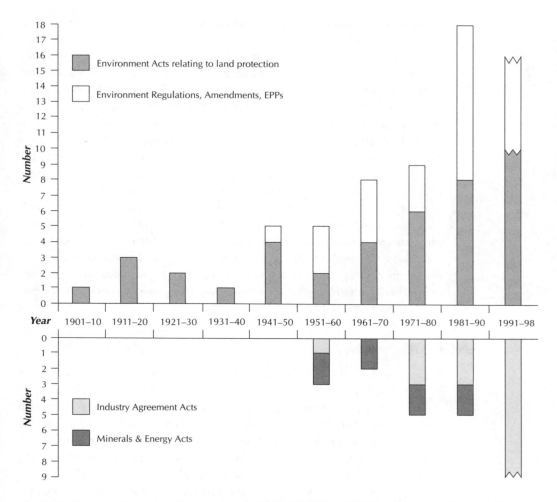

Fig. 15.1   Statutes relating to environment protection in WA 1900 – 1998 in relation to land, minerals, energy and industry agreement Acts. Drawn from data in Clement and Bennett (1998), Bates (1995) and www.erin.gov.au. Excludes Commonwealth legislation.

(NSCP) for research and community projects to be administered in conjunction with the States. Some developed their own soil conservation strategies. Also around this time, the federal government embarked on a series of sectoral *Ecologically Sustainable Development* (ESD) investigations, including one on agriculture (ESD Working Group 1991), and in 1992 the Commonwealth adopted ESD as a national policy (Commonwealth of Australia 1992c).

The guiding principles of ESD included the uncertainty principle—namely that the lack of scientific certainty should not be used as a reason to postpone measures to prevent environmental degradation; and also the principle that decisions and actions must provide for broad community involvement in issues which affect it. These principles conform with the spirit of the Brundtland report (WCED 1987) and the Rio Declaration and Agenda 21 from the world environment conference in Rio de Janeiro in 1992 (UNCED 1992). Land degradation problems and policy recommendations were included in the national ESD policy statement, no doubt influencing a subsequent Australian Senate Standing Committee report on land degradation policies and programs (Senate Standing Committee on Rural and Regional Affairs 1994) and the later Industry Commission (1997) inquiry into ecologically sustainable land management. The ESD program is now administered through a national committee which reports to COAG (Appendix 3), and ESD principles are explicitly incorporated in various environment policies and programs, including the *National Landcare Program* (NLP). These national (and also State) legislative and policy initiatives were reinforced by a range of other inquiries and reports into aspects of land degradation or related issues during this period (Table 15.1).

State of the Environment (SoE) reporting (Table 8.2) has provided valuable background assessments and a deeper perspective on the condition of the biophysical environment nationally and in different States. The reports were intended to be produced at regular intervals to assess changes and trends. An important development was the proposal to adopt a consistent set of environmental indicators throughout Australia to better assess land conditions and changes over time (Chapter 8).

LANDCARE

Detailed discussions of Landcare and its origins can be found in Campbell (1994), Alexander (1995), Baker (1997) and Lockie and Vanclay (1997).

It was in response to an unprecedented, joint National Farmers' Federation and Australian Conservation Foundation proposal for a national land care program, that the Prime Minister's *Statement on the Environment* announced the *Decade of Landcare* (Hawke 1989). The program was to focus on and implement ecologically sustainable land use around Australia by the year 2000. The program established a number of important initiatives and there is little doubt that these provided a major impetus to improving awareness, research and action on land degradation problems around Australia. Former NSCP and Federal Water Resources Assistance programs were amalgamated with the *National Landcare Program* (NLP) and Waterwatch was added in 1995–96.

The Prime Minister's *Statement* also set up major revegetation and replanting programs, which included: *The One Billion Trees* and *Save the Bush* remnant vegetation programs (with particular encouragement given to the formation of Landcare groups, operating at the community level, and an emphasis on education and community projects (see further below)); a *National Reforestation Program*, encouraging hardwood plantations on private land to relieve logging pressure on native forests; the establishment of the *Land and Water Resources Research and Development Corporation* (LWRRDC), which was to examine soil, water and forestry issues and promote an integrated approach to land and water research; and additional funding for the Murray–Darling Basin (MDB) Natural Resources Management Strategy, to implement land, water and vegetation programs. Agricultural chemicals legislation was to be reviewed. Total funding for all the programs was to deliver $320 million over 10 years, upgraded in 1992 by

Prime Minister Keating (Keating 1992) and again in 1996. A decade after the 1989 *Statement*, many of the National Landcare Programs and objectives were being implemented, although not without some problems (discussed below).

The plan placed the contribution of revegetation into context at the farm and district level through community Landcare groups. In 1992, a national office was established, with a national co-ordinator. Groups seeking community grants from the National Landcare Program, MDB Natural Resources Management Strategy, the One Billion Trees Program or Save the Bush could apply to the program using a standardised application process. Guidelines were expanded in 1996–97 for funding to stimulate wider development and sustainable natural resource management as part of catchment or regional plans.

As noted above, the term 'Landcare' was formally adopted by the Commonwealth Government's *Decade of Landcare* program in 1989. However, the word was borrowed from a Victorian government program set up during the mid-1980s. That program involved statutory land conservation groups which enjoyed a measure of support from government agencies.

But the idea of self-help community groups formed to deal with land degradation and conservation issues had earlier origins. There had been some limited farmer participation in soil conservation groups over the 1930s to the 1960s in several States, supported by agency extension services, and the Soil and then Land Conservation Districts (LCDs) of NSW and WA developed from them in the 1970s and 1980s. With agency guidance, farmers grouped together (with local governments) in districts to deal with local and regional land degradation problems, particularly salinity and soil erosion. During the 1970s a parallel development took place in Victoria, where landholder revegetation groups were set up to address rural tree decline through Farm Tree Groups (Oates *et al.* n.d.). Such programs were subsequently taken up under the *National Tree Programme* (1982) and administered by a non-government organisation, Greening Australia. This program was later incorporated into the NLP.

'Landcare' is not based solely on its most popular and tangible expression of community groups. It is also a value system—one based on an ethic or sense of caring for, or stewardship of, the land for the benefit of present and future generations. It also has a more practical or structural component in its technical and funding support through government resourcing. This has occurred most notably through the National Landcare Program (NLP—formerly the NSCP) and, in the late 1990s, the Natural Heritage Trust (NHT—discussed further below). In this regard the movement retains, often unfavourably, a bureaucratic focus insofar as community groups rely on expertise and funds through a 'top-down' administrative process. On the other hand, the framing and implementation of projects is predicated on grass roots or community input and support—but as later discussions show, this is also sometimes a flawed process.

Landcare is not solely a rural movement. Increasingly, schools, local and conservation groups not wholly located in rural, non-metropolitan areas are engaging in various community Landcare projects. Many urban and rural schools, for example, are involved in programs such as Frogwatch, Streamwatch, Bushcare, Saltwatch, and Ribbons of Blue; and numerous other

community groups are involved in revegetation, monitoring, vegetation retention, bushland management and wetland conservation. All fall within the
overall land, water, vegetation and biodiversity guidelines of the government
program. Alexandra *et al.* (1996) estimated that the membership of such groups
throughout the country could exceed 150 000 people.

The timing of the foregoing national and State initiatives on soil conservation
and other land-use strategies could not have been better in terms of capturing the
mood of rural communities. They had become increasingly aware of serious problems in their areas and motivated to do something about them. Within four years
of setting up Landcare programs, more than 2000 Landcare groups had formed
around the country (Mues *et al.* 1994; Campbell 1994). National numbers
increased to 3250 groups by 1997, and 4250 a year later (Industry Commission
1997:187; *Australian Landcare* September 1998:18) (Fig. 15.2). The ABS (1996a)
estimated that over 30 000 farm business were involved in Landcare or similar
groups. Significantly, membership was greatest among broadacre farmers, and in
1992–93 about a third of all broadacre farmers belonged to a Landcare group.
State differences were notable, from a high of 44% in Western Australia to a low
of 12% in Tasmania. The comparable figures for Victoria and South Australia
were 38% and 21% respectively (Mues *et al.* 1994), proportions which have
increased subsequently. However, a range of difficulties relating to management
and funding of Landcare programs has emerged.

How Successful is Landcare?

Landcare's success has been remarkable, most notably measured through its
near exponential growth (Fig. 15.2). It might also be measured through the per-
ceptible shifts in bureaucratic attitudes and behaviour towards community
involvement in relation to natural resources and land-use planning and management. Another performance measure can be found 'on the ground', for
example in extensive farm forestry projects, streamline rehabilitation, fencing
of remnant vegetation, restoration of wetlands or erosion control measures. In
South Australia, for example, it was estimated that between 1989 and 1995,
76 million trees and shrubs were planted through various rehabilitation programs
(which would have included Landcare groups but excluding natural regeneration
and mine rehabilitation) (EPA SA 1998:185–6). Heightened levels of activity
were also exemplified from a 1994 ABS survey of more than 12 000 establish-
ments in Western Australia. The survey recorded significant revegetation and
fencing activities (Landcare Review Committee 1995:Table 32). Other areas
had been planted to farm forests, shelter belts and fodder shrubs. By 1998, WA
had over 100 000 ha planted to Tasmanian blue gums on previously cleared
farmland—from virtually nothing a decade or so earlier (*Australian Landcare*
June 1998:47). Although Landcare members were not identified and no earlier
data were provided for comparison, there is general agreement that the above
practices are significantly different to those in previous decades, when natural
vegetation was there to be cleared rather than conserved.

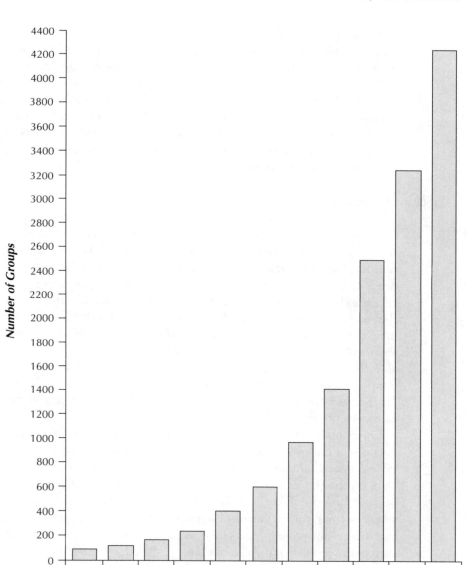

Fig. 15.2 The growth of Landcare groups in Australia, by State and Territory, from 1985 to 1998. Sources: Conacher and Conacher (1995, Fig. 7.2); Lockie and Vanclay (1997); Industry Commission (1997, p. 187); *Australian Landcare* (September 1998, p. 18).

Then there are the unsung heroes of Landcare. These are not the highly visible policy makers, but the thousands of individual landholders and community volunteers who, with an immense amount of goodwill, give their time, energy and resources; and also the important and changing role of women in rural communities. They are frequently key players in Landcare, both amongst the landholders and volunteers, but also as paid Landcare co-ordinators (Beilin

1997; Raynor 1997a). Another interesting community benefit or change engendered by Landcare has been the encouragement for farmers to walk over each others' properties—something that was unthinkable a decade or so previously—with the effect of breaking down barriers and building community trust. There has been an important shift of the broader community's attitudes and commitment towards a 'caring for the land' ethic (Lockie and Vanclay 1997). Finally, government agencies are increasingly adopting a 'whole of government' approach to land management, notwithstanding the competitive frictions amongst them for resources and power.

Problems in Landcare

But, as already alluded to, the successes have not always been unequivocal. A range of concerns has affected the nature and performance of Landcare, and still does. 'Success' should not be measured solely in terms of numbers in groups or trees planted, impressive though these numbers may be. Despite initial evidence of rapid growth in Landcare *groups* (Fig. 15.2), *membership* growth rates appear to have slowed, at least in the agricultural sector. Numbers in 1995–96 showed only a modest increase from those in a previous 1992–93 survey; little more than a third of broadacre and dairy farmers belong to a Landcare group; and there is evidence of reduced participation by long-term members (Mues *et al.* 1994, 1998). In South Australia, the number of Landcare groups declined from a peak of 300 in 1996 to 287 in 1998 (EPA SA 1998:182). This is not to say that non-Landcare members are ignoring the introduction of better land-management practices.

On-ground evidence of improvements in land quality or reduced land degradation is often weak. This refers to evidence such as lowering of groundwater tables, reduced area of salt-affected land, or improvements to water quality. Data are patchy or not analysed comprehensively (to be addressed in the *National Land and Water Resources Audit*). The lack of evidence may reflect the absence of monitoring, the long lag times needed before improvements become evident, or inappropriate targeting and co-ordination of programs. Another slant to this problem is the need for some re-direction of research effort to more management-related knowledge. For example, a recent review of Australian dryland salinity research found that from more than 2000 references screened, a little over 50 produced findings which could be used directly for farm management decisions, policy formulation or bio-economic modelling (Schilizzi and White 1999). On the other hand, without curiosity-driven (or basic) research, it could be argued that there would not have been even 50 useful references. Nevertheless, in this context implementation of benchmarking and use of land-quality indicators are crucial as measures of program performance and for justification of major expenditure.

There have been several formal reviews of the Landcare program since the early 1990s (Government of WA 1995; ANAO 1997; summaries in Industry Commission 1997, Appendix E, and individual evaluations such as those by Raynor 1997a, 1997b; and Curtis *et al.* 1998). An Australian National Audit

Office report found that despite the passage of eight years of the National Land-care Program decade, there had generally been no clear outcomes (the methods used in the audit are described in Chapter 8). Considerable overlap occurred amongst some programs whilst monitoring, review and reporting of perfor-mance and financial administration were inadequate for managing potential risks. It was not clear whether funds allocated had been spent for the intended purposes. Constraints of poor baseline data were recognised, but there was con-siderable room for improvement in the funding process (ANAO 1997).

Taking another example, Curtis *et al.* (1998) assessed the Murray–Darling Basin *Riverine Environment Investigation and Education Program* (covering 48 completed projects) and the Commonwealth's DPIE *Farm Forestry Program*, which had attracted Commonwealth and State funding totalling more than $11 million. The researchers found that while there were significant outcomes, about a quarter of the projects in both programs performed poorly. There were particular concerns over project logic and program implementation. Some of the problems identified included:

- weak links between attitudes/behaviour and adoption of practices;
- poor co-ordination amongst agencies;
- gaps in socioeconomic research (biophysical research dominated);
- considerable overlap amongst projects;
- poor program management (managers were often over-committed);
- an inadequate early screening process of projects, possibly partly due to the limited training and experience of the evaluators;
- insufficient field auditing, and
- limited peer review of findings.

Some improvements were noted subsequent to the survey.

The importance of the above kind of monitoring for Landcare programs has been acknowledged (*Australian Landcare* December 1998:10–11). Government bodies rely heavily on self-monitoring or surveys to assess funding targets, but approaches must be to a standard and reliable if they are to be useful. Longer-term monitoring, however, is needed to assess matters such as the effects of groundwater changes on vegetation or ecosystem recovery.

Most reviews of Landcare have noted the intense frustration of Landcare participants in having to deal with excessive paperwork and administrative mindsets in the process of securing funding for programs. Frustrations include the difficulties involved in determining what is required, where to go for assis-tance, the hours of work involved, and perceived or real bias in allocating funds (refer NHT below, and Curtis 1999). Some reviewers queried an excessive emphasis on education and awareness-raising programs, particularly when behavioural changes towards better land management were still not being achieved. It has been suggested that the focus perhaps should be on finding out *why* the desired changes in awareness and management have not occurred.

Natural resources and land-use management is extremely complex in the nature of the problems encountered and their treatment. Towards the end of the Decade of Landcare, it became clear that despite a decade of program focus and huge injections of public funds and public reforms, management approaches still often fell short of program objectives and operated in disjunct frameworks (ANAO 1997). The value of public participation might also be queried. Is the process simply a heavily disguised transfer of public costs to the private sector in local communities? As the following chapter on regional planning shows, it is evident that the heavy hand of bureaucracy persists—although it needs to be recognised that agency involvement is essential for its skills, data bases and guidance.

The foregoing in some respects seriously challenges some aspects of the much-vaunted successes of Landcare. Has it been, as veteran Greens Senator Bob Brown alleged, 'a triumph of publicity over outcomes' and 'advertising over action' (reported in Lockie 1997:34)?

Senator Brown's comments may be a little harsh since, on balance, Landcare has been responsible for major shifts in attitudes and practices in land management in Australia. To achieve this was a challenging and difficult task—which is by no means complete. This is well illustrated in a study by Johnson *et al.* (1998) of the complex and regulated Australian sugar industry. As a high-input industry, environmental impacts are often serious and include: severe erosion; sedimentation of coastal environments; agricultural chemical residues in soils, groundwater and marine environments; disturbance to riverine and estuarine environments, and heavy demands on water resources for irrigation. Historically, there was little evidence of a sustainably managed industry until the *Sugar Industry Act 1991*, which introduced significant changes in management attitudes and practices. However, a 1996 environmental audit of the industry found that, despite the introduction of a Code of Practice, Landcare and ICM, sustainable resource use was slow to develop. At the time of the audit, only 5% of growers had property management plans and none had environmental management plans. This indicated the need for better information transfer, incentives and co-operative arrangements, amongst a range of suggestions put forward by Johnson and his co-authors. As is true for most agricultural industries in Australia, future policies need to be proactive, locally relevant, and operating within an ESD and regional planning framework.

The Natural Heritage Trust (NHT)

Following the part sale of the public communications utility, Telstra, which raised $1.25 billion, the NHT was set up as a natural environment program at the end of 1996, subsuming the National Landcare Program. The Trust provides for the protection and rehabilitation of Australia's natural environment. It aims to integrate environmental protection, sustainable agriculture and natural resource management with principles of ESD. A major initiative is to foster partnerships amongst the community, industry and all levels of government. Its enabling legislation (the *Natural Heritage Trust Act 1997*) is administered jointly by the Commonwealth's AFFA and Environment Australia (Fig. 15.3). Funds raised were to be used in a five-year environment rescue program which has five major foci: vegetation; rivers; biodiversity; land care, and coastal/marine. Table 15.2 shows the allocation of funds as proposed in 1996. Sustainable agriculture, and natural resource management and conservation, were the key objectives of the program.

Programs of the NHT

Fig. 15.3   The Natural Heritage Trust. * Indicates programs through which the community may obtain funds. Adapted from *Australian Landcare*, April 1998, p. 12.

Table 15.2   **The Coalition Government's Environment Programs through the NHT: $1.15 billion over five years. Source: the Coalition's *Saving our Natural Heritage* policy document.**

Budget ($ million)	Program
318	Natural Vegetation Initiative to tackle land and water degradation
32	National Land and Water Resources Audit to provide a national appraisal of the extent of land and water degradation
163	Murray–Darling 2001 Project in partnership with relevant States
85	National Rivercare Initiative
100	Coasts and Clean Seas Initiative
80	Development of a National Reserve System to preserve Australia's biodiversity
8	National Wetlands program
16	Additional assistance to the Endangered Species program
12	Additional assistance to manage Australia's World Heritage Areas
16	Implementation of the National Strategy for Feral Animal Control
19	National Weeds Strategy formulation and implementation
16	Various measures to address city air pollution
5	Community-based waste management awareness
164	National Landcare Program

Natural Vegetation

Supported by $318 million, the aim was to see that the rate of revegetation exceeds the rate of clearance. The program absorbs the One Billion Trees, Save The Bush and Corridors of Green Programs and has strong links with the NLP.

Rivers

More than $250 million was to be allocated to an ESD program and projects which support the Murray–Darling Basin, national river care and wetlands programs.

Landcare

A range of environmental and sustainable agriculture programs were to be funded by more than $300 million through various Landcare, farm forestry and pest management strategies. Included in the funding was $32 million for the establishment of a National Land and Water Resources Audit (Chapter 9).

Coastal/Marine

A $100 million coasts and clean seas initiative was to tackle problems of pollution, water quality and marine biodiversity.

Funding Arrangements under the NHT

Under the NHT, separate agreements or partnerships between the Commonwealth and the individual States and Territories were to be struck. Commonwealth funding was contingent on State/Territory contributions and alignment of the programs with NHT objectives. The funding was not to be used for State government activities unless explicitly agreed to (such as WA's State Salinity Strategy). Provisions were included for an annual reporting process. Funding was to act as a catalyst for significant, long-term, environmental, economic and social improvements. Most Landcare, river and vegetation initiatives would be delivered through regional mechanisms, particularly large integrated projects, unlike the narrower focus of earlier Landcare projects.

Government approval of the sale of a further 16% of Telstra will add $250 million to the NHT fund and extend the program to 2002. Some programs, such as Murray–Darling Basin 2001 and National Rivercare, are to receive reduced funding, while others, such as the NLP, pollution, waste, farm forestry and audit programs, will receive increased support or have the duration of their programs extended.

A *Living Cities Program* was a new Commonwealth budgetary measure for 1999 – 2000. Waste recycling, air quality, chemicals and natural resource issues are addressed through urban water management and coastal pollution schemes. These build on existing programs (such as Wastewatch and the National River Health Program) and introduce a new one—Urban Stormwater Management. This type of program acknowledges that natural resource management issues are not confined to rural areas and that integrated approaches for environmental management are also needed in urban areas. As discussed below, catchment or regional programs help to achieve such integration.

Such a large injection of funds into the management of Australia's natural resources was a great boost to environmental management efforts. But four important concerns have emerged.

1   How Secure is the Longer-Term Funding Process for Australia's Environment?

In real terms, there was in fact a *reduction* in core environmental budgets for the third year in a row, with the shortfall being made up by the NHT (Telstra) funds (*Habitat*, June 1998:10). The Australian Conservation Foundation (ACF) claimed that over half the $1.25 billion allocated to the NHT had been effectively depleted by cost shifting by Commonwealth and State governments. A total of $654 million of forward expenditure had been used to fund cuts at various levels of government, or had gone to projects of little or no environ-mental value. Only about half the NHT funding was additional to that which already existed at the time the Coalition Government came into office. The sale of a further 16% of Telstra will barely cover the cost shifting that has taken place to date (*Environment Institute of Australia Newsletter* No. 18, February 1999:4). Given that expenditure on the environment is already low on the Commonwealth Government's expenditure agenda (less than 1% of GDP—incidentally, a similar proportion as that spent on the environment by farmers), the chances of successfully arguing a strong case for restoring government fund-ing after the Telstra money has run out in 2002 are not strong. There has been an assumption that projects will generate their own momentum once set up; but experience suggests otherwise.

It is now evident that governments are giving some serious thought to the post-NHT period. The following examples demonstrate the growing momen-tum for change, although the precise nature and extent of commitments are not yet clear.

An Industry Commission Inquiry into the use of agricultural land and other natural resources and the need to achieve sustainable management, was set up by the Commonwealth Government in January 1997 and was to report to Par-liament within 12 months. A draft report was released for public comment in September 1997 (Industry Commission 1997) and the final report was produced in January 1998 (Industry Commission 1998). However, the latter was not tabled in Parliament until April 1999, more than 12 months later, with Cab-inet's response due late in 1999 (S. Stone, Parliamentary Secretary to Minister for Environment and Heritage, pers. comm., July 1999). The reasons for the delay are not clear, although a hint of some possible reasons lies in the body of the report. It is evident that there was considerable unease within several State natural resource agencies as well as farming bodies over some of the Commission's recommendations. To take an example: one of the Inquiry's major recommendations was that existing environmental protection and natural resources legislation in the Commonwealth and each State and Territory be replaced by a single, unifying statute, supported by an independent administering agency. Each statute would set out principles for land management and environmental protection around a central principle of 'duty of care' for the environment, and would be complemented by mandated and voluntary standards. While Queensland appears to be one of the few States to have advanced such a statute in its proposed Natural Resource Management legislation, consolidating six pieces of natural resource legislation into one (State of Environment Advisory Council 1996:6.45), others have been more reluctant, although WA has drafted an Agriculture Management Bill that will integrate 16 Acts into one. Several States made it quite clear to the Inquiry that they feared this type of reform would be disruptive, costly and counter-productive. In any event, it was argued, a considerable amount of legislative and structural reform had already been implemented and was considered to be functioning adequately. The arguments were acknowledged by the Inquiry, although it was noted that New Zealand had demonstrated that this type of statute was technically achievable (although also not without its problems, as discussed in Chapter 20).

An alternative proposed by the Inquiry was the creation of an over-arching statute (not unlike Tasmania's RMPS model—Chapter 20) built around principles of ESD and duty of care. But the Commission of Inquiry feared that under this less formal arrangement, the desired degree of rationalisation and integration would not be achieved. Despite the Commission's reservations, it appears that this is the course of action which has been adopted by the Commonwealth.

Thus, the Commonwealth Government's response to the Inquiry has been to set up a Taskforce to develop a national, natural resource management policy statement in close consultation with all governments and key stakeholders. The draft policy was to be open to public comment in late 1999. The policy is to provide strategic direction for the next 10–15 years, focusing on better management practices and decision-making, regional management, community empowerment and acceptance of shared responsibilities. It is seen as an overarching framework which will include agreed actions and timelines. Whether there will be any legislative or regulatory actions is, as yet, unclear (T. Ryan, Commonwealth Minister's Office, AFFA, pers. comm., July 1999).

A parallel development to the framing of a national, natural resource management policy was the setting up by the House of Representatives Standing Committee on Environment and Heritage of an Inquiry into Catchment Management in Australia. This was in response to the Standing Committee's review of the Department of Environment and Heritage Annual Report (House of Representatives Standing Committee 1999), in which concerns were raised over the condition of Australia's water resources. A more co-ordinated approach was called for, by rationalising programs. The inquiry is to investigate specifically the value of catchment management, best practice methods, and the roles of governments in planning, resourcing and monitoring issues (www.aph.gov.au/house/committee/environ).

2   Project Evaluation

Funding under NHT particularly targets on-ground activities and larger projects. But questions persist over the effectiveness of targeting, implementation and other issues. It is clear that many Landcare-associated problems identified in the past have not been resolved. In 1998, Curtis (1999) surveyed half of Victoria's 890 Landcare groups. He found that while there was evidence of important accomplishments in on-ground and community work, a significant decline of on-ground activities had occurred despite increased funding to the groups. Some of this decline was attributable to a severe drought; but other causes included member 'burnout' in delivering NHT outcomes and program management deficiencies—including late delivery of NHT funds, inadequate training for group leaders, insufficient group co-ordinators, and an absence of priority setting and catchment planning by many groups. Curtis believed the focus should move beyond a pre-occupation with on-ground matters to providing improved support in order to remedy these deficiencies.

Such concerns have not been ignored by government: the Industry Commission (1998) advised that an evaluation framework for NHT was being developed by a working group of representatives from Environment Australia, AFFA and the Department of Finance. Action was taken by mid-1999, when the NHT Ministerial Board directed that a mid-term review of the Trust be undertaken to evaluate the effectiveness and appropriateness of its programs and administration (NHT/AFFA advertisement, *Weekend Australian*, 8–9 May, 1999:42).

3   Political Bias

Accusations of political bias in the disbursement of NHT funds were raised early in 1998, a federal government election year. It was alleged that the incumbent Coalition Government had spent 90% of electorate-specific funds from the first tranche of grants in Coalition seats. The Government denied the charge and suggested that the apparent bias was due to a large proportion of Coalition seats being located in rural areas where most projects were needed. Political opponents countered that it was little more than a 'gigantic pork barrelling exercise'; while some skewing towards country seats could be expected, it was the dimensions of the bias that were of concern. The heads of at least half the State assessment panels overseeing disbursement of NHT funds were Coalition party members (*The Australian* 16/2/98; 20/2/98). NHT funding has also been seen as a big 'subsidy' to rural areas.

This poor publicity for NHT was countered subsequently by the National Landcare Co-ordinator, who drew attention to the number of success stories in Landcare and the important environmental changes taking place (Polkinghorne 1998). Whatever the arguments, a bipartisan approach is highly desirable.

4   Vegetation Clearance

However laudatory and valuable the extensive revegetation schemes around Australia are, the fact remains that in general, more land is being cleared of its native vegetation than is being replanted (Hawarth 1997). There are excep-tions: in South Australia, the number of trees planted in 1995 actually exceeded clearing—although the removal of remnant/individual trees in viti-cultural and horticultural regions is causing concern (EPA SA 1998:185). But the irony and futility of government agencies and policies pulling against one another remains unresolved. In early 1999, WA's Minister for Environment upheld several farmer appeals over clearing bans, allowing several hundreds of hectares of land to be cleared despite contrary advice from relevant agencies.

The foregoing concerns have been raised not to diminish the considerable achievements of the various land management programs: these are acknowl-edged and fully supported. Rather, they point to the difficulties which accom-pany such major and innovative initiatives and highlight those areas which might benefit from closer scrutiny in order to realise effective resource manage-ment goals.

SOIL CONSERVATION LEGISLATION AND LAND CONSERVATION DISTRICTS

Important amendments to soil conservation legislation have been made in recent years, with Victorian and New South Wales examples being discussed in the following section under 'Catchment Planning'.

In **Western Australia**, the *Soil and Land Conservation Act (1945–92)* has been revised on several occasions to deal with soil and land resource conserva-tion and to mitigate the effects of erosion, salinity and flooding (Clement and Bennett 1998). The Act provides the Commissioner for Soil Conservation with powers to serve soil conservation notices and to enter into agreements with landholders on the protection of remnant vegetation. Importantly, the legislation includes a facility for establishing Land Conservation Districts (LCDs) and committees (LCDCs) (formerly *Soil Conservation Districts*, established in 1982 (Select Committee into Land Conservation 1991:130)). The original LCDCs were to report to the Soil Commissioner on land degradation problems, but there was little strategic co-ordination. By the late 1990s, the 150 LCDCs covered most of the agricultural and pastoral areas of the State. Their primary role is to develop co-operation between land users, local governments and agri-cultural agencies in order to implement sustainable land management systems and to solve local and regional land degradation problems. They were also ini-tially the main unit to receive NSCP funds (now the National Landcare Program administered through the Natural Heritage Trust). By the late 1990s, LCDCs were increasingly involved in integrated catchment management, regional planning and Landcare. The first rural/urban LCDC, linked to the Swan–Avon catchment management program, was approved in 1999. It covers three local government areas on the fringe of the Perth metropolitan region.

Some LCDCs in WA have been very successful, benefiting from good lead-ership, well-defined goals, and the energetic involvement of farmers, community groups and agency personnel. However, as with Landcare, there has been a range of problems (which are not unique to WA). Many of the LCDCs are effectively 'defunct'. Failures have arisen for a variety of reasons. These include: lack of focus and consensus; poor leadership; disillusionment or 'burnout'; anti-bureaucratic attitudes; conservatism; the inability of some farmers to recognise the existence of the less visible degradation problems, and difficulties in accessing information, technical assistance or funds (see further, Select Committee into Land Conservation 1991:132–3; Government of WA 1995). Moreover, funding for projects which have an impact on the ground is now directed primarily through Landcare and the other NHT programs discussed above, with the effect of downgrading the perceived importance of the LCDCs—although some groups remain strong. This shift has been reinforced by the establishment of staffed Landcare centres (offices) in several country towns, providing a valuable, tangible focus for that initiative. Additionally, some LCDCs have been combined within larger catchment groups, benefiting from better resourcing and a broader focus.

In **South Australia**, the equivalent of the Western Australian Land Conservation District Committees are *District Soil Conservation Boards*, with 27 (in 1997) covering most of the State's agricultural areas except Kangaroo Island and the southeast. The Boards are based on broad community membership. Their role is to promote the use of land within its capability and to conserve and rehabilitate degraded land. Under South Australia's *Native Vegetation Act 1991*, the Boards must be consulted about the development of management guidelines for heritage agreements and about applications for minor clearance of vegetation. South Australia's *Soil Conservation and Landcare Act 1989*, in which land degradation is comprehensively defined, required the Boards to develop district plans by 1995 and to develop three-year Action Plans. Over two-thirds of the groups had prepared plans by 1998, with the remainder being in draft form or in preparation (EPA SA 1998:18–19). District plans identify:

* classes into which land falls;
* capability and preferred uses of the land;
* land uses;
* nature, causes, extent and severity of degradation;
* measures that should be taken to rehabilitate each type of degradation, and
* land management practices best suited to preventing degradation of the various classes of land (Community Education and Policy Development Group 1993:93).

The object of the *Soil Conservation and Landcare Act* is for land to be used within its capability without damage or reduction of future productivity. There is a statutory obligation on all landholders to take reasonable steps to prevent land degradation. The Act was under review in 1998, with one recommendation being that the Soil Conservation Boards also consider water and vegetation.

In 1992, South Australia's Department of Primary Industries (now PIRSA) commenced a Property Management Program (PMP) to assist landholders with sustainable land management. By 1997, 157 groups had participated, and it was intended that the program would be expanded over the following three years, under NHT funding, to help secure a State-wide coverage (EPA SA 1998:175).

CATCHMENT PLANNING AND MANAGEMENT— BROADENING THE FOCUS

The focus of land management in Australia increasingly is away from individual farm, district or sub-catchment levels and towards integrated catchments and regional scales. Overall objectives of integrated catchment management are to manage resources in a sustainable manner, and to minimise land degradation within river or drainage basins or other appropriate spatial frameworks. These objectives are achieved through land management practices designed to: control land degradation; protect and conserve biodiversity, and recognise the economic, social and cultural needs of people. Policies and strategies are developed through co-operation amongst agency, community and industry representatives.

The basic elements of integrated catchment management are:
- identify the problems;
- involve landholders, community and government agencies;
- set goals and objectives;
- collect base data and maps;
- identify vulnerable areas, natural features, land use;
- delimit management units;
- incorporate risk management strategies (for example, droughts, commodity price fluctuations);
- prepare plans;
- make plans available for public comment;
- implement plans, and
- monitor and review.

The concept is not new, with well-known examples of river basin management including the Tennessee Valley Authority in the USA and others in the UK, France and New Zealand (Mitchell 1990). However, its acceptance reflects a global trend. National, integrated programs such as the four-States Murray–Darling Basin management strategy and interstate agreements for the management of the Lake Eyre and Great Artesian Basins have bolstered this acceptance. It should perhaps be noted that there is often confusion (and overlap) between catchment management and Landcare groups, and that increasingly the functions or groupings are being drawn together.

ICM in Australia

The notion of jointly considering land and water management received increasing interest following publication of *Water 2000* (Steering Committee/DRE 1983). Eight major issues were identified, one being the need for co-ordinated management of land and water. Several integrated approaches have emerged since then, of which the most notable was the establishment of the Murray–Darling Basin Commission to deal with salinity and other land degradation and environmental problems in the Basin (see Chapter 5). The first national workshop on integrated management was held in 1988 and New South Wales passed its ICM legislation on total catchment management (TCM) shortly after in 1989. The national workshop did much to further ICM in Australia, with support from three peak water, environment and soil conservation councils. Its aims were to establish major ICM objectives, develop the best ways of implementing ICM and bringing people together, improve knowledge and skills, and promote State ICM strategies. A report on recommendations for action from the first workshop was submitted to the Australian Water Resources Council and subsequently fully endorsed by the Council with ministerial agreements to promote State strategies.

ICM groups vary in size and power around Australia but generally they possess a common intent of integrated management. The following sections explore ICM as a lead in to integrated regional, environmental planning in subsequent chapters. Unless otherwise indicated, details have been drawn from the second National ICM Workshop (1997): refer also to Williams *et al.* (1998). New South Wales and Victoria have had the most extensive experience in Australia with formal ICM arrangements, although they have differed in detail. However, by 1998 most States had established formal ICM policies or legislation.

New South Wales

New South Wales was the first State to pass ICM legislation in 1989 (under review in 1999). Significantly, the focus is not just on water, but under its TCM (total catchment management) program includes 'land, water, vegetation and other natural resources'. Management takes place through Catchment Management Committees (CMCs) which in turn are co-ordinated by a State Catchment Committee. The legislation also provides for the establishment of Catchment Management Trusts with powers to raise revenue from resource users within the catchment (a strategy which increasingly has been adopted in other jurisdictions). Few Catchment Management Trusts have been set up: two examples are the Hawkesbury–Nepean (Druce 1999) and the Upper Parramatta Trusts. The CMCs, which have broad representation, are required to develop regional strategies with action plans to address resource management and environmental issues. Any individual or government agency may set up a CMC with eligibility for funding and technical support from government departments. By 1995 there were 38 TCM groups covering 80% of the State.

Victoria

In Victoria, relevant legislation has gone through several changes. Its *Catchment and Land Protection Act 1994* replaced the former *Soil Conservation Act*; it addressed land degradation issues and embraced the concept of sustainability.

A framework for integrated catchment management, which encouraged community participation, was established. Regional Catchment and Land Protection (CaLP) Boards, covering the entire State, were set up to correspond with areas covered by existing Landcare groups. A CaLP Council was also established to advise government and the Boards. A key feature was the requirement for Boards to develop Regional Catchment Strategies which set strategic directions for catchment and natural resource management and sustainable agriculture. Local action plans function at a more specific operational level. Land owners are required to take reasonable steps to conserve soil, protect water and prevent the spread of feral animals and noxious weeds.

Following a major review in 1997, all existing resource management groups and CaLPs within a catchment were amalgamated into Catchment Management Authorities (CMAs) with significantly increased responsibilities. The aim was to overcome duplication and to rationalise planning and service delivery in one community-based authority, with a strong regional focus (see the Port Philip Bay example in Chapter 19). CMAs took over responsibility for salinity control, floodplain and river management and have strong links with Landcare groups. The CMAs are to review and co-ordinate implementation of regional (catchment-based) natural resource management strategies.

Western Australia

In Western Australia, management and protection of the State's natural water resources are the responsibility of the Water and Rivers Commission (WRC), set up under the *Water and Rivers Commission Act 1995*. The Commission has powers to protect, conserve and manage the State's waterways. The powers exist under various laws in relation to water extraction, use and pollution, as well as catchment and conservation areas. However, there is no ICM legislation in WA. Efforts have been largely voluntary but with strong State agency and community support.

In 1987, the State Government adopted an ICM policy to link the activities of local governments and State agencies through community involvement in local and regional structures. Land, water, vegetation and other elements of the natural environment were incorporated in a management system. An ICM policy group and secretariat were established as an umbrella group, the Office of Catchment Management (OCM). The findings of an ICM review carried out by Canadian geographer Bruce Mitchell in 1991 were only partly implemented and the OCM's role was weakened, eventually lapsing following the creation of the Water and Rivers Commission (WRC) in 1995. Several subsequent attempts to strengthen the ICM strategy failed. There is still no proper statutory or institutional backing and strong agency resistance has persisted.

Agency roles often overlap. AgWA is responsible for sustainable rural development in co-operation with local governments within regional strategies, assists Landcare Co-ordinators and, especially, LCDCs, and is involved in considerable rural research including secondary salinity and other problems of land degradation. Somewhat strangely, CALM oversees the four-agency salinity action plan. The WRC is responsible for riverine catchment management, and the Department of Environmental Protection (DEP) also plays a role in water management. It has issued several EPPs covering the protection of metropolitan groundwater mounds, coastal wetlands and estuaries.

Two important early developments in ICM were the formation of management Authorities (under the *Waterways Conservation Act 1976* such authorities have the power to declare conservation and management areas) to plan and manage several large drainage basins and estuaries. These include the Swan–Avon and Blackwood river basins, the two largest (of five). They are severely degraded drainage basins in the State's southwest, with areas of 124 000 km^2 and 28 000 km^2, respectively. The Authorities operate through private landholders, Land Conservation Districts, catchment groups, State government agencies and local governments to help improve land management in the drainage basins through sustainable agricultural methods and care of the river systems. Farmers are advised of impacts of their practices and river water quality is monitored to assess the effects of the management initiatives. Preparation of management plans is encouraged; and penalties apply to certain activities which adversely affect the conservation areas (Government of WA 1993; Clement and Bennett 1998).

By 1995, 12 catchment groups had been established in WA, but in 1998 four major regional groupings dominated (the Blackwood, Swan–Avon, Murchison–Gascoyne and the South Coast), with further such developments anticipated. Swan–Avon Catchment local groups committed $9.3 million and state agencies $8.3 million to projects for 1998–2001, with NHT funds expected to double the total amount. Regional strategies for the Swan–Avon catchment are in preparation (Swan–Avon Integrated Catchment Management Group 1997) and a strategy for the Blackwood catchment was being revised in 1999 (Saan Ecker, Blackwood Catchment Group Project Manager, pers. comm. 1999; Masterton 1999). Regional strategies are also being employed to implement the State's Salinity Strategy. Initially, focus catchments were used to test the program, with the possibility of eventually adopting a partnership amongst farmers and agency in a services for-implementation approach. In other words, in return for carrying out good management practices, farmers would receive technical and economic support from the agency (Government of WA 1996b). However, under CALM's guidance, 'good management practices' seem most often to translate into 'establishing Tasmanian blue gum plantations', with all the attendant risks of a monoculture.

Northern Territory

There is also no single ICM legislation in the Northern Territory. However, the *Water Act 1992* provides for a whole catchment approach, enabling declaration of regions and the establishment of advisory committees for water management purposes. Other land management legislation cannot be overridden, however. In general, land management is approached on a case-by-case basis in a co-operative environment. One problem is that when conflict arises, there is no statutory basis for resolution.

Queensland

In Queensland, ICM has operated under a State strategy since 1991, with over 30 Catchment Co-ordinating Committees (CCCs) in existence and almost 80% coverage of the State. Committees operate on a voluntary basis, with State and local government support. By 1997, nine CCCs had produced management strategies; a further seven had strategies in draft form, and there was one regional strategy. Five case studies across different environments had been set up for review, due in 1997–98. Queensland became a signatory to the Murray–Darling Basin Agreement in 1992 (discussed in Chapter 5) and, later, to agreements for the Lake Eyre and Great Artesian Basins.

The Lake Eyre Basin is one of Australia's largest catchment management projects. It covers an area of more than 1 million km² of east-central Australia and is one of the world's largest endoreic drainage basins, extending over parts of the Northern Territory, Queensland, New South Wales and South Australia. An arid to semi-arid region, the basin has a range of bioregions and land uses. Key problems relate to the degradation of rangelands and ground- and surface-waters, the management of which is hampered by the region's size, isolation and multiple jurisdictions. In 1995, a Steering Committee was set up to represent the diverse stakeholder interests and to identify the main environmental management issues. The Committee's recommendations were endorsed in 1997 and a *Lake Eyre Basin Co-ordinating Group* established to assist the development of catchment groups. Amongst its objectives are development of a shared vision in the ecologically sustainable management of the Basin's resources. In mid-1999 a discussion paper was released for public comment, as the first stage for finalising an inter-governmental agreement for the Basin. It outlines various options for institutional arrangements (see further, Andrews 1999).

Queensland is also party to the interstate agreement for the Great Artesian Basin. A Consultative Council was set up in 1997 with a co-ordinating role to facilitate improved management of the groundwater resources, which have been under increasing threat as a result of mismanagement. The Council has representation from industry, government, traditional landholders, conservation interests and the general community (S. Stone, Parliamentary Secretary to the Federal Minister for Environment and Heritage, pers. comm., July 1999). A strategic management plan for the Basin is to be developed.

New natural resources management legislation for Queensland was under preparation in 1997 to provide for the formation of statutory bodies or trusts to manage catchments (State of the Environment Advisory Council 1996:6.45). A Landcare and Catchment Management Council is to replace Queensland's Landcare Council and Catchment Co-ordinating Committee following a review. The new Council is to include representatives from Landcare, industry, local and State governments, the Conservation Council, Greening Australia and the Great Barrier Reef Marine Park Authority. The Council will play a key role in approving strategies developed by Regional Strategy Groups formed to integrate

Landcare groups and Catchment Co-ordinating Committees. The State has been divided into 12 bioregions based on catchments, land, vegetation or socioeconomic type, and regional environmental strategies are being prepared for each region. These are to be part of broader, State regional initiatives (Chapter 19).

Tasmania

Tasmania produced an ICM strategy paper in 1988, and various attempts have been made subsequently to formalise a policy. However, progress has been thwarted by changes of government and abandonment of the process in the face of disagreements and other reform priorities. The State *Resource Management and Planning System* (RMPS) was established in the early 1990s. This is an integrated package of statutes which oversee comprehensive natural resource management and environmental control within a government, industry and community framework (discussed in Chapter 20). Under the RMPS, ICM is acknowledged as part of comprehensive natural resource management. In 1998, a new draft ICM policy to cover all of Tasmania and with links to the RMPS, was released. However, the policy raised numerous issues regarding its position in relation to existing policies, programs and legislated objectives. It has not been given a high priority under the current Labor Government (Pretty, pers. comm. 24/3/99).

South Australia

ICM development tended to be slower in South Australia than in the other States, partly because of the subdued landscape and aridity, and because most of the catchment of the major river (the Murray) lies outside the State. In 1995, the *Catchment Water Management Act* was passed. This set up Catchment Water Management Boards (CWMBs) which are required to develop and implement catchment plans in consultation with communities. The plans must be consistent with the State's Water Plan and they are subject to five-yearly reviews. Boards are semi-autonomous and have powers to levy Councils. In 1997, the Act was subsumed under the *Water Resources Act 1997*. A feature of the Act is that it allows management plans to combine flexibility with legal enforceability and strong policy direction (Dyson 1997). Other resource issues covering soils, vegetation and pollution are to be taken into account. By 1998, several Boards had been formed and were implementing plans. The Patawalonga and Torrens CWMBs included major rehabilitation programs. Three other Boards were preparing plans for the Adelaide/Barossa, Onkaparinga and River Murray catchments.

The Mount Lofty Region Catchment Program (MLRCP), set up in 1995, has continued to be active (EPA SA 1998:98–9). The MLRCP is jointly supported by PIRSA, DEHAA and the NHT. There is a board of management with community and agency representation. The catchment program is to address natural resource issues of the region. Its vision is:

> to achieve major institutional change to facilitate the development of land-use planning based on the principle of land capability adopted by all tiers of management (State, local governments, landholders) which ensures the

improved management of soil, water and vegetation in the Ranges for the benefit of existing landholders and the wider community (see further, www.mlrcp.sa.gov.au).

South Australia is also party to the interstate agreement on the Lake Eyre Basin (discussed above, under Queensland). Additionally, the SA Government has announced that it will adopt a more regional approach to natural resource management, using the nine regions identified for NHT funding purposes (including the Lake Eyre Basin, Mt Lofty Ranges, Murray River region and the southeast: refer Chapter 18).

FUTURE NEEDS

Above all, and as is true for all good planning frameworks, ICM and Landcare need to be proactive, comprehensive and flexible. The Australian experience with ICM raises a number of questions and issues: initially written in relation to ICM, many of the following points are relevant to Landcare and LCDCs as well as to integrated regional and environmental planning and management generally:

- *legitimacy*—ICM needs a statutory basis which provides formal recognition by governments and agencies. But it has been argued that this is unnecessary when the process is happening anyway, and that statutory arrangements could be restrictive. Legislation therefore needs to be flexible;
- *clear goals, strategies and roles*—there tend to be too many overlapping roles, introducing unnecessary complexity. ICM needs to be focused, simple, co-operative and achievable;
- *effective leadership*—too many groups have failed without it. Effective leadership includes good guidance, interpersonal skills and mediation but also a strong commitment to the program and good network links;
- *momentum*—there are serious problems if momentum is lost through 'burnout' or loss of interest by ageing and economically constrained rural communities;
- *central office*—a central office is needed for use as a focus for action and clearing house, providing access to information, for meetings, interaction, continuity and technical exchanges. Assistance is currently too dispersed;
- *are river basin boundaries always appropriate?*—clearly not, in more arid environments and also in some extensive coastal dune or perhaps urban environments. Additionally, bioregional groupings are being increasingly favoured, and there are also significant benefits to be gained from using existing administrative and census boundaries for infrastructural or data purposes. Perhaps different boundaries or regions could be used for different priorities?
- *are Landcare and ICM attacking symptoms, not causes*, through 'band aid' or *ad hoc* approaches? Long-term, strategically-directed management is needed within integrated frameworks;

- *skills*—many people involved have not been trained appropriately. This is especially a problem in relation to the devolution of environmental and other responsibilities to local governments, which are now finding themselves confronted with land management issues for which they have neither experience nor expertise (for example in GIS, land suitability evaluation, environmental engineering, conflict resolution, integrated regional/environmental planning and management). Similarly, community participation is excellent for many matters, but expertise is often also required: a balance is necessary;
- *are we merely creating another layer of bureaucracy or government* (regional, as in New Zealand) or substituting one bureaucracy for another in ICM? In either case, decision-making still tends to be 'top-down' with community participation being paid lip service only. Is the so-called 'bottom-up' process really a 'top-down' strategy to elicit local support (and resources at least cost) rather than a genuine attempt to obtain grass roots involvement?
- *flexibility is needed*—application of the precautionary principle requires the ability to respond to future change and uncertainty—which perhaps implies that incremental (or 'responsive') planning is still the most appropriate model;
- *monitoring and auditing*—there is still too much administration and paper shuffling, which is frustrating and wasteful. Stronger accountability on program performance in relation to meeting clearly defined goals, accounting for expenditure and demonstrating outcomes is required;
- *openness*—this has generally improved, but deceit and manipulation are still encountered. Agencies or individuals are reluctant to relinquish their power, thus non–co-operation and old mindsets are barriers to improved land and environmental management. A free flow of good information and data is essential;
- *who pays?*—are local levies appropriate? To what extent should volunteers be expected or asked to bear the costs? Are Landcare funds just a private subsidy with little public gain?—which introduces issues of public versus private rights: who 'owns' the commons? Should offsite costs be internalised? Should polluters be penalised? How does one achieve social equity?
- *for how long and how far can community goodwill be relied upon?*—elsewhere we have discussed the devolution of central government costs and the shedding of responsibilities to local communities and governments: is this a failure of central government to govern, or the outcome of a genuine desire by governments to move closer to people's needs? Elected local government councillors operate on a voluntary basis, as do members of community groups. There is a conflict here between empowering local communities on the one hand and relying on them to do work which perhaps should be done by paid public servants on the other. The 1998 Commission of Inquiry into ecologically sustainable land management recognised some of this responsibility, in developing 'duty of care' and partnership approaches to management;

- *better management practices*—there is a danger of confusion or duplication amongst EMBs, BMPs, EIAs, EPPs, NEPMs: all have similar goals and could perhaps be harmonised within a natural resources and environmental management context. Is it satisfactory to rely on voluntary compliance with best practice? Lines of least resistance are taken that may favour managers' 'bottom lines' more than environmental needs.
- *external factors*—market forces, new technology, social changes and government policies may adversely influence the capacity of rural communities to embrace or implement Landcare activities. Overcoming these constraints requires good community leadership, strategic regional planning and appropriate assistance.

This chapter has considered the moves towards more integrated approaches to natural resource and environmental management in non-metropolitan areas, from early soil conservation efforts to integrated catchment management. Chapter 16 now looks at the next level of integration in natural resource management and land-use planning, tracking the change towards the integrated regional planning frameworks of the 1990s.

CHAPTER 16

INTEGRATING ENVIRONMENTAL PROTECTION WITH REGIONAL PLANNING AND MANAGEMENT

INTRODUCTION

The previous chapter explored the nature of Landcare and catchment management in Australia. These initiatives developed crucial links between rural and urban planning in recent years, and addressed some of the serious gaps which existed previously in approaches to natural resources management. By incorporating the various linked elements of the natural environment into land management strategies, and by considering processes and consequences of land-use practices, managers developed an appreciation of the effectiveness of comprehensive, integrated approaches to planning and management. However, in non-metropolitan areas, and notwithstanding the major impacts of Landcare and catchment management, the latter remain essentially *de facto* planning methodologies. A major deficiency is the absence of statutory backing and formal structural arrangements—with some exceptions.

This chapter looks at some of the ways in which traditional planning (mainly urban-centred) is being transformed to integrate environmental and natural resources factors with planning objectives. Of particular interest is the development of and changes to regional planning. These changes are providing appropriate models for comprehensive, integrated approaches to environmental management and land-use planning. A State-by-State assessment follows in subsequent chapters, to illustrate various changes in planning and land management and the shift towards integrated planning— though the latter is not without its flaws. Part 5 concludes with a brief evaluation of some of the benefits and problems associated with integrated, regional planning and environmental management and what is needed to overcome the problems.

Buckley (1988) reviewed the phases in progression towards more integrated environmental planning and management in Australia. They broadly cover the sequence followed in the preceding chapters as set out in the introduction to Part 2 and in Table 4.0. The four phases described by Buckley are:

Phase 1—the 'squatter' or scramble phase where resources were used with little discretion or central control.

317

Phase 2—the activities of different users began to overlap, leading to the introduction of planning and resource controls; but still with a strong emphasis on individual resource rights.

Phase 3—particular resources became so severely degraded (for example, salinity) that land and water users were forced to change regardless of individual rights or social controls.

Phase 4—the experience of Phase 3 leads to the public perception that resource management should not be left to the individual or unregulated market mechanisms if they jeopardise national assets. This results in calls for more integrated planning approaches based on TCM and similar strategies.

A fifth phase can be added to Buckley's four. This integrates natural resource management and TCM with comprehensive land-use planning and environmental management.

CHANGES IN CONVENTIONAL PLANNING

Town planning in Australia served the country well, at least until the late 1940s, when population and economic growth placed new stresses on the system (Chapter 5). The primary role of town planning is to ensure orderly land use and the provision of infrastructure and services through zoning, incentives and development controls. This is achieved at the local government level by means of town planning schemes and other regulations. In recent years, local authorities have also become responsible for implementing various State laws governing environmental matters such as protection of native vegetation and pollution control. In general, planners are required to have 'due regard' to the environmental effects of a development or plan, with significant proposals being subject to EIA (Chapters 7 and 11). Requirements are greater in some States or Territories than in others.

While most urban areas have quite stringent land-use controls, the same has not been true of non-metropolitan regions where a less formal approach to planning has applied. In general, responsibilities in country areas have resided with lands departments or natural resource agencies (such as those responsible for water, agriculture, mineral resources, forests or national parks). As discussed in Part 2, their main roles focused on controlling land release and developing infrastructure and natural resources. The emphasis was nearly always an economic or practical one: managing the environment *per se*—with the possible exceptions of national parks and wildlife departments—was never a prime concern of the managing agencies. The agencies operated in general isolation from one another with little integration within regions. Overlapping responsibilities on the one hand, and the lack of an authority responsible for integration on the other, were (and still are) common problems in resolving problems of land degradation and land-use conflicts in particular places. This lack of inter-agency integration was illustrated much more recently—and with potentially serious consequences—in the *sectoral* approach that was taken to the ESD process, in an apparent and unfortunate contradiction of the holistic intent of the national ESD policy.

The focus on economic development strongly coloured planning imperatives after 1945. Grandiose schemes for dams, irrigation, hydro-electric power, land releases, mining towns and transport systems, along with decentralisation policies, preoccupied politicians and planners during the 1950s and 1960s. Whether as an effort to rein in some of the excesses or to further promote development, some key changes in planning were initiated during those years.

In major urban areas throughout Australia, as well as in New Zealand, the UK, Europe and the USA, the outer boundaries of cities and their immediate hinterlands were incorporated into more comprehensive city regions and regional planning schemes to deal with pressures of growth (Dickinson 1964). One example was the *Perth Metropolitan Region Scheme* (Stephenson and Hepburn 1955), adopted in 1959. A particularly bold and interesting development in the regional scheme was the provision of a special fund to acquire land for public open space. This was an explicit acknowledgment of the intrinsic value of land for recreational and aesthetic purposes. The later *Corridor Plan* (MRPA 1970) reinforced this provision with its large, intervening wedges of rural and naturally vegetated land between the urban corridors (regrettably since abandoned, at least in part). Another related development was the Conservation Through Reserves scheme, which provided a blueprint to guide planners on key areas to be set aside for conservation and recreation throughout the State (DCE 1981). System 6, one of 12 'systems' or regions, covered a broad region centred on the metropolitan area, and this report is still influential in metropolitan region planning.

Outside the main metropolitan areas, regional economic development engaged planners in the early 1970s, fostered by the former federal Department of Urban and Regional Development (DURD—later the Department of Housing and Regional Development (DHRD), and now the Department of Transport and Regional Sciences (DTRS)). Regional planning, with a focus on employment potential, infrastructure development and social needs, was actively encouraged throughout Australia during this period (example: Pilbara Study Group 1974). While this type of planning was more 'comprehensive', and to some degree 'integrated', than traditional models, environmental matters were still largely ignored.

Various political and social changes since the 1980s were to reshape planning policies, covering matters such as growing public concerns over environmental deterioration, conflicts over large private developments, lack of accountability and increased calls for public participation in decision-making processes. Some of the key changes which influenced a more comprehensive and integrated style of regional planning included:

- greater inter-governmental co-operation;
- broad recognition of ESD and environment in planning policies;
- clarification of government roles and responsibilities in planning;

- merging of natural resource management agencies into single, major, administering bodies (Table 6.1);
- consolidation or revision of natural resource management legislation to include environmental protection;
- strengthening of environmental protection legislation, often with mandatory requirements;
- development of environmental planning strategies (conservation, wetlands, coastal zone, biodiversity strategies, ICM, state planning strategies, planning protocols: Appendix 3);
- broadening of regional plans with an economic/urban focus to include non-metropolitan regions, environmental factors and other externalities, and
- broad community consultation to help shape policies.

Within the foregoing context, national, State and local governments of the 1990s have worked to improve planning at regional and local levels. Common objectives and protocols have been developed (refer Chapter 7: COAG, IGAE, Ministerial Councils and various national environmental strategies). These objectives were greatly enhanced by the national Decade of Landcare Program, and subsequently the NHT, while at the State level, State Planning Strategies, ICM policies and regional initiatives have furthered the process.

Many of the benefits of regional co-operation and development were redefined by a Commonwealth Taskforce on Regional Development (Taskforce 1993). Under a Federal Labor Government's employment and development program (Commonwealth of Australia 1994a), funds were allocated in 1994 over four years to Regional Environmental Employment Programs (REEP). Although still with economic foci, these particular regional strategies differed from their predecessors in that environmental management was accorded a high priority in the plans. Employment programs were to be linked with regional environmental plans. Several such strategies were prepared around Australia between 1995 and 1997: with Commonwealth funding, local government Councils were selected for pilot programs by the Australian Local Government Association (ALGA) under the direction of a national co-ordinator. The South West Environmental Strategy in WA and the Southern Region of Councils Environmental Strategy in SA are two examples of the project (discussed in Chapters 17 and 18). The Federal Coalition Government withdrew Labor's regional development program in 1996 on the grounds of inefficiencies. It has subsequently launched its own regional policy, which emphasises innovation, leadership and capacity building. However, a plethora of regional programs persists (refer Dore and Woodhill 1999, and Chapter 13 in Productivity Commission 1999), some of dubious value.

Local governments did not ignore the transforming processes of changing priorities and the public mood of the 1980s and 1990s either. Hundreds of local government reports relating to integrated planning and policies were produced (Greening Australia 1995:127–40). They are strong indicators of the changes referred to, covering topics such as: roles of local government in land conservation; cultural development; environmental management; integrated local area planning; coastal management; regional development; local state of environment reporting; public participation in local government; greenhouse responses; sustainable living; 'greening' cities; biodiversity; open space and green plans; flora and fauna protection; rural strategies; environmental management strategies; roadside management; conservation and heritage, and environmental indicators.

By the end of the 1990s, it was becoming apparent that regional planning was being enlivened in several States. While the economic focus is still apparent, greater State responsibility, community involvement, inter-agency co-operation, broader and stronger planning principles, the incorporation of environmental protection and their greater complexity distinguish these plans from their predecessors. This mood change was reflected in newspaper advertisements for environmental officers in the Queensland Main Roads Department (*The Weekend Australian* 10–11/4/99). Regional teams have been set up in district centres to meet the Department's environmental management commitments and to ensure a consistent approach to environmental management across the regions.

WHAT IS REGIONAL PLANNING?

A regional plan is a package of land-use and development strategies which provides a basis for establishing preferred present and future land uses and influencing land-use practices. It is proactive, long-term and strategic. The plans are based on a recognition of sustainable use, environmental impacts, the capacity or suitability of land to support uses, and the effects of land use on socioeconomic development: they integrate economic, ecological, social and cultural values in accordance with State or national environment, development and social policies and strategies. Regional plans or strategies in turn set out policies and physical frameworks to guide the development and implementation of local plans. They reflect the particular present and future problems and needs of the region and find ways of better using local resources.

Planning Regions

Regions can be delimited on the basis of various criteria: by former regional economic development boundaries; through an amalgamation of local administrative areas (Councils, Shires, Counties); according to particular needs (community services, employment, ICMs, Landcare, or particular land degradation problems); as sustainable rural regions, or on a bioregional basis (below). Detailed plans, specifications and regulations can be set down in a local planning strategy, district scheme, rural strategy or local management plan. Regional offices may be established to provide a link between regional and local areas, and between regions and central or State offices and agencies. Environmental components of regional strategies are supported by other State and federal documents such as Conservation, ESD, Greenhouse and Biodiversity Strategies, or ICM, protection of agricultural land, and wetland policies.

Bioregions

According to the National Biodiversity Strategy (signed by the Australian Prime Minister, Premiers of all the States and Chief Ministers of the Territories—DEST 1996), environmental characteristics are a principal determinant of boundaries for bioregional planning. Bioregions are based on ecological parameters, vegetation types, catchment areas and climatic factors, but are also combined with the interests of those living and working in the region. DEST notes that the Murray–Darling Basin Commission plans on an environmental basis, using catchment boundaries as well as existing local, State and Commonwealth structures. Several State and Territory governments were also beginning to plan and manage on a bioregional basis in 1996 as part of their land management responsibilities (refer also the *agroecological regions* proposed by Williams and Hamblin, discussed in Chapter 8 under 'Ecological Approaches').

Principles for establishing bioregional planning units include (DEST 1996:8):

- identifying the biological diversity elements of national, regional and local significance, the extent to which they are protected or need protection;
- identifying the major human activities in and beyond the region and the effects of those activities on the region's biodiversity, to ensure that the activities are ecologically sustainable;
- identifying areas which are important for biological diversity and require repair or rehabilitation;
- identifying priority areas for biodiversity conservation and for ecologically sustainable use, and their relationship with essential community requirements such as infrastructure and urban and industrial development;
- providing mechanisms for 'genuine, continuing community participation and proper assessment and monitoring processes';
- co-ordinating mechanisms to ensure ecologically sustainable use of biodiversity (*sic*), with particular reference to agricultural lands, rangelands, water catchments and fisheries, and
- incorporating flexibility to allow for changes, multiple and sequential land uses, and new developments and knowledge.

An example of bioregional planning involving local governments can be found in the Upper Yarra Valley and Dandenong Ranges Authority of Victoria. Extending over four Councils, a comprehensive strategy was being developed in 1994 to conserve the biodiversity of the region (Christoff and Wishart 1994).

PRIME OBJECTIVES OF REGIONAL PLANNING

The objectives of regional planning may be summarised as being to:

- identify appropriate planning regions;
- adopt ESD principles and the sustainable use of land, resources and the environment for the betterment of the people living in the region and the

maintenance and improvement of environmental quality (in the broadest sense, as discussed in Chapter 1);

- adopt equity principles: social equity (cultural values, access to basic services such as health and education); intra- and inter-generational equity; spatial equity (between richer and poorer regions); and balancing environmental, economic and social issues;
- comply with relevant State and national policies;
- retain and improve the 'character' of the region and identify specific needs;
- achieve efficient allocation of land and resources;
- protect the region's natural resources;
- seek co-operation amongst government agencies, local authorities and other bodies;
- achieve public participation and develop ways of dealing with conflict;
- assist (guide) local governments to develop and implement local plans consistent with the regional plan;
- encourage greater acceptance of environmental values;
- encourage the adoption of better management practices in the use of land and resources;
- have the capacity to respond to future uncertainties;
- develop plans which reflect land capability and suitability for particular uses and protect land of high value;
- ensure periodic reviews of plans, achievements, problems, accountability and openness.

DEVELOPING A REGIONAL PLAN

Once the region has been delimited—generally by State and/or national bodies—plan development follows a fairly well-established series of steps. For any plan to be accepted by a community, the people must have a feeling of 'ownership' of the plan. This is achieved by involving the community as much as possible (Part 4) in implementing the following steps:

- a Steering Committee is appointed by government in close consultation with the community. Its membership includes representatives from the regional community, local governments, industry and relevant government agencies;
- the region's needs and 'vision' are identified within broad ESD principles using 'scoping' and 'screening' methods involving the community;
- achievable, key, short- and long-term objectives are identified, again with active community involvement;
- information relevant to the objectives is collected, data bases and inventories built up, reports gathered—avoiding duplication, relevant policies and legislation noted;
- the needs, concerns and aspirations of neighbouring regions are identified and acknowledged;
- objectives are refined, reviewed and re-prioritised;

- a draft plan (or strategy) is produced using a range of transparent planning methods including land suitability evaluation, Siroplan-type procedures, cost-benefit analysis and environmental impact assessment, as appropriate;
- net social, economic and environmental costs of implementing actions are identified and quantified and objectives revised accordingly;
- the draft plan (or strategy) is made available to the community and responses actively encouraged;
- the strategy is finalised, action plans developed and costed and bodies responsible for implementation identified and consulted;
- measurable performance targets and time scales are identified, budgets drawn up by implementing bodies, sources of funds identified and approached;
- plan implementation commences with an initial focus on high-priority and/or affordable actions, and
- regular reviews, SoE reports, monitoring and auditing of performance targets and environmental impacts, and community feedback lead to *ongoing* revisions of the plan, with the revisions being implemented as required.

(Adapted from O'Riordan and Turner 1983; Greening Australia 1995; Wright 1995; EMRC 1997; Sandi Evans, EMRC, pers. comm.; ALGA Website www.alga.com.au—*How to develop a strategy that works*; amongst others, with modifications.)

Local Environmental Planning

Local governments are being encouraged to adopt Local Agenda 21 (see Website in Appendix 3), SoE reporting (including the use of environmental indicators) and EMS into policy and planning. Greening Australia's (1995) *Local Greening Plans* sets out a detailed strategy to assist in the development of local environmental plans. Topics covered include: landscape value; wildlife corridors; buffers for fire, water, vegetation and air protection; rehabilitation of degraded areas; identification of significant habitats; aesthetics (building styles, fencing, landscaping, infrastructure design); erosion/stormwater runoff control; road verge protection/rehabilitation; streamline protection; fire, weed and feral animals management strategies; landholder obligations, and recycling. Although EIAs are not generally required, Councils are encouraged to develop a checklist matrix and follow up intersects and take appropriate actions as necessary. Use of Codes of Practice is seen as desirable, particularly for contractors undertaking Council works. Public participation is encouraged (through open meetings, public comment, feedback and consultation) and education through workshops.

Barriers

In relation to these new responsibilities in local government planning, Wright (1995) identified several barriers. They include: low priority accorded to the environment by local authorities; lack of skills, training and resources to prepare and implement plans; lack of control or influence over broader policies and strategies as agents of the State, and poor government agency interest in local affairs—indicating a need for greater recognition. Historically, of course, environment and land planning and management were not the responsibility of local Councils, especially rural ones: instead they dealt with the three Rs— roads (many local governments originated as Roads Boards), rates and rubbish. Each local government is principally concerned with the problems and issues within its own boundaries, which take up all their energies (John Pretty, pers. comm.). From there to comprehensive, integrated, environmental and land-use planning and management is a huge step.

CONCLUSION

Previous chapters have outlined legislative and structural reforms in environmental and natural resource management and planning around Australia, but it is evident that much remains to be done. In general, an abundance of legislation remains with few unifying structures or laws to draw them together in a rational and more efficient framework, such as attempted in New Zealand's *Resource Management Act 1991* (Chapter 20).

The State-by-State discussion in the following chapters illustrates the many positive moves which have been taken towards more integrated forms of regional planning and environmental management. The evidence, nevertheless, is sometimes suggestive of a 'cobbling together' of laws and agencies in a rather unwieldy manner, hinting at a process of putting old wine in new bottles. On the other hand there seems to be an increasing use of some common methods discussed in Part 3 (including environmental, land capability and land suitability assessment, land evaluation and GIS), but less agreement on the application of actual planning methodologies such as Siroplan. Indeed, there is a feeling that the allocation of land uses to land is still being done in an *ad hoc* manner. But there also appears to be a welcome increase in public participation, particularly the use of workshops to identify planning issues and objectives.

CHAPTER 17

THE INTEGRATION OF ENVIRONMENTAL PROTECTION WITH REGIONAL PLANNING AND MANAGEMENT IN WESTERN AUSTRALIA

INTRODUCTION

Western Australia is discussed in some detail, partly in recognition of the authors' greater knowledge of this State, to exemplify the nature of developments relating to the integration of environmental protection with regional, non-metropolitan land-use planning. These developments have their parallels in the other jurisdictions, but there are also differences which are discussed in subsequent chapters. Additionally, the State comprises one-third of Australia's land area and a significant proportion of its mineral and energy resources, though only about 10% of its population.

A quiet revolution has taken place in Western Australian rural policy and planning. Since 1988, and more particularly in the 1990s, the Ministry for Planning (MfP) (previously the Department of Planning and Urban Development—DPUD) has been preparing regional plans, strategies and discussion papers which have a major rural component. These plans were also explicitly seen as vehicles for integrating environmental protection with land-use planning.

Regional planning in Western Australia has a relatively long history. The first, the *Plan for the Metropolitan Region: Perth and Fremantle*, was prepared in 1955 by Stephenson and Hepburn for the State Government, and the first Bunbury Regional Plan was produced by the Town Planning Department in 1972. Both plans, however, were essentially *urban* plans, and the latter was focused on roads. In 1974 a joint Commonwealth/State study was prepared for the Pilbara region, but its emphasis (in line with national regional policies at the time) was on resourcing, industrial development, wealth enhancement and competitiveness (Pilbara Study Group 1974). MacRae and Brown (1992) noted that a regional study produced for the southwest region in 1965 was also framed around development objectives, but included some protection of environmental features. A later Peel–Preston planning study warned more clearly of development pressures and ways in which proposals should be assessed in order to retain regional character and function.

According to MacRae and Brown (1992), government concerns over *ad hoc* rural development had been evident since the 1950s. Broad policies were established to manage this, ranging from a moratorium on subdivision in 1965 to a broader approach to land-use planning in the 1970s. With relatively rapid growth in regional centres, interest in regional matters gained considerable momentum. For instance, Harman (1987) identified 61 non-metropolitan studies/strategies or profiles undertaken between 1970 and 1986. By the 1980s, the emphasis in regional planning had shifted—reflecting a new government policy which supported State and regional planning strategies based on community participation. Greater value was placed on rural planning to protect productive land and rural values from growth pressures. Regional plans now included economic, social and biophysical environmental objectives, and were complemented by a land-use strategy (WAPC 1995a).

The South West Development Authority (with four planning units) was established in 1983 and three others followed in 1987–90 in the South and Mid-West. Planning offices were established in three major towns. However, these and other plans were severely constrained by the outdated provisions of the *Town Planning and Development Act 1928*, which was not amended significantly until the new planning Amendment Act was passed in 1996. Further reforms to planning legislation are proposed (Ministry for Planning 1998). A number of the economic development regions persist as Commissions, set up under State legislation in 1993 but relics of former Commonwealth programs. Two (of the nine) notable ones are the South West Development Commission and the Great Southern Development Commission. They focus on partnership development with other organisations and facilitate regional projects. A voluntary regional organisation of local governments was also established in the southwest in 1996, aimed at resource sharing and regional development.

RURAL LAND-USE PLANNING

In the post-war years rural lands came under increasing pressures, from rural/urban subdivisions, loss of prime agricultural land, land fragmentation, alternative lifestylers, new crops and other demands, and from land degradation. These pressures showed up the inadequacies of existing planning legislation. One consequence was the introduction of a rural smallholding policy in 1977 as the first government attempt to recognise the demand for small rural holdings. Subsequently, town planning policies were modified to guide rural subdivisions, but they have been criticised for being too complex and rigid.

In 1989, a broader Rural Land Use Policy was adopted by the State Planning Commission (SPC) (now the WA Planning Commission (WAPC)). This required all local governments experiencing development pressure to prepare a local or limited rural strategy as the basis for rural subdivision and zoning. Local strategies are proposed within the context of regional strategies, where they exist; but the introduction of rural strategies has tended to outstrip the latter. Various guidelines have also been introduced by the WAPC to guide local governments on the preparation of local strategies, land capability assessment and other matters. The prime objectives of these strategies are to protect areas of agricultural significance, manage the fragmentation of land and prevent land-use conflicts in identified areas.

(As an aside, it is of interest that the Industry Commission's inquiry (1998) into land management in Australia noted in its Chapter 20 that some concerns had been raised over the effectiveness of policies protecting agricultural land around Australia. Inappropriate or inefficient actions by governments (such as tax relief measures, rezoning or granting of exemptions) often undermined the policies. However, the Commission made no formal recommendations to deal with the problems. It was recognised that land-use conflict in rural/urban areas was likely to continue and that buffer zones and separation areas might offer a means of reducing such conflict. In this context, the draft Peel Regional Scheme released in early 1999 is notable in that it has attached to it a strategic agricultural resource policy which is to be incorporated into the statutory scheme (discussed further below).)

The 1989 Rural Land Use Policy was under review in 1997 (WAPC 1997). Only one regional rural strategy had been completed (for Albany) and most local/regional strategies were not being implemented. Some of the reasons are considered in the following paragraph. However, there have been other regional developments which are discussed under the next heading.

Stronger links between rural strategies and town planning schemes, and between natural resource management, conservation policies and local/regional policies, are seen as desirable (refer Taskforce 1993). An adequate level of responsibility at local level is necessary to achieve the implementation of regional and local plans. However, local governments have been critical of the cost of preparing strategies and the skills and costs associated with undertaking land capability assessments. A revision of policies is sought for speedier approvals processes and greater consistency and co-ordination in policies. Other Australia-wide problems cover: uncertainties over the effects of external influences such as markets, government policies and technological change; the impacts of socioeconomic changes in rural areas, such as rural depopulation and loss of services (refer Productivity Commission 1999); and the tendency for local strategies to be used for the benefit of developers and not local communities. Given the large area of WA (2.5 million km²), regional plans also need to be appropriate or tailored to the areas covered: there are numerous, significant differences between the remote, pastoral areas in the north and the (relatively) intensive agriculture of the populated southwest. Local governments are also seeking more involvement with State agencies. But they need better access to professional assistance and provision of land capability and suitability information at no extra cost and in a timely manner. Many local government areas have only a few hundred ratepayers and therefore very limited resources.

Land and water degradation is of particular interest to rural and agricultural regions. In response, a number of government agencies have established committees and systems, some with statutory powers, to manage these issues. The committees and systems include: integrated catchment management (ICM) under the auspices of the Water and Rivers Commission; environmental protection policies (EPPs) (by the Environmental Protection Authority); Land Conservation District and Landcare Committees under the guidance of Agriculture WA and local governments, respectively; land capability assessment, also by Agriculture WA; the State Salinity Strategy (with CALM the lead agency), and various environmental management plans and controls introduced by some local governments (such as Mundaring, Jarrahdale–Serpentine and Denmark Shires). The Ministry for Planning stated that it would use the strategic frameworks provided by regional land-use plans and policies as an integral part of ICM and management of land and soil conservation. Land capability assessment 'is now an integral part of land use planning' (DPUD 1994).

STATE PLANNING STRATEGY AND REGIONAL PLANNING

Prior to the above developments in rural land-use planning, however, the *State Planning Commission Act 1985* had created the State Planning Commission (SPC—since renamed the WA Planning Commission: WAPC). The SPC/WAPC is an executive body comprising a small number of Commissioners and served by the planning agency, the Ministry for Planning (MfP—previously the Department of Planning and Urban Development (DPUD)). The MfP is responsible for statutory and regional planning and providing advice to local government authorities (LGAs). Problems of land degradation also bring other agencies and supporting legislation (including the *Local Government Act*) into play (Chapter 3 in Clement and Bennett 1998). The WAPC is responsible for planning at the State level, and it initially identified twelve regions, later reduced to ten (including Perth), based on ward boundaries of the Country Shire Councils Associations (Fig. 17.1). However, regional plans produced to date are for realistically smaller regions than these ten State Planning Regions. In the South-West Region, for example, regional plans have been produced for the Peel, Bunbury–Wellington and Leeuwin–Naturaliste regions (Fig. 17.2).

A draft *State Planning Strategy 1996* (see below), which focuses on strategic planning and policy outside the Perth metropolitan region and is a requirement of the *Planning Commission Act 1985*, was preceded by the WAPC's release in 1995 of eight discussion papers for public comment. These included papers on *Environment and Natural Resources, Managing Growth* and *The Regions*. Each paper discussed key issues and invited comment. The *Environment and Natural Resources* discussion paper recognised the need to take into account the long-term impacts of development on the environment and the importance of finding a balance between competing demands on resources. Key issues included air quality, land degradation, coastal pressures and biodiversity. While open space was considered to be generally secure in the metropolitan area (a debatable assumption—as recognised by the 1998 Perth *Bushplan*, which seeks better protection for significant areas of urban bushland—WAPC 1998), securing areas of remnant vegetation was regarded as a high priority in non-metropolitan

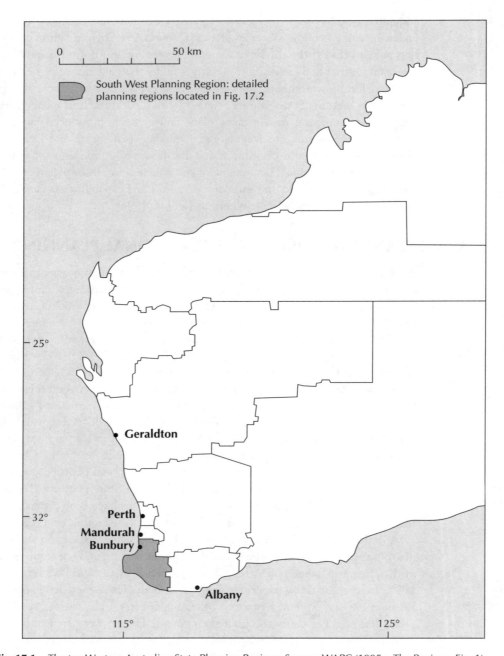

Fig. 17.1   The ten Western Australian State Planning Regions. Source: WAPC (1995a, The Regions, Fig. 1).

areas. A co-ordinated approach to environmental management was sought to avoid the mistakes of the past. Issues such as waste minimisation, recycling, energy reduction, protection of flora and fauna and the co-ordinated management of land degradation were also addressed. The draft Planning Strategy's *Summary* discussion paper stated that:

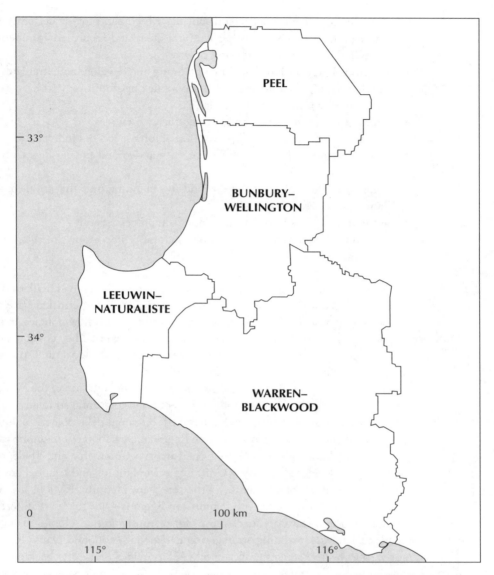

Fig. 17.2   Planning regions in the southwest of Western Australia. Source: DPUD (1993a, Fig. 1).

planning must take a more active role in protecting water catchments, agricultural land, aquacultural areas, forestry areas and natural resources generally.

Regional plans or strategies are expected to conform with the draft *State Planning Strategy 1996*, which is also intended to guide decision-making authorities and the community. The *Strategy* has a number of recommendations which apply to the entire State and which are of significance to rural land-use planning policies (WAPC 1996a, 1997). They show clearly the intention of balancing development with environmental and resource protection and the incorporation of these objectives in the land-use planning process:

- introduction of statutory region schemes which can provide certainty for strategic land uses such as horticultural areas and secure environmental areas and resources such as water reserves;
- continued preparation of regional planning and development strategies to support social, environmental and economic opportunities;
- protection of water and soil quality;
- promotion of management and protection of resources;
- protection of landscape and prevention of further loss of biodiversity;
- ensuring land and soil resources are safeguarded and further degradation does not occur;
- improving the linkage between land-use planning and the provision of human services;
- minimising delays in government approval processes, and
- promoting the development and optimal use of important strategic infrastructure.

Further policies are adopted by the Commission following the circulation of drafts for comment. The policies are contained in a document entitled *Development Control (Including Subdivision)* which is updated with new policies and is available to the community, other agencies and developers. They cover a variety of topics including coastal planning, the preparation of a local rural strategy, land capability assessment and bushfire protection.

After 1996, when the *Planning Amendment Act* came into force, the WAPC was given the mandate to integrate and co-ordinate planning strategies even though other State agencies may be involved. Regional development policies are prepared and reviewed through State Regional Development Commissions (set up under 1993 legislation), the Department of Commerce and Trade and the Regional Development Policy Unit. Regional planning and regional development are seen as complementary—the ten State Planning Regions are the same as the State Development Commission Regions—but there are different statutory and non-statutory mechanisms for planning and developing regions, including land-use planning strategies and regional economic plans. Importantly, the 1996 *Planning Amendment Act* provides statutory backing for regional planning strategies in non-metropolitan areas (the first of these statutory regional strategies, the Peel Regional Scheme, was released for public discussion early in 1999—discussed below). Economic and social considerations are to be developed within sound environmental objectives, and economic development and environmental management are to be considered in conjunction with land releases, regional open space and regional road reserves. However, unlike the metropolitan region (with its Metropolitan Region Improvement Fund), there is no tax base in country areas for the purchase of land for public open space and other purposes.

TOWARDS A REGIONAL DEVELOPMENT POLICY

Following a series of stakeholder meetings in the nine, non-metropolitan Development Regions of WA, a discussion paper on a *Regional Development Policy for WA*, was released in 1999 (Government of WA 1999). Community views were summarised and priorities identified.

The central message of the regional consultations was the belief that regional development should be seen as an investment and not a cost. Emphases were placed on developing community strengths, encouraging governments to maintain access to services and infrastructure, and on protecting and enhancing natural environments and heritage. Stakeholder concerns over the environment covered matters such as the rate of degradation of the natural resource base and the reduction in the quality of physical, biological, cultural and heritage conditions. Investment to arrest and reverse this decline was seen as '...critical for maintaining the productivity of the land and the amenity of regional, rural and remote areas' (Government of WA 1999:6).

Although environmental matters (as is usual in such reports) tend to be low on the list of priorities, their importance is nonetheless acknowledged in the discussion paper. In setting regional development goals which encourage more vital and enriched communities and responsive governmental intervention, the need for sustainable natural resource management and environmental protection is recognised. This is to be achieved through improved and integrated land-use planning practices and management.

REGIONAL STRATEGIES/PLANS

As alluded to above, environmental issues became the catalyst for a number of planning studies for the first time in the 1980s. The first non-metropolitan regional environmental plan (as distinct from regional economic development plans) was produced in 1988 (SPC and CALM 1988). However, the *Shark Bay Region Plan* differed somewhat from its successors, in that it was produced as a direct result of the nomination by the Commonwealth Government of the region for listing on the World Heritage Register—which subsequently came to pass (Chapter 6). The plan was a response to a 'State's rights' issue, an ineffectual assertion by the State of its capability of running its own affairs—ineffectual in that the attempt to avoid Shark Bay's listing on the World Heritage Register failed. Partly as a consequence of the idiosyncratic objectives, and also because it was the first, the plan has a number of deficiencies. It was produced too quickly, community input was inadequate and the data base was deficient (Macklin 1987). These kinds of deficiencies were largely corrected in subsequent plans.

A later and serious environmental problem of eutrophication in a major estuary south of Perth was the subject of a second environmental study and plan (SPC 1989). Prescribed land uses, better management practices, the use of land capability assessment and several other considerations were used to define zones, each with its own land-use and management guidelines. The approach was further developed and used in a number of regional environmental and land-use plans.

Four regional plans (Bunbury–Wellington, Leeuwin–Naturaliste, Albany Region and Central Coast) gave significant coverage to rural planning. They addressed rural land-use planning on a catchment basis combined with land capability, land use and other administrative/geographic boundaries to form planning units or precincts. Within these units, predominant land uses were identified and broad planning guidelines recommended and a land-use strategy formulated. Common elements of the regional strategies include: protecting good agricultural land and water resources; minimising land-use conflict; providing for other appropriate land uses; protecting significant environmental values, and using natural resources in a sustainable manner.

Methodology of Non-Metropolitan Regional Strategies/Plans in WA

The more recent plans have settled into an interesting and sound methodology. Even though the regions are often very different from one another, there is a recurring pattern. All were carried out under the direction of a steering committee appointed for the task and incorporating public participation. The work was done by or with the assistance of the Department of Planning and Urban Development (DPUD—now the MfP). Those regional plans discussed here are the *Albany Regional Strategy* (SPC 1994), the *Bunbury–Wellington Region Plan* (WAPC 1995b), and the *Central Coast Regional Strategy* (WAPC 1996b). These (and other) plans, or strategies, have a major rural component; they were also explicitly seen as vehicles for integrating environmental protection with land-use planning. Final strategies are adopted as policies and may be translated to State planning schemes. They provide guidance for local rural strategies and town planning schemes.

With regard to the planning process, all Reports comment to the effect that:

> the process used to prepare the land use plan included elements of intuitive decision-making and is not necessarily based on scientific analysis. The values and opinions of the community, the Study Team and Steering Committee input, have contributed to the outcome (DPUD 1994:49).

Thus there was no formal method of allocating land uses to land units, as (for example) employed in FAO's land suitability framework or CSIRO's Siroplan (Chapters 9 and 10).

Contents of the Reports

All Reports are similar in their structure and contents, which are summarised below. Nevertheless there are differences. The Bunbury–Wellington and Albany regions each focus on an urban area (relatively small, regional central places), whereas the Central Coast region has no urban focus. Specific problems and issues reflect the characteristics of the regions. Thus, the rural parts of the

Albany region are mostly agricultural, with some tourism; Bunbury–Wellington has a large State forest component as well as mining (coal, bauxite, tin, mineral sands) and secondary industry which is to some extent decentralised outside the city of Bunbury; and the Central Coast region has a sparse population of a little over 4000—yet with a significant problem of more than 1000 squatter shacks, whose inhabitants are not counted in the census—and also mineral sand mining, an important marine component (particularly the export-orientated lobster industry), and relatively large areas of unallocated State-owned land.

Underlying Principles

Each plan sets out a number of underlying principles, with all stating that a fundamental objective is 'to provide a link between State and local planning which is based on a balance of economic, social and environmental considerations' (from DPUD 1994:13). These statements are then followed by 'Strategies', 'Issues', 'Objectives' and 'Key Actions'.

For example, the *Albany Regional Strategy* sets out to (SPC 1994:4):
- provide for long-term sustainability of land use;
- ensure the quality of soils and water including marine waters is maintained or improved;
- protect land uses which contribute to the economy of the region and the State;
- protect the natural, human and environmental heritage;
- enable land uses to respond to changes in the economy;
- guide the expansion of urban areas to avoid adverse impact on valuable rural land and waterways;
- ensure an adequate level of government services and facilities, and
- plan for development which enhances, or at least minimises, any adverse impacts on the values of places of cultural heritage significance.

It will be noted that environmental principles are high on the list.

Strategies

'Strategies' are then outlined under five headings:
- natural and human environment;
- rural land;
- housing and residential land;
- industry, transport and infrastructure, and
- commercial and other economic activity.

Under each of the above headings, the *Albany Regional Strategy* lists separately Issues, Objectives, and Key Actions. For the first strategy, 'Natural and Human Environment', 21 Issues are identified, of which the first four are:
- land degradation;
- water degradation;
- sustainability, and
- loss of and need for retention of natural vegetation.

Nine Objectives follow, and then the crucial, 24 Key Actions. The Objectives and the Key Actions clearly match the identified Issues; and importantly, each Key Action identifies the agency(ies) responsible for implementation. The first four Key Actions are (SPC 1994:29):

- establish the location, nature and extent of land and water degradation, and implement actions for their improvement;
- prepare Integrated Catchment Management Plans for each catchment, to guide the formulation of sub-catchment/farm plans, to address land and water degradation and ensure long-term productivity of the rural environment;
- investigate the need for an Environmental Protection Policy to clarify requirements for land and water conservation and sustainable rural land use, and
- support and expand current programs encouraging the retention of remnant vegetation and the reforestation of private land holdings.

Ten Issues, four Objectives and 12 Key Actions are identified under the 'Rural Land' Strategy heading (above), with some overlap with the Natural and Human Environment section. Key Action Number 2 refers explicitly to the preparation of local rural strategies and rural planning guidelines, as part of integrated catchment management plans. The local rural strategies and rural planning guidelines are to be reflected in local authority planning schemes, produced by Local Government Authorities (LGAs) with the assistance of the State Planning Commission, the Office of Catchment Management and Land Conservation District Committees. These 'location specific strategies' refer to the Planning Units referred to previously, the smaller spatial units nested within the catchments. An example of such strategies for a Planning Unit is reproduced in Fig. 17.3, from the draft *Central Coast Regional Strategy*.

The Bunbury–Wellington Region Plan was the only one which paid significant attention to Aboriginal requirements (in addition to noting the need to protect significant sites, which is also done in the other Plans). Thus, under Issues and Values relating to 'Heritage and Culture', the *Bunbury–Wellington Region Plan* noted that:

> There are numerous places such as burial sites, distinctive physical features of mythological significance, trails and special areas for hunting and camping which have Aboriginal heritage and cultural value. There are various types of connection between individuals, families, the land and sites in Aboriginal society. These are spiritual, mythological or religious, historical and social associations. An important part of Noongar culture is 'cultural revitalisation' where people are rediscovering their heritage. The Noongar language and Culture Centre serves the whole of Noongar territory from an office in Bunbury and has produced language materials and a dictionary to revive the Noongar language (DPUD 1993a:57).

Accordingly, the second of three Specific Objectives of the *Bunbury–Wellington Region Plan* (under the 'Heritage and Culture' heading') is 'to pre-

Planning Unit E9

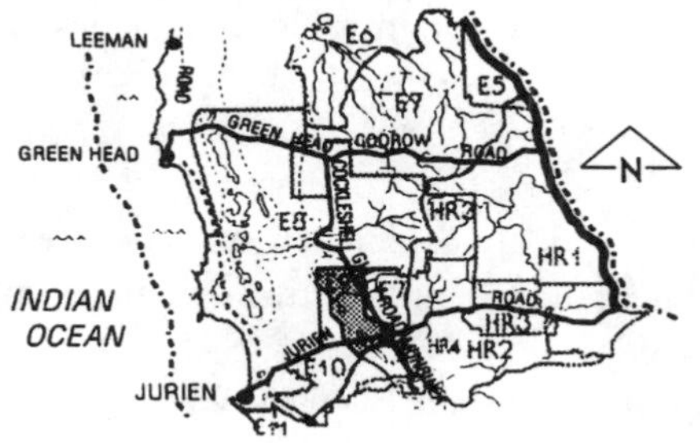

LAND USES TO BE PROMOTED OR ENCOURAGED

Continued 'broadacre' agricultural use involving:

- grazing of improved pastures; and
- limited cropping to lupins and cereals.

Horticulture possible in limited areas with suitable soils and availability of water; and with due consideration of the need to protect against nutrient losses to drainage.

Floriculture - based on intensive cultivation or on sustainable harvesting of existing natural areas for dried flower market and possibly also for plant extracts - volatile oils, medicinal compounds.

MAJOR ISSUES & PLANNING CONSIDERATIONS

Drainage from upland areas collects in swamps and evaporates (increasing salinity) or percolates to groundwater passing through Tamala limestone entering the sea as sub-surface springs.

Soils variable over short distances; all have low capability for most agricultural uses.

Lower areas saturated in winter and can become saline if water not used. These lower areas are most productive when sown to improved pastures. Has some vegetable growing potential due to abundance of water.

Supports attractive wildflowers.

Sandy rises are infertile, have low ability to store water and are prone to wind erosion.

Swamps mainly fresh 'blackwater' types, are common.

Ironstone/ferricrete, associated with areas of regional or local groundwater discharge near dissected laterite of Yeeramullah land system (planning unit E7) limits many agricultural uses.

Continued access to mineral and energy resources and basic raw materials.

PLANNING UNIT AND FEATURES

Bassendean Backplain (dunes & wetlands mainly south of Hill River.

Diverse landscape of predominantly grey sands comprising low dunes, sand plains and poorly drained plains and swamps.

Private land which contains clumps of remnant vegetation.

Covered with Mining and Petroleum Act tenements.

PLANNING & MANAGEMENT GUIDELINES

- Increase water usage by establishing deep rooting perennials (pastures or trees).
- Improved crop sowing techniques to overcome non-wetting soil limitations.
- Permit subdivision for agricultural diversification into horticulture or floriculture subject to adequate supportive studies.
- Wetland protection in accordance with principles of EPA's EPP for Swan Coastal Plain Lakes (including adequate buffers or management controls for horticulture).
- Reduced or minimum tillage, careful stubble management, conservative stocking and establishment of windbreaks to reduce risk of wind erosion.
- Application of fertiliser and water in accordance with plant requirements and the need to protect water resources.
- Protect landscape amenity areas over important landscapes and areas with remnant vegetation.
- Mineral resource development in accordance with State Government policy and Mining Act.
- Basic raw material extraction in accordance with approved plans and strategies.
- Rehabilitation of quarry and mine sites.

Fig. 17.3 Example of recommendations relating to a Planning Unit in WA's Central Coast Regional Strategy. Source: DPUD 1994, p. 63.

serve places of significant Aboriginal cultural and heritage value', and the relevant Actions are to:

- establish a process for identifying and verifying with Aboriginal people information about places of Aboriginal heritage and cultural value;
- establish a process for Aboriginal people to be directly involved in decision-making about the preservation of places with Aboriginal heritage and cultural significance, and

- establish a regional body for Aboriginal heritage matters which is comprised of Aboriginal elders connected with the region, and is responsible for bringing together Aboriginal people for interaction and discussion (DPUD 1993a:58).

Implementation

Each Report concludes with a section headed 'Implementation'. As noted, the Central Coast and Albany Region reports identified the specific agencies responsible for implementing their recommended Actions.

However, it is interesting to note that the draft *Bunbury–Wellington Region Plan* and the draft *Albany Region Plan* both recommended that adoption of the plan as a whole should have statutory backing. Despite this, the final *Albany Regional Strategy* noted that the State Planning Commission preferred to adopt the Strategy as policy. Similarly, the Bunbury–Wellington Region Plan simply recommended the adoption of a Statement of Planning Policy under the *Town Planning and Development Act*, which would provide a statutory framework which decision-making authorities must 'have regard to'. The SPC (1994:50) noted that amendments to planning legislation which were to be introduced to Parliament proposed to give the Commission power to produce statutory regional planning schemes beyond the Perth metropolitan region (now implemented). But although there would, therefore, be the option of adopting a statutory Regional Planning Scheme, the SPC considered that it may be appropriate for the urban areas of Albany but would not be 'necessary' for the whole Albany Regional Strategy area. Reasons were not given.

The SPC (1994:51) concluded by noting that the LGAs which comprise the Albany Regional Strategy area had formed a voluntary regional organisation of Councils, known as the Rainbow Coast Regional Council, and commented that 'this organisation could provide an excellent forum for the Councils of the region to implement many of the actions of the strategy'. Similar, voluntary associations of LGAs are also appropriate for the other regions.

Comment

The Reports briefly reviewed here were particularly significant in a number of respects. These include: the integration of environmental protection with land-use planning; the considerable effort taken to involve the community in the preparation of the plans, or strategies; the improving methodology, routinely incorporating land capability evaluation and GIS; the recognition of Aboriginal values, especially in the Bunbury–Wellington Plan, and the explicit nesting of the regional plans within State planning policies.

Of course, there are also some areas where improvements could be made. The attention given to Aboriginal requirements in the Bunbury–Wellington Plan should be carried through in all regional planning as appropriate for the region. Actions need to be more precise and detailed (whilst it is recognised that the actions specified in these plans are intended only as guidelines for LGAs and other agencies, the more specific they are, the better). GIS could be used more effectively than merely as a means of storing and overlaying spatial data—in particular to assist in allocating land uses to land (compare the steps in land suitability evaluation and in Siroplan, for example: Chapters 9 and 10) and to assess the spatial consequences for other land uses in the region of such allocations, as well as in carrying out 'what if' analyses.

It is still too early to determine the 'success' of these plans/strategies in terms of identifying improvements on the ground—although, in a number of instances, the strategies were formalising actions that were already being under-taken. Two major areas are cause for congratulation: the commitment to mean-ingful public participation, and the high priority given to environmental considerations. On the other hand, an unpublished investigation by Throne in 1997 came up with some concerns. With reference to the Albany *Strategy*, she first identified the various bodies responsible for implementing the actions as specified by the State Planning Commission (SPC 1994). Senior representa-tives of each body were then interviewed to determine the extent to which the actions were being implemented. Somewhat alarmingly, a number of intervie-wees were unaware of the existence of the Albany *Strategy*! This reinforces the need for statutory backing of such plans. A mitigating factor is that, as noted by Throne, many of the actions were in fact being implemented, though often not with reference to the *Strategy*. It has been said that good planning essentially facilitates developments or changes which are already nascent within society: perhaps Throne's findings indicate good planning.

There has been some apparent progress since Throne made her observations. The South Coast Region (which covers 14 local government areas, including Albany and Esperance) is involved in a regional initiative which originally focused on natural resource management in farming areas. A regional report (*Southern Prospects*), intended as a framework for various projects and better natural resource management, was produced by a community-based planning team in 1996—the South Coast Regional Initiative Planning Team (SCRIPT). SCRIPT receives support from the NHT, State natural resource agencies and the regional development commission. This report is now being updated and expanded to a regional natural resource management strategy, with sustainable development of the region becoming the focus. Socioeconomic issues are also considered (Dore and Woodhill 1999).

The Peel Region Scheme

The Peel Region Scheme (WAPC 1999) was released in March 1999 for a three-month public discussion period. In brief, this is the first *statutory* regional planning scheme prepared outside the Perth metropolitan area under the 1995 WAPC Act. It differs from previous schemes in that it will have the force of law and can exercise control over land use. It is also the first regional plan subject to an EIA under the *Planning Amendment Act 1996* (Fig. 14.1).

The area covered lies south of Perth, incorporating some important coastal recreational and agricultural areas and internationally significant wetlands. Embracing three Shires, the region is experiencing considerable population growth, with numbers expected to double to more than 100 000 people from 1996 to 2021. Thus a strategic regional plan was considered necessary to provide appropriate zoning and services for future land uses (requiring some major government land acquisitions), to ensure adequate environmental protection and sustainable management, and to provide certainty in the development processes.

A regional structure plan (the Inner Peel Region Structure Plan) was prepared in 1997. The outcome of extensive public consultation, the plan identified sufficient land to support a population in excess of 200 000. It was to guide planners on various land uses, but it did not have the statutory basis of the Region Scheme.

The Region Scheme is to implement the structure plan within the context of the 1997 State Planning Strategy, promote nodal settlement patterns and protect regional environmental features. Incorporated in the Region Scheme are some key strategies and policies relating to coasts and wetlands, minerals and basic resources, protection of agricultural land, and floodplain management. Provision is made for additional policies to be incorporated in the scheme. Where there are inconsistencies between the regional and local schemes, the regional one will prevail.

As noted above, an EIA has been prepared for the Region Scheme. The Scheme was deemed by the EPA to require a level 3 assessment (an ERMP); consultants prepared the Review which has been submitted to the EPA. This Review is also open to public comment during the regional scheme's advertised period. The environmental Review focuses on zoning, which differs from local schemes—and which has the potential for environmental impact. Several key areas are covered, including: impacts of future development on the nationally and internationally significant wetlands in the region (including a Ramsar site); the protection of reserves, identified under a previous study of the need for conservation and passive recreation; impacts of development on groundwater resources, and the need to control point and non-point pollution (excess nutrients mainly from agricultural land have caused severe eutrophication in the Peel–Harvey estuary, leading to the development of an EPP to control nutrient inputs through better land management). Other areas covered include air quality, solid waste, and risk and hazard management.

OTHER REGIONAL PLANS

It should be noted that regional plans have also been produced for parts of the State's north and northwest, including the Gascoyne, Pilbara and Kimberley regions. Another is under development for the northeast Goldfields. They also assess regional land use but mainly in relation to pastoral land uses. They incorporate environmental repair through the application of management plans, and explore some alternative land uses.

Integrated Environmental Planning at the Local Government Level

The integration of environmental protection with land-use planning has also occurred at the local government level, as this tier of government increasingly recognises the need and as State-based agencies and governments delegate some of their responsibilities. Western Australia provides three good, recent examples.

The Shire of Mundaring Environmental Management Strategy

The *Environmental Management Strategy* (EMS) for the Shire of Mundaring (1996) offers an insight into some enlightened and proactive planning in an outer metropolitan local government area. Located in the combined urban fringe and rural setting of the Darling Ranges east of Perth, the Shire has a population of 35 000 in an area of 645 km². There are significant areas of natural bushland and forest in both private and public ownership, with some relatively extensive areas protecting public water catchments. Thus major priorities are to protect the natural vegetation cover for its own sake and as wildlife habitat, as well as for water supply and quality. The Shire co-ordinates the activities of local (community) conservation, Landcare and catchment groups, and the Shire Council has access to an environmental officer and a catchment groups co-ordinator, and several other supporting personnel and environmental committees (which include community representation). Community education is a specific objective in matters such as bushland and streamline management—whereas some of the community groups consider education of Shire personnel (Shire officers and contractors) as being high on their lists of priorities.

An implementation strategy for the EMS has been developed. It incorporates a priorities rating and time frames for completion developed from extensive community consultation and other information. A number of strategies were under way by 1998, incorporating an erosion and sedimentation control policy, an ICM plan, recycling, urban drainage, Landcare, Friends (of bushland reserves) groups, reserves management, revegetation, wildlife corridors, integrated pest and weed control, the control of domestic and feral animals, and a compliance audit. Whilst there has been considerable community enthusiasm for the environment program, there have also been some difficulties relating to community goodwill, which were mostly in the process of being resolved. There have been few precedents for this project, as a result of which all players are learning on the job (refer also comments in Wright 1995). Thus, the various people involved in the Shire's management strategy were enormously encouraged in 1999 to learn that the Shire had won the State's Landcare award.

The Eastern Metropolitan Region—Regional Environmental Strategy

A boost to the Shire of Mundaring's *Environmental Management Strategy* was the establishment of a major regional environmental strategy which draws the Shire into a six-members Council partnership (EMRC 1997). Covering almost 20 000 km²—one-third the area of the Perth metropolitan region—and with a population approaching 250 000, the *Eastern Metropolitan Regional Environmental Strategy* aims to develop a co-ordinated and structured approach to environmental management in the region. Funding is contributed jointly by the NHT and member Councils. Many of the environmental problems of the region are shared by the Councils—stormwater and drainage management, eutrophication of wetlands, loss of native vegetation, groundwater protection, weed invasion, fire management and the spread of exotic pests and diseases. The objective is to establish a framework of action for the region and provide benchmarks for Councils to assess their own environmental programs. Member Councils work in close consultation with the community in setting priorities. Local Agenda 21, ESD and best practice objectives are being incorporated into policy. Links have been established with other environmental management groups such as various integrated catchment and land management programs in the region. The *Strategy* proposes to proceed along the lines set out in Chapter 16: 'Developing a Regional Plan', and was due to be completed in 1999.

The South West Regional Environmental Strategy

This final example from WA differs somewhat from the above examples, and also returns to a predominantly rural environment. The region was one of nine pilot programs selected by the Australian Local Government Association (ALGA) under Commonwealth regional development funding: refer also to the Southern Region of Councils in Chapter 18. The *South West Environmental Strategy* (Bradby and Pearce 1997a, b) sets out in a strategic format the environmental conditions and needs of the region, which comprises 12 member LGAs. The Strategy provides a basis for future environmental management in the region through a co-ordinated action plan linked to local labour markets. It incorporates separate papers on the State of the SW Environment, Green Jobs (the potential for environmental employment in the region), and an Environmental Indicators Project (as part of a national pilot program). The objective is for local governments to use the information and recommendations to rehabilitate the environment within their normal operations. Based on regional consensus through public consultation, the Strategy covers areas such as environmental education, biodiversity protection, openness in planning, 'best practice for integrated regional environmental actions', targeted rehabilitation works, and farm forestry. Thirty-nine action proposals are presented. The status of the Strategy in 1999 was uncertain, as other regional initiatives take place in the southwest.

OTHER FORMS OF REGIONAL PLANNING

It should be noted that, as in other States, government agencies other than planning departments also undertake various forms of regional management and planning—and there is no shortage of such regions. To some extent this is a hangover from the regional development thrust of DURD in the 1970s, which in turn represented an attempt to reduce the primacy of the capital cities in most States and Territories—with Queensland the notable exception. Thus governments attempted to encourage the growth of regional centres by decentralising a number of government agencies (ironically while country towns were simultaneously losing railways, hospitals, schools, post offices and branches of privatised banks). In other respects the changes were administrative, reflecting the 1980s and 1990s push for greater efficiency and devolution.

In WA, CALM and Agriculture WA continue to have regional centres. In 1996, Agriculture WA set up six Sustainable Rural Development Regions (SRDRs) to achieve future sustainable rural development. This is being done by developing 'regional partnerships' with industry, other agencies and the community, identifying key issues, building a framework for natural resources management and regional economic development, encouraging sustainable farming systems through Landcare, ICM and property management, rural adjustment, conserving natural resources and recognising the environmental and social impacts of agricultural activities (*Primary Focus*—Agriculture WA Newsletter July and November 1996).

Despite an apparent abandonment of the 1997 Taskforce report into sustainable agriculture (Taskforce, 1997), it appears that the role of SRDRs is being strengthened. A discussion paper (Anon. 1999) was released following twelve months of community participation and agency input. Regional strategies incorporating natural resource management (NRM) goals are to be developed for each region and are to be consistent with other strategies, such as the State Salinity Plan. They are to cover a full range of land, water and biodiversity issues and should be capable of integration into an overall State NRM strategy. NRM is viewed as a shared responsibility, although land managers are seen as having prime responsibility. Partnership agreements between communities and agencies are envisaged with roles and responsibilities spelt out to assist in the promotion of NRM principles and achievement of regional goals. Such regional groupings are also expected to enhance State/Commonwealth funding allocations and delivery processes.

Other regions have been defined by, for example, the Water and Rivers Commission and WA's Department of Environmental Protection (1998). The latter regions, used in the State's SoE Report, were adapted from interim bio-geographic regions proposed by Thackway and Cresswell 1995: each SoE region is based on geology, landforms, vegetation and climate, with some boundaries varied to accommodate local factors.

LEGISLATIVE CHANGES

During the latter half of the 1990s, the Coalition Government's reforms and amendments to environmental protection and planning legislation essentially brought it into line with the other States, with implications for regional planning. Amendments were proposed to the *Environmental Protection Act 1986* which would increase penalties for causing environmental harm, incorporate a form of strategic environmental assessment and revise rights of appeal (Edwardes 1997: refer Chapter 7). In addition, numerous changes not requiring formal amendments to the Act had already been made arising out of an inquiry held in 1992. Additionally and separately, the *Planning Legislation Amendment Act* of 1996 linked planning approvals to the environmental assessment process

(Fig. 14.1), although it has not been without its difficulties. Scheme amendments for some regional plans now proceed with comprehensive instructions on environmental issues to be addressed in statutory planning processes. The EPA also has the power not to assess some amendments and the public cannot appeal—a matter of concern to some conservationists who interpret this as a reduction of public openness.

In 1998, the Ministry for Planning released a discussion paper on a proposed consolidation of WA's planning laws (the *Town Planning and Development Act 1928*, as amended; the *Metropolitan Region Town Planning Scheme Act 1959* and the *Western Australian Planning Commission Act 1985* plus subsidiary legislation including Regulations and By-Laws) under a new *Urban and Regional Planning Act*, broadly in line with changes in other States. In essence, the proposal:

- streamlines planning and development decisions;
- extends planning beyond the metropolitan region, and
- seeks to achieve inter-agency and government co-operation (Ministry for Planning 1998).

The proposed legislation aims to increase statutory planning powers in relation to Statements of Planning Policy (SPPs), to which local governments are currently required only to have 'due regard'. Thus, statutory and other methods are proposed to ensure that State powers are properly reflected in regional and local schemes and new developments. Planning processes are to be streamlined, the legislation simplified and enforcement powers increased. But it would appear that the role of local government is to be reduced and ministerial discretion increased. The environmental aspects of the *Planning Legislation Amendment Act 1996* are not included but may be the subject of a later review.

A *Planning Appeals Bill* introduced in State parliament in June 1999 proposes to set up a new Planning Appeal Tribunal with an independent Director of Appeals. However, the Planning Minister may intervene in an appeals process if a matter is considered of State, regional or other importance. The move has been seen as granting the Minister unprecedented power over the planning appeals system (*The West Australian*, 25 June 1999:1).

COMMENT

By and large, developments in regional planning since the 1970s were favourable in their increased recognition of environmental needs. But there are still concerns: environmental priorities still tend to be relegated to an inferior position in policy statements; plans are made and then ignored; ministerial discretion may be invoked inappropriately; agencies still tend to work independently of State planning structures, and there may be room for further improvements to the statutory bases of regional planning.

CHAPTER 18

THE INTEGRATION OF ENVIRONMENTAL PROTECTION WITH REGIONAL PLANNING AND MANAGEMENT IN THE NORTHERN TERRITORY AND SOUTH AUSTRALIA

THE NORTHERN TERRITORY

Town and regional planning in the Northern Territory (NT) is undertaken by the Planning Division of the Department of Lands, Planning and Environment (DLPE), under the terms of the *Northern Territory of Australia Planning Act 1993*. The DLPE was created in 1997 by merging the Conservation Commission with land and planning agencies. Environmental impacts must be considered under the provisions of the Act. The planning Act allows for the establishment of Land Use Objectives which are developed into Control Plans administered by planning authorities. Land Use Objectives may relate to the whole Territory, to smaller geographical areas or particular issues.

There is no single, integrated, natural resource management legislation for the NT, although the DLPE is responsible for the overall management of land and water development, with input from other agencies. A whole catchment approach, integrating land, water and vegetation, is taken under the 1992 *Water Act* which declares water regions for management purposes. However, this legislation cannot override other land management legislation, nor does it have statutory backing should conflict arise. A Territory Plan sets out principles for an 'attractive, safe and efficient environment', and the *Environmental Offences and Penalties Act 1996* and the *Waste Management and Pollution Control Act 1999* consolidate environmental protection legislation. EIAs are conducted under the *Environmental Assessment Act 1982–94*, administered through the Division of Environment and Heritage in the DLPE.

Regional plans incorporate environmental protection and management with land-use planning. Several land-use structure plans were produced during the 1990s, including regional structure plans for Alice Springs, Darwin and Katherine (the Gulf Plan is discussed below). Regional plans are based on integrated and/or total catchment management; but regional boundaries may also be influenced by biogeographical, administrative or cadastral considerations. Plans vary in content but generally incorporate recommendations to reduce land degradation, protect and enhance remnant natural vegetation and wildlife habitats, improve water quality, reduce or prevent conflicts between competing land uses, and encourage development. However, separate bodies have not been created to implement any non-metropolitan regional plans. The most recent strategic land-use plan is the Darwin Regional Land Use Structure Plan 1990. This, with several more detailed sub-regional and locality plans, are subject to ongoing review under the *Planning Act*. Other agencies have produced strategic plans for the Darwin region in relation to tourism, parks and transport, and the NT government is involved with the WA government in strategic planning for Stage 2 of the Ord Project.

Land capability assessment or land suitability assessments are undertaken for regional planning areas. Many factors influence the assessment, including land-use structure plans, land units, topography, conservation issues, land claims and sacred sites. Various GISs are used. Environmental assessments are usually in the form of an EIS by consultants and/or environmental assessment reports/recommendations prepared by the Environmental Protection Unit of the DLPE.

The main methods used to involve the community in the planning process are displays, questionnaires, public notices in the newspaper, and advertising periods for submissions. Additionally, local government has numerous opportunities to participate in the planning and implementation process. In particular, local government is expected to play an important role in establishing Land Use Objectives in making and amending Control Plans, Policies and Guidelines and in reviewing the effectiveness of these outcomes.

A major problem in undertaking and implementing regional plans is the cost of research, which can be prohibitively expensive in remote areas of the Northern Territory.

The Gulf Region Land Use and Development Study 1991

The Gulf Region Land Use and Development Study 1991 (Northern Territory Department of Lands and Housing 1991) is an example of a non-metropolitan region plan in the Territory. In the Foreword, the Minister noted that land-use proposals and development directions envisaged in the document may change as further ecological, cultural and land resource discoveries are made and cultural conditions change. Land-use objectives need to allow for flexibility and review. He therefore invited people or organisations to submit comments on the document, and provided a full postal address.

The objectives of the Study were to provide land-use policies in the Gulf Region, with proposals for achieving specific development goals in keeping with objectives included in the Northern Territory Economic Development Strategy of 1988, and complementing the 1986 Holmes Report. There are three parts to the document: a *Regional Profile*, summarising existing resources, land use and land tenure; a *Land Use Structure Plan*, outlining broad principles of government policy with respect to future land use and tenure; and *Land Use and Development Proposals*, which itemise specific actions to achieve development objectives in keeping with the land-use structure.

The region spans environments ranging from monsoonal to near desert and from steep rocky escarpments to the sea. Eighty-four land systems were identified and assessed, but soils are generally shallow and of extremely poor quality. The many rivers which flow into the Gulf of Carpentaria provide habitats for a wide range of plants and animals and also support the pastoral industry. Sea grass beds and mangroves are important nurseries for prawns, barramundi and mud crabs, and tidal flats may be suited to aquaculture.

Population barely exceeds 1000 and there are only two small gazetted towns. The cost of servicing these and the pastoral stations is high. The pastoral industry occupies 70% of the land, though fishing is the largest single economic activity ($24.5 million in 1990 in wholesale values). Unfortunately, most of the incoming benefits of this industry accrue outside the region, partly due to the lack of infrastructure such as ports. Although there are no mines in the region, deposits of several minerals are known. There are areas with a high potential for tourism but none has been developed.

Given this difficult and somewhat unpromising context, the objectives of the *Land Use Structure Plan* were the culmination of research into the natural values of the Gulf, detailed review of current land use and economic activities, canvassing the long-term land-use requirements of residents, Government departments, other organisations and interest groups, and investigation into possible growth areas within the local economy.

There are five main objectives:

1   encourage development which will have long-term benefits for the region, the economy of the Northern Territory and for employment;
2   retain the lifestyles and character of the Gulf;
3   protect the important natural and cultural values of the Gulf;
4   ensure that land-use proposals are in keeping with political and social constraints in the Gulf, and
5   allow flexibility in planning for future discoveries and consequent development.

Pastoralism is expected to continue essentially unchanged. Although 14% of the Gulf region is considered to have high conservation and recreation value, with some being of national and international significance, there are no declared recreation or conservation reserves in the region (it is interesting that recreation and conservation are considered together: cf. Chapter 4, national parks legislation). It is envisaged that Parks (national parks or other types of parks and reserves) will be established gradually by a variety of specified means; and that public access to foreshores and inland waterways can be achieved by a number of options dependent on the level of usage and management controls required. Environmental Management Areas for rare and endangered plant and animal habitats and areas of land which contain one or more land types which remain relatively undisturbed, have been identified. All such areas are proposed for management by the Northern Territory's Conservation Commission.

Mining and aquaculture will be encouraged. Development zones are identified for recreation and leisure facilities, including accommodation which will fit unobtrusively within the remote character of the region. In relation to Aboriginal enterprises, the study proposes to support Aboriginal participation, provide advice on grazing and tourism, encourage cultural centres, and provide infrastructure support and training in construction and maintenance. Existing roads will continue to be upgraded and new ones constructed to parks and new developments.

Part 3 of the Study Report contains fairly specific proposals designed to achieve the above objectives. They are presented under eight headings under the broad umbrella of 'Initiatives for Developing the Gulf Region'. The headings are: pastoral development; conservation and recreation; conservation and management of river frontage and foreshore areas; mining development; aquaculture and fishing; Aboriginal enterprise; transportation, and tourism. Thus it can be argued that environmental values are fully incorporated and integrated in the overall regional plan; although the discussion under the development-orientated sub-headings contains little cross-referencing to environmental objectives.

Under the 'Conservation and Recreation' sub-heading, the Study proposes that the following initiatives should be undertaken:

1   existing pastoral lease holders to be given the option of privately using recreation and tourism potential. Separate titles could issue, with suitable development covenants. Plans for development and management would be approved by the Ministers from Land and Housing in consultation with the Conservation Commission;

2   where pastoral lease holders do not wish to take up the above option, or where the formation of a park is dependent on bringing together separate parcels of land so that features of conservation significance are within a single holding, consideration should be given to acquiring this land and vesting it in the Conservation Land Corporation as a national park, nature park or reserve;

3   the Conservation Commission to commence negotiations with Aboriginal owners of land identified in the Structure Plan for conservation and recreation purposes under a joint management arrangement;

4   in the case of the Mountain Creek Nature Park, Aborigines should be offered joint ownership under a plan of management with the Conservation Commission;

5   the Conservation Commission to compile plans of management for each Environmental Management Area within two years, and

6   the Conservation Commission to identify and provide advice on management of areas of high conservation and recreation value on pastoral or Crown leases beyond those areas shown on the Structure Plan, to enable pastoralists and other land managers to take appropriate conservation management actions, including development for tourism purposes.

Thus pastoralism and conservation appear to be integrated; and under the 'Pastoral Development' sub-heading there is reference to 'protecting ecological, cultural and environmental values of significance'. But under 'Mining Development'—which is considered to be 'the most important economic consideration in the Gulf Region' with far-reaching regional and national implications—there is no such caveat. This is also the case with the discussion under the 'Aquaculture and Fishing' sub-heading.

SOUTH AUSTRALIA

Houston (unpublished manuscript) wrote that of all the Australian States, South Australia is probably the most centralised, urbanised and spatially unicentric. Within its 'settled areas' (covering roughly the southern 20% of the State's territory and 99% of its population), there is a well-marked dichotomy between a central core area of population growth or stability, within about 1–2 hours' driving time from central Adelaide, and the rural periphery. As in some other parts of non-metropolitan Australia, the latter is marked by general population decline, low population density, selective loss of the younger age groups, loss of services, and increasing concentration of the remaining population into some of the small country towns, with socioeconomic consequences (Smailes 1993). Thus rural planning in South Australia is designed to deal with growth on one hand, and containing or combating decline on the other.

The urban origins of the planning profession have spawned a physical planning mentality adapted to a situation in which people, capital and economic activities are plentiful and competing for space; land is the scarce factor, and the main object of planning is to conserve, ration and allocate it. In the remoter rural areas, however, land is the only cheap and plentiful factor and meaningful planning must be aimed at preserving community structures and service provision, encouraging capital investment, population growth and employment-generating activities. 'This basic fact has been grasped and acted upon extremely slowly, not only in South Australia, but in rural Australia generally' (Houston, unpublished manuscript).

Developments in Land Use and Environmental Management and Planning: 1980s – 1990s

A number of important changes took place in South Australia's natural resource management, environmental protection and planning/development legislation, structures and processes during the mid-1980s to the mid-1990s. In 1983, a working party on land use for primary production recommended that all land (including freehold) ultimately be subject to land management legislation (Environmental Protection Council of South Australia 1988:83). There were also reforms in and consolidation or introduction of natural resource legislation—soils, vegetation, water and environmental pollution (referred to in Chapters 4, 5 and 7). The setting up of State-wide soil conservation and catchment management boards was another important development. These changes have been enhanced by the introduction of Landcare and ICM policies and the establishment of peak natural resource management policy bodies, which aim to integrate natural resource management into local, regional and State planning processes. Some of the legislative deficiencies have been addressed in the *Water Resources Act 1997* (which subsumed the *Catchment Water Management Act 1995*) (Chapter 15) and in a proposed new *Land Management Act*, which will replace the existing soil conservation and Landcare legislation.

In line with developments in other States, the 1990s was also a time of some transformation of planning legislation and agencies in SA. A metroplan was developed for Adelaide during the 1960s and revised in the 1990s. Following a major policy review (mainly urban but with some reference to the need for improved non-metropolitan planning) (Government of SA 1992), several regional reviews were undertaken (Flinders, Mt Lofty, Murray Valley, Barossa Valley). The *Development Act* was passed in 1993 (amended in 1996), replacing the former *Planning Act*. This legislation adopted a comprehensive approach to planning (including a requirement for EIA) through non-statutory, State-wide planning policies at local, regional and State levels. Local Councils, for example, are empowered to develop Council Development Plans, which are to be consistent with relevant metropolitan, regional and State plans and address environmental issues. They are subject to three-yearly reviews. Planning policies acknowledge the need for sustainability and environmental protection while recognising economic, social and cultural needs. DEHAA (the Department of Environment, Housing and Aboriginal Affairs) is the State's main environmental manager and administers the *Environment Protection Act*. It comprises six divisions, covering: environmental policies and protection, heritage and biodiversity, resource information (spatial data sets) and Aboriginal affairs (www.dehaa.sa.gov.au). The *Development Acts* are administered by Planning SA (within the Department of Transport, Urban Planning and the Arts (DTUPA)) but must be consistent with the requirements of the *Environment Protection Act*. Natural resource management—agriculture, primary industries, mining and energy—is the responsibility of PIRSA. Although adopting a sectoral approach to NRM, PIRSA is examining ways of developing more strategic spatial methods integrated with regional development plans. It has established a sustainable resources group to manage natural resource sustainability in co-operation with other agencies and the community.

The South Australian government's policy for development was presented in its 1994 Planning Strategy. However, the (re)published planning strategies for the individual regions are standardised, repetitive and still couched predominantly in terms of physical planning—focusing heavily on matters such as services, open space and the appearance of land and buildings. With few exceptions, proactive development planning provisions for the periphery are conspicuous by their absence (a 1976 government report recommending the protection of prime agricultural land had not been acted on a decade later: Environmental Protection Council of South Australia 1988:80). To remedy this, the rural regional plans (apart from the two completed for the urban pressure regions of the Mount Lofty Ranges and the Barossa Valley, discussed below) were to be reviewed. A Regional Development Task Force, comprising the Chief Executives of six major State Government Departments and located in the Department of Premier and Cabinet, completed a review of the planning strategy for country regions by the end of 1995 (refer to: www.pir.sa.gov.au/SARDT-report). The emphasis of the Task Force (and of the Regional Development Board for each region) was and is on economic development, as is evident for the six non-metropolitan regions (Spencer, Murray Lands, Riverland, South East, Kangaroo Island, Mid North and Outer Metropolitan, administered by DTUPA) shown on the government Website www.sa.gov.au. However, the newly created Office of Regional Development and Regional Development Council intend to broaden the focus of the regions beyond an economic one and seek greater co-operation between government agencies and communities (D. Hockey, Office of Minister for Primary Industries, Natural Resources and Regional Development, pers. comm., July 1999). A Regional Development Statement is being prepared.

Regional Planning

Strategy Plans for two regions adjoining metropolitan Adelaide—the Mount Lofty Ranges and the Barossa Valley—were prepared by the then Department of Housing and Urban Development (DHUD) by 1993. The renamed agency (DTUPA) continues to be involved in development and review of other development policies for non-metropolitan areas. This work principally takes the form of studies for particular areas such as the Murray Valley or the coast, and preparation of Ministerial amendments to Development Plans and review of Council-initiated amendments to Development Plans. South Australia is also a signatory to two interstate management agreements—for the Murray–Darling and Lake Eyre Basins. The Mount Lofty Ranges and Barossa Valley Strategy Plans integrate measures aimed at environmental protection/management with land-use planning. A similar approach was taken with the Murray Valley Management Review and is also adopted in reviews of Development Plans. Non-statutory committees also operate in six NHT regions and it is possible these will evolve into regional NRM bodies if the over-arching legislation under consideration becomes operational (A. Johnson, PIRSA, pers. comm., August 1999).

Planning Methods

No particular technique is used to evaluate ecosystems or to assess land capability and suitability. The Department operates ESRI's Arc/Info GIS software, which is used extensively by the planners to store environmental and socio-economic data sets for integration with administrative, planning and development proposal boundaries. They also use it to overlay and combine data in various ways such as for land suitability and land evaluation, hazard assessment, habitat analysis and environmental sensitivity assessment. A wide range of data sets is audited and maintained using the GIS, with data manipulation varying from single edits and updates to carrying out 'what if' scenarios.

Methods of involving the community in the planning process vary. Public notice is required to be given of Amendments and the Plan Amendment Report has to be made available for public inspection. Interested persons are able to make a written submission and a public hearing is held. A range of methods was used with the Barossa Valley and Mount Lofty Ranges Reviews, including a consultative committee of key stakeholders and public meetings. However, there is no standard approach.

The two regional plans include measures to: reduce land degradation; protect/enhance remnant natural vegetation and wildlife habitats; improve water quality (a fundamental tenet of the Mount Lofty Ranges Strategy—an area which encompasses 60% of Adelaide's water supply catchments); reduce conflicts between competing land uses, and encourage appropriate development. A wide range of complex issues to address included: rapid expansion of local townships; hobby farms; commuter belt development; loss of prime agricultural land; water catchment protection; tourism and recreation activities; plantation forestry; protection and enhancement of primary production; high fire risk, and protection of natural areas. Management zones were established for water protection, primary production and rural zone development.

Implementation

Implementation is achieved by various means. One of the principal measures is control of development, which is administered by local Councils. At the regional scale, the Mount Lofty Ranges Regional Strategy Plan initially proposed to create a Regional Management Authority to oversee the implementation of regional planning policy and to co-ordinate actions identified in the Plan. However, it was considered that the Region was too large (containing 15 local authorities or Councils), possibly requiring two Regional Management Authorities. Eventually a centre was established and made responsible for funding and technical support to local groups and agencies—a 'one-stop shop'. A separate catchment management program, jointly administered by PIRSA, DEHAA and the NHT, has since been set up. A Board of Management operates with community and agency representation. The program is to address NRM issues in the region through integrated approaches.

The functions of the Authorities were envisaged as being to:

- monitor and review the Regional Strategic Plan policies (noting that a 'policy' is defined as being a 'specific course of action to achieve the objective');
- advise on and co-ordinate regional planning policies for the region;
- administer development control for prescribed classes of development within the region;
- advise on and co-ordinate natural resource management policies within the region;
- promote public awareness of planning and land management aims within the region, and
- investigate specific issues as they arise (DHUD 1993).

Nevertheless, the report noted that 'it is not intended to create a fourth tier of government, as the role of the Regional Management Authority would involve the transfer of relevant State and Local Government powers by delegation'.

South Australia's Regional Strategy Plans: Mt Lofty and Barossa Compared

The *Mount Lofty Ranges Regional Strategy Plan* (DHUD 1993) and the *Barossa Valley Region Strategy Plan* (Barossa Valley Review Steering Committee 1993) are similar in structure. Both have Executive Summaries, a contextual chapter, and then chapters on: environmental protection; water resources; primary production; mining and mineral resources; residential and urban development; tourism; recreation and sport; bushfires, and management. The Barossa Valley report additionally has chapters on landscape protection, rural development, economic development and human services.

Four principles, upon which the development of regional priorities and policies should be based, were identified in the Government's terms of reference in establishing the Mount Lofty Ranges Review in 1987. The principles are:
- the opportunity for viable agricultural pursuits should be retained;
- the quantity and quality of water production from water catchments should be protected;
- the allocation, use and management of land should be based on land capability principles, and
- there should be strong community involvement in all facets of the Review.

The aims and objectives of the Strategy Plan conform to these principles, although the support for public participation referred to the process of developing the plan and not its subsequent implementation. Not surprisingly, given the region's much smaller area and emphasis on vineyards and wine production, the Barossa Valley Region Strategy Plan has more focused objectives, namely that the region:

> should maximise its economic potential while retaining integrity as a Region
> of cultural significance and as a Region with an environment of high quality
> due to its unique blend of natural and man-made landscapes.

The integrity referred to in the quote is to be retained through a commitment to:
- support viticulture and wine making as a major basis for the Valley's economic viability and landscape qualities;
- maintain and improve the variety of landscapes which include vineyards, other horticultural and farm land, natural and built form heritage, native vegetation and towns and buildings;
- promote value-added tourism, compatible with the Region's character, and
- increase economic activity in a manner sensitive to the existing environment.

Under the heading 'Environmental Protection', the Mount Lofty Plan aims to enhance and protect the natural and cultural characteristics of the Region through the management and protection of places of conservation value, cultural significance, scenic beauty and scientific interest. The inclusion of 'cultural significance' is of interest in the light of the earlier discussion concerning the meaning of 'environment' in Chapter 1 and, in relation to environmental impact assessment, Chapter 11. Twelve specific objectives are listed under the above aim. They concern remnant native vegetation, watercourses and wetlands, native fauna, resolution of land-use conflicts 'in a manner that affords the highest priority to environmental protection', coasts, scenery, visual aspects of the built environment, Aboriginal and European heritage of the Region, solid waste disposal, and 'promotion of public appreciation and understanding of the natural and cultural resources of the Region'.

The Barossa Region Plan has similar but perhaps more focused objectives under the 'Environmental Protection' heading. The overall aim is the 'maintenance and enhancement of all aspects of the existing natural and modified physical environment', followed by five objectives. Interestingly, the first is that environmental protection is 'to be given priority in the assessment of land management practices'. Another refers to the establishment of a Regional Authority for the purpose of co-ordinating natural resources management in the region. There are also six objectives under the 'Landscape Protection' heading (which is not present in the Mount Lofty Ranges Plan, although the objectives arguably are included in the latter Plan's longer list of objectives under 'Environmental Protection'). Two of the six objectives, however, are specific to the Barossa Valley; namely to discourage outward metropolitan Adelaide expansion into the southern edges of the Region, and attempts to confirm the Barossa Valley's 'sense of place' by constructing identifiable gateways to the Region.

Interestingly, objectives under the other, more development-orientated objectives, appear to incorporate environmental objects more clearly and consistently than was the case with the Northern Territory's Gulf Region Plan. Thus both South Australian Plans refer to the development of water resources and primary production in a 'sustainable' manner; and the aim of developing mineral resources is to be done 'in an environmentally sensitive manner' (Mount Lofty) and 'consistent with the environmental quality and rural character of the Barossa Valley Region'. Environment is not mentioned in relation to the extraction of mineral resources in the Northern Territory Plan. Thus the South Australian Plans arguably are more truly integrative in combining environmental with development objectives.

The question then arises as to how these objectives are to be achieved.

The Mount Lofty Ranges Regional Strategy Plan has a chapter devoted to each of the main headings (Environmental Protection, Water Resources, Primary Production and so on). Within each chapter, the objectives (as discussed above) are first listed and then elaborated upon under separate headings. For example, the first objective in Chapter 3, 'Environmental Protection', of the Mount Lofty Plan, is 'Conservation of remnant native vegetation and its protection from further clearance, disturbance and fragmentation'. This is followed by Policy, Action and Responsibility statements, which are defined respectively as: policy—a specific course of action to achieve the objective; action—a specific mechanism or procedure required to implement a policy, and responsibility—indicating the agencies responsible for implementing an action, with the 'lead' agency marked in bold type. Thus, continuing with the remnant vegetation objective, 14 'policies' are identified, each with an associated background statement, followed by 'action' and 'responsibility'.

Again to illustrate, the first of the 14 remnant vegetation policies reads:

> No further clearance of remnant native vegetation, except for essential clearance of a minor nature as set out in Part II of the Regulations under the *Native Vegetation Act (1991)* or other minor clearance approved by the Native Vegetation Council.

The policy background provides a context and further explanation for the policy, followed by the Action (Responsibility) statement:

- Implement *Native Vegetation Act (1991)*. (**DENR**)
- Review adequacy of Native Vegetation Act Regulations in the context of the policies set out in the Regional Strategy Plan.

In comparison, there are six objectives listed in Chapter 6, Mining and Mineral Resources. The listing of the six objectives is followed by an italicised Note stating that 'all actions regarding environmental management of mining operations and mineral resource planning are of High Priority'.

The first of the six Objectives is: 'Management of mining operations to minimise environmental impact and to maximise the extraction of the resource'. This is followed by nine Policies, the first of which is: 'proprietors and operators of mining operations to be responsible for maintaining a high level of environmental management in their operations', and the Action (Responsibility):

> - Responsibility for the management of mining operations to be specified in mine development programs, especially in regard to the management of external impacts and rehabilitation. (**DME**)
> - Prepare guidelines on how to manage the environmental aspects of mining operations. (**DME**, EWS, DENR)

The final Chapter 11 discusses the setting up of the proposed Regional Management Authority, commented on in the first section of this chapter.

Implementation in the Barossa Valley Region Strategy Plan is very similar, although there is perhaps more introductory, explanatory material. Within each chapter (devoted to the various headings discussed previously) there are 'Management Guidelines', which comprise a number of Objectives, with each Objective further elaborated upon with statements of Policies and Action/Responsibility.

For example, the first Objective in Chapter 3, Landscape Protection, is: 'minimisation of landscape alteration in visually significant areas'. Below that, the first of the seven Policies is to 'protect the quality of natural and rural land-scapes including native vegetation, vineyards and roadside vegetation along approach roads to the Barossa Valley from significant disturbance'; and the Action/Responsibility statements are:

- incorporate into Regional SDP through setback requirements;
- guidelines to be prepared for use by local and service authorities and landowners.

(DHUD, DENR, DOT, Regional Authority)

The emphasis on water resources is indicated by the number of Policies (41) and associated Actions/Responsibilities. As with the Mount Lofty Ranges Strategy, environmental objectives are incorporated in the more development-orientated objectives. For example, under Primary Production, the first Objective is to use rural land for 'sustainable commercial production of food and other products', and the fifth Policy under that Objective is:

Primary production land uses to be defined as consented to or prohibited depending on the degree of:

- water usage,
- environmental impact,
- visual impact on the landscape.

Under Mining and Mineral Resources, the *first* Objective is to contain 'the environmental impacts of mining and quarrying to acceptable levels' (with DME to prepare guidelines on how to manage the environmental aspects of mining operations). There are nine Policies under this Objective, covering areas such as proprietors and operators being responsible for maintaining a high level of environmental management in their operations, assessing landscape values against the value of mineral production, the requirement for compre-hensive rehabilitation programs to be developed before operations commence, the establishment of buffer zones, inspection procedures, avoidance of water pollution and avoidance of damage to stands of 'significant' remnant native vegetation. As is common, 'significant' is not defined.

The Barossa Valley Region Strategy Plan strongly recommends the estab-lishment of a statutory body, the Barossa Region Planning authority, to 'oversee the implementation of regional planning policy' within the region and 'to co-ordinate ongoing actions identified in the Strategy Plan'. Many of the recom-mended Actions specify the Authority as the body responsible for their implementation. It follows that failure to establish such an Authority would require considerable re-assessment to identify alternative responsible bodies.

The Schedule to the Regional Strategy identifies the circumstances which will determine whether an application is of regional significance. They include, for example, bus, fuel and public service depots, road transport terminals, extension to industry of warehouse buildings resulting in a total floor area exceeding 500 m², and new wineries with floor areas exceeding 1000 m². Part B of the Schedule identifies other development proposals previously dealt with by particular bodies which are to be the responsibility of the new Regional Planning Authority. A flow chart indicates the procedures whereby applications of regional significance will be handled, including public notification procedures and provisions for appeals. This part of the Schedule also outlines the scope, formation, structure, membership and resourcing of the Authority (the latter from member Councils and Government Departments), and presents an organisational chart showing the relationship of the Regional Authority with other organisations.

Some of the problems with the Mt Lofty Plan have been identified by Rolls (1997), namely: delays in the program; the complexity of issues, which has made management difficult, and the large number of Councils and players involved. As a consequence, partnerships sought have been slow to develop, resourcing has been inadequate (some NHT funding assisted the program in the late 1990s) and leadership has suffered as a result. Its strength, on the other hand, has been in having a co-ordinated and focal resource centre. Partnership experience has laid the foundation for a comprehensive approach to natural resource management in the Ranges.

The Southern Region of Councils (SROC)

Social, economic and environmental factors were also placed in a regional context in the *Southern Adelaide Plan* (DHUD 1995), in line with the State's Planning Strategy. The plan was developed amongst five local Councils—termed the Southern Region of Councils, or SROC—of the southern Adelaide region, in consultation with State government representatives and stakeholders of the region. SROC is a voluntary amalgamation of Councils in a region extending over 680 km^2 and with a population of 240 000. The region takes in part of metropolitan Adelaide as well as a coastal zone, old townscapes, Aboriginal heritage sites, conservation parks, vineyards, some industry, and other rural land uses. Particular pressures on the region relate to population growth, land degradation, loss of biodiversity and water resource issues.

Under the *Southern Adelaide Plan*, the Councils committed themselves to the following guiding principles:
- integrated planning using land-use capability and total catchment management (ICM is strongly supported in law under the 1997 *Water Resources Act*);
- achievement of best practice in environmental management;
- adoption of ESD in decision-making, and
- recognition of social justice in connection with quality of environment and lifestyle in the Council communities.

The Southern Region of Councils (SROC) Environmental Management Strategy

A subsequent environmental management strategy (SROC 1996) was seen as part of an ongoing planning process set out in the *Southern Adelaide Plan*. The non-statutory environmental strategy further developed key environmental themes identified in the 1995 Plan and incorporated the guiding principles outlined above. It does not override the former plan but offers guidance and consistency in environmental management. The environmental strategy was one of seven national pilot projects sponsored by the Australian Local Government Association (ALGA) with Commonwealth funding (cf. WA's South West Regional Environmental Strategy in Chapter 17). Key objectives were to develop an integrated regional framework to enhance ESD in the region, to develop best management practice environmental management, and to identify environmental projects which would generate employment in the region. Following a consultative process with key stakeholders, several major foci were identified. These were: catchment management; open space management; biodiversity; coastal management; ecotourism/sustainable tourism, and sustainable industry/green jobs. Transport and waste were also important regional issues but were dealt with under other portfolios in SROC.

SROC has been disbanded following changes to Council boundaries and some amalgamations. However, the Strategy has been used by the newly created City of Onkaparinga and its Water Catchment Board in the preparation of local environment and catchment plans.

Other Regional Initiatives

In parallel with the SROC environmental strategy, a pilot project on developing SoE indicators, also funded by the Commonwealth, was reported in SROC (1996:Vol. 2). Attention was focused on monitoring within the Southern Region as part of a strategic environmental planning process. The objective was to develop a core set of environmental indicators at local levels as inputs into regional, State and national SoE reporting. Consistency and good standards in monitoring were seen as key issues.

Eyre Peninsula Regional Development

West of Adelaide, the Eyre Peninsula covers 55 000 km^2 and is home to 33 000 people. In 1993 an *Eyre Regional Development Board* (ERDB) was established with ten participating local governments. Enjoying the support of local, State and federal governments, the ERDB focuses primarily on economic development and employment for a region which has been disadvantaged by its size, isolation, limited potable water, small population base and a high dependency on agriculture. Nine Target Teams cover issues such as infrastructure, business and local government, as well as primary industries and the environment. With agency participation, the Environment Team brings together various environmental matters under one regional umbrella. The body is also seen as a potential avenue for NHT funding.

An *Eyre Peninsula Regional Strategy Committee* oversees a project to enhance the environmental sustainability of agriculture in the region. With Commonwealth and State support, the program is largely run through PIRSA (refer Parts D and E in Dore and Woodhill, 1999).

Local Agenda 21

Several Southern Councils have also adopted a *Local Agenda 21* as an environmental management framework. The end result will be a Local Environmental Plan to improve local environments and integrate environmental management in all Council operations, with community involvement. Some other, separate Council initiatives cover revegetation and wetland projects, and management of stormwater and wastewater.

Local Government Act Reform

Under the requirements of the *Local Government Act (Boundary Reform)*, local governments are to identify the 'communities of interest' in their areas. These refer to physical, economic and social systems which are central to the interaction of communities and their local environments. The boundary reform process offers opportunities to rethink administrative boundaries (such as catchment boundaries) as a basis for local government reform. Three Councils in the SROC region discussed above, have since been amalgamated into the City of Onkparinga.

CHAPTER 19

THE INTEGRATION OF ENVIRONMENTAL PROTECTION WITH REGIONAL PLANNING AND MANAGEMENT IN QUEENSLAND, NEW SOUTH WALES AND VICTORIA

In addition to published sources of information as cited, and government Web-sites for the three States (follow planning and environment links), we have drawn on unpublished papers written in 1994 by Daryl Low Choy (Qld), Roger Epps (NSW) and Colin Leigh (Vic.) and, additionally, material kindly provided by several people as acknowledged in the Introduction.

QUEENSLAND

The Planning Context

The historical context for planning in Queensland can be summarised by Minnery's (1987) observation that Queensland is not only geographically on the fringe of Australia but also that planning in the State was on the fringe of planning. According to Robinson (1994), this view referred to both the nature of Queensland's location and pattern of economic growth compared with New South Wales and Victoria, and the State Government's attitude towards planning and the management of economic development, which Mullins (1980) described as politically authoritarian, socially repressive, fundamentalist and full of hillbilly panache.

Minnery and Mullins were referring largely to the era of the Country/National Party State Government from 1968 to 1987 under the Premiership of Johannes Bjelke-Petersen. Robinson (1994) noted that this was a period of very rapid growth, especially in the southeast corner of Queensland where the traditional planning processes were frequently circumvented or 'fast-tracked' to facilitate large-scale development projects. The powers of the *State Development and Public Works Organisation Act 1971* and the *Integrated Resort Development Act 1987* were particularly 'useful' in this regard (coincidentally, the Federal Labor Government of the 1980s associated this process with its deregulation policies). Flooding, erosion, silting, pollution and adverse effects on coastal ecosystems resulted, and the State was notorious for sales of tidal mangrove 'real estate' around Moreton Bay to unsuspecting overseas buyers. Additionally, the lack of a regional approach to strategic planning was a crucial weakness, making it impossible to develop rational, consistent and coherent development policies for tourism, pollution control, water supply and environmental protection in relation to the growth of the coastal region and the Brisbane metropolitan area (Kozlowski 1990; Robinson 1994). Yet ironically, the badly-planned Brisbane of the 1960s was under the control of a single municipality, while better-planned cities in other parts of Australia and New Zealand, governed by numerous LGAs, considered that the solution to their planning difficulties lay in the amalgamation of their local authorities. A further weakness was that the State itself was not bound by strategic plans, especially for Crown Land. Neither had it been particularly interested in the Federal Government's regional initiatives of the 1970s, apart from a brief period late in the decade when Regional Co-ordinating Councils and Regional Planning Initiatives were set up and then dissolved. Only a few voluntary regional bodies, such as the Moreton Regional Organisation, survived this demise of regional planning. One gain of this period was an extensive coastal management survey which looked at the urgent need for statutory planning schemes and regional plans to control environmental and other problems in sensitive coastal areas.

Developments After 1989

Bjelke-Petersen was replaced as State Premier in December 1987 and two years later the National Party, after 32 years in government, was succeeded by Labor. During its first years in power substantial changes were made in the way in which the State Government operates (Robinson 1994). The introduction of the *Local Government (Environment and Planning) Act* in 1990 heralded a period of major policy reform for Queensland. Parallelling changes in other States, the Act consolidated and rationalised planning procedures and strengthened environmental protection measures. However, the devolution of statutory planning functions to local authorities remained unchanged and has been one of Queensland's planning strengths (as in New South Wales). A considerable number of local authorities were already displaying strategic plans, and more were expected in response to increasing development pressures in rural and coastal areas. Other changes included the introduction of a more efficient land administration system through the creation of a Department of Land Management, which initiated an inquiry into the State's land tenure system. The National Parks and Wildlife Service was reinstated within a new Department of Environment and Heritage (becoming a separate agency in 1999); a number of reviews were initiated, and more land was targeted for conservation. The State Government also sought to control urban sprawl; and the newly created Department of Housing and Local Government (now the Department of Communication and Information, Local Government and Planning) proposed to co-ordinate the actions of State, local and federal agencies. In effect, the new regime proposed a more rational and balanced planning system, incorporating more community involvement and less opportunity for exploitation and manipulation by individual Ministers or business interests. In parallel with these reforms, major trials were under way to deal with corrupt public officials of the former government (Briody and Prenzler 1998).

At the State-wide level, the State government issued planning policy and support documents which indicated the intention of co-ordinating environmental protection with regional planning. For example, the *State Planning Policy 1/92: Development and the Conservation of Agricultural Land* was supported by a second draft set of guidelines in 1993 dealing with the separation of agricultural and urban land uses, issued jointly by the Department of Primary Industries and the (then) Department of Housing, Local Government and Planning. Reflecting similar developments in other States, the guidelines dealt primarily with measures to ameliorate potential conflicts between urban and agricultural land uses, particularly new rural residential developments, and were designed to assist local authorities, developers and the agricultural sector. A revised set of guidelines was issued at the end of 1994.

Also in 1994, the Department of Primary Industries released a Discussion Paper on *The Sustainable Use and Management of Queensland's Natural Resources*. Policies and strategies were outlined on the management of natural resources for which the Department has legislative responsibility; namely, land, water, forests and fisheries. The following policy directions were emphasised in the Discussion Paper:

- ecologically sustainable use;
- voluntary action supporting the Landcare approach;
- integrated catchment management approach;
- statutory natural resource management plans for water and forest resources and for critical land areas;
- community participation in the planning process, and
- clearer definition of entitlements to natural resource use.

The Discussion Paper was the prelude to proposed natural resource management legislation intended to integrate the nine existing natural resource Acts for which the Department was previously responsible (the proposed legislation was still under consideration in the late 1990s—refer Chapter 15). It should be noted, however, that despite the positive nature of the above policy directions, there is some evidence that their application is sometimes less than the ideal. In relation to the Landcare objective, for example, Morrisey and Lawrence (1997) found that less than 5% of the farmers in a central Queensland region were members of a Landcare group, in contrast to the 30%+ figure for the rest of Australia. Half the Landcare group members studied gave a somewhat less than laudable reason for joining—namely, membership was seen as a means not of combating land degradation but of counteracting an increasing 'greenie' influence over rural environments: but there are parallels elsewhere.

Various government reviews covered land management issues related to rural planning, ICM, land releases, water allocation rights, heritage protection, air pollution and coastal management, among others. Serious deficiencies in

planning were identified in a major planning review in 1992, with various recommendations made for reform (QDHLGP 1993). Major areas of concern
included excessive fragmentation, lack of co-ordination and complexity of
planning arrangements; unnecessary delays for developers and lack of flexibility; inadequate public participation, and incomplete procedures for EIA. The
review led to the release of a public discussion paper on the proposal for new
planning and development legislation as a precursor to the PEDA Bill, which
subsequently was to become the *Integrated Planning Act 1998*.

The Planning, Environment and Development Assessment Bill 1995 (PEDA)

PEDA had its origins in a Planning Protocol which had been signed between
the State and the Local Government Association in 1993. This protocol committed both spheres of government to integrated State, regional and local planning (also reflecting the recommendations of the IGAE). It incorporated
biophysical, economic and social environmental considerations in development planning, clearly defined governments' roles, ensured public participation
in all stages of the planning process, provided appropriate mechanisms for dispute resolution and for law enforcement, and simplified legal procedures. Local
governments were to retain their primary responsibility in local planning matters but could also have an interest in the development of State and regional
planning policies. The State, on the other hand, assumed responsibility for the
development and implementation of regional planning, the co-ordination of
major infrastructure and public works, and the formulation of State planning
instruments and their integration with local government planning schemes.
The State also has interests in developments of major economic or environmental significance. The PEDA Bill was aimed not so much at consolidation of
laws as at retention of the 'best' features of existing laws, and incorporating new
features such as fast tracking for development consent.

England (1996) reviewed some key elements of the PEDA Bill. She found
that the problem of fragmentation and complexity in existing legislation was
addressed to some extent and the approvals process simplified under the new
Integrated Development Assessment System (IDAS). However, there were
inconsistencies in some of the procedural details and, although clearer language
was used, some obscurities remained. The most serious concern was related to
what England saw as potentially draconian powers. The authority of State agencies and ministers would be strengthened without regard to other principles of
reform, such as enhancing public participation. Few mechanisms were included
to achieve the goals of integrating economic, environmental and social needs.
These concerns were shared by McDonald (1995), who feared that there was a
grave risk of undermining what some would argue was an already weak EIA
regime. England (1996) considered that in some instances, the above principles—including public participation—appeared to take second place to the goal
of defining and safeguarding the role of State agencies. She also found in other
cases that lofty statements of intent were poorly translated to operational mechanisms, leaving much to the discretion of individual development managers.

The question was posed: did the PEDA Bill amount to a radical new change of philosophy, or was it really only a change of form which did not seriously threaten the old philosophy of unaccountable, centralised State government? The following question could be added: and did it differ in any way from similar processes occurring in other States?

Briody and Prenzler (1998) echoed some of the foregoing concerns. They challenged the degree to which the legislated environmental reforms of the 1990s in Queensland address some of the outstanding environmental management problems, such as pollution, and noted that, despite the various reforms, environmental problems appeared to be worsening. They provided clear evidence of a reluctance by government agencies to use their increased punitive powers to break the dominant paradigm of development at any cost, and suggested that the non-enforcement of environmental laws 'must include a notion of partial capture of the Department of Environment'. Possible factors responsible for this reluctance include political influence, over-familiarity with industry and a tradition of under-enforcement (it is also likely that, in line with trends in the other States, regulatory agencies prefer to encourage voluntary compliance rather than using mandatory approaches). Establishing a well-funded EPA insulated from the foregoing was recommended, and an EPA was eventually set up in March 1999.

Nevertheless, environmental concerns and the need to integrate management processes have not been completely ignored by local authorities. In 1998, the Brisbane City Council produced its second *State of the Environment Report* (Brisbane City Council 1998b) and a *Biodiversity Strategy* (Brisbane City Council 1998a). One of the key objectives of the latter is to 'improve the conservation outcomes of future development through integrated urban planning'. In this regard, recognition is given to the intentions of the new *Integrated Planning Act* (see below). Efforts are to be made to create conservation zones and incorporate biodiversity protection into the City and local plans. There are strong imperatives for this since the City has over 30 000 ha of natural bushland and wetland vegetation, more than 400 bird, 100 plant and 31 mammal species, and important international wetland sites. But until the late 1980s, natural areas were being lost at a rate of 550 ha per year—down to less than 100 ha per year a decade later.

The City's strategic plan and environmental policy both adopt ESD principles, and EMS is used by the Council to assist it to comply with the *Environmental Protection Act*. A restructuring has made possible a better integration of policy and planning for the natural environment in areas such as bushland protection (which includes the successful use of Voluntary Conservation Agreements), green space plans, water management and support for community involvement in environmental protection.

The Integrated Planning Act 1997

The PEDA Bill was passed, with amendments, in 1997 as the *Integrated Planning Act* (IP Act), repealing the *Local Government (Planning and Environment)*

Act 1990. The following discussion outlines the new Act but needs to be read in the light of the above comments on its PEDA precursor.

As discussed by Allison and Keane (1998), the *Integrated Planning Act* recognises the need to incorporate economic, social and biophysical environmental objectives. Its centrepiece is the Integrated Development Assessment System (IDAS) which ensures streamlined, fast-track approvals for development in a one-stop shop, assessment process. More than sixty State/local approval processes and thirty Acts have been consolidated into a single process. (It should be noted that this Act goes further than the equivalent NSW legislation, in that it creates one legal system to deal with approvals, whereas the NSW legislation is concerned with co-ordinating arrangements within the existing system.) An essential notion of IDAS is its comprehensive assessment of environmental impacts. Unlike its predecessor, the 1990 *Local Government (Planning and Environment) Act,* the IPA Act does not prescribe a formal structure for planning schemes. The Act is aimed at introducing performance-based controls which assess a development against its environmental impact, removing the necessity for a planning scheme's regulatory function to be based around land-use zones. This supposedly offers greater certainty and flexibility to developers without prescriptive mechanisms to hinder growth and investment. But it also raises questions concerning the effectiveness of self-regulation and performance-based measures (refer EMS in Chapter 8).

England (1999) has conducted an early review of the IP Act, noting that it introduced some sweeping reforms which other States have watched with interest. As noted earlier, one such reform was the impetus given to performance-based, outcome-orientated planning in place of traditional prescriptive zoning practices. Another was the duty of planners and other decision-makers to have regard to the overarching requirement of incorporating ecological sustainability and the precautionary principle into the planning and development process. Planning remains largely the responsibility (and at the discretion) of local governments who are guided by certain core criteria which give weight to environmental matters. Thus, environmental goals are seen as pervading the whole planning process rather than being peripheral to it, as was previously the case. The overall approach is designed to be more strategic, flexible, responsive and holistic.

However, England once again raises some concerns over aspects of the new legislation. She asks whether the Act is in fact any 'greener' than the *Local Government Act* it replaced. Rather, the new Act would appear to be more focused on the procedural streamlining of planning processes (as has been the case in other States)—and indeed she considers there are some aspects of the Act which are regressive. While the discretionary approach, for example, may work well for more environmentally conscientious or progressive Councils, others might well fall short in their environmental planning obligations. The IP Act also offers greater opportunities for the State government (and Minister) to intervene in local planning matters. The use of such discretionary powers may ultimately work in favour of business and developer interests, with concerns for better environmental planning taking second place.

Regional Planning

The IP Act formalises the role of regional planning by providing a framework for Regional Planning Advisory Committees (RPACs) (see below). Although lacking statutory powers, these committees can advise on regional (and sometimes State) issues of importance. This will permit more effective planning co-ordination at all levels of government and public participation in the planning process. Government departments are to be more involved with local planning schemes through a co-ordinated process. Infrastructure development is to be made more efficient through improved strategic and comprehensive planning. By involving local authorities in regional planning, planners will be forced to integrate economic, social and environmental issues.

The Department of Communication, Information, Local Government and Planning supports regional planning projects in Queensland. These include: Wide Bay 2020; Far North Queensland 2010 and South East Queensland 2001 (below); Townsville/Thuringowa; Whitsunday, Hinterland and Mackay 2015; Gulf Region; CYPLUS (Cape York Peninsula Land Use Study (below)), and Central Queensland. Other projects are under development through the Department of Natural Resources (DNR) (for example, the South West Strategy, discussed below).

The South West Strategy

To all intents and purposes, the *South West Strategy* is a regional project within a bioregional framework. As part of a broader State initiative, Queensland has been divided into 12 bioregions to attract and aid preparation of regional environment strategies. Launched by the Queensland government in 1994, with federal support, the Strategy evolved out of extensive community consultation and the publication of position papers investigating the problems of the mulga rangeland region of southwest Queensland and northern New South Wales (Department of Lands 1993). Land-use studies of the 1980s had identified that the area suffered from extensive land degradation. This was associated with vegetation clearance, the invasion of woody shrubs, changed fire regimes, reduced water flows from the Great Artesian Basin and overgrazing by cattle and feral animals. Some severe economic and social difficulties (such as uneconomic pastoral holdings, high indebtedness and withdrawal of community services) exacerbated the problem. Following the announcement of the 1994 Strategy, a *Mulga Land Use Advisory Group*, with broad community and government representation, was set up with the objective of implementing an integrated approach to the various environmental, economic and social problems of the region. Several key strategies were to be developed, including enterprise reconstruction, natural resource management and regional development. Property management services were to be made available to landholders and regular resource monitoring was to be undertaken. A regional support team (now under DNR) was established in Charleville to deliver and co-ordinate the Strategy (Hoey 1994).

By the late 1990s, implementation of the Strategy was well under way with the release of a *Strategic Plan* to the year 2003 (Macarthur Agribusiness 1998), and a major injection of funds from the NHT. While the earlier focus had been on economic and structural matters such as rural reconstruction and leasehold arrangements, later emphasis centred on improvements to natural resource management (South West Natural Resource Management Group 1999). This has been supported by several targeted programs covering feral animal control, safe carrying capacity, bore drain replacement and nature conservation (South West Strategy *Annual Report 1997–98*, South West Strategy Office, Charleville).

The South East Queensland Project (SEQ 2001)

Another serious attempt by the Queensland State Government to engage in regional planning was the SEQ (South East Queensland) 2001 regional planning project, which was completed early in 1994. It was a co-operative planning exercise involving all three levels of government with representatives from the business sector and community organisations. The project was overseen by a Regional Planning Advisory Group (RPAG). The main output of the exercise included fourteen policy papers, including ones on agricultural land, nature conservation, open space and recreation, rivers and coastal management, and water and waste water. The *Final Report* was released in March 1994 after a further period of public review. The report provided a regional framework for growth management, a regional outline plan, recommendations for co-operative and co-ordinated planning, and proposed institutional arrangements. It contained 16 objectives, 94 principles and 144 proposed actions, including recommendations on environmental quality and proposals to establish a regional open space system. Sub-regional structure plans for the four sub-regions of the SEQ Region were then produced.

The SEQ 2001 project was seen as a model for others in an integrated planning framework, incorporating economic, environmental and social criteria. The RPAG had government recognition; but although this and other RPAGs are recognised in the new IP Act they still have no statutory powers and rely on the support of interest groups to address issues.

Far North Queensland and Cape York Peninsula Projects

A further major regional planning initiative by the State Government was the FNQ (Far North Queensland) 2011 Project, centred on Cairns. The project had a similar Regional Planning Advisory Group supervisory structure to the SEQ 2001 project, and worked towards the production of four regional strategies, namely:

- urban growth and transport management strategy;
- regional economic development strategy;
- conservation strategy, and
- integrated catchment management strategy.

A final Regional Growth Management Report was produced in 1996 and further information papers on future strategies have been published.

There is also a joint State–Federal planning initiative which commenced in 1992 entitled CYPLUS (the Cape York Peninsula Land Use Study). It is a process to guide ESD in land management of the Cape Region, an area where environmental issues, development proposals, land rights and other socioeconomic factors have generated conflict over the years (Mobbs and Woodhill 1999). Stage One (1992 to 1996) included the collection and analysis of data. Stage Two (1995–97) addressed key issues of natural and cultural resources, conservation, economics, land tenure, infrastructure and lifestyle in a Strategy Report and made recommendations. Stage Three (due at the end of 1998) develops State/Commonwealth policy responses to the Stage Two report.

Regional plan developments for other areas include Fraser and Bribie Islands, Moreton Bay, the Great Barrier Reef and the Great Sandy region. There are also voluntary arrangements amongst Shires; for example, the seven Councils of the Sunshine Coast have banded together for mutual benefit, although the emphasis is on economic development.

Finally, integrated catchment management (ICM) continues as State Government policy (Chapter 15). More than three-quarters of the State has catchment management coverage. Former catchment and Landcare committees now operate under strengthened arrangements with broad representation. There are also interstate arrangements in partnership with the Commonwealth for the Murray–Darling, Lake Eyre and Great Artesian Basins.

Thus it can be seen that Queensland has been moving, albeit schizophrenically, towards integrated environmental protection policies, principles and measures in regional planning, and it is perhaps still too early to assess the effectiveness of these developments. Encouragingly, the Labor Government elected at the end of 1998 appears to be continuing support for regional initiatives.

NEW SOUTH WALES

The Department of Urban Affairs and Planning (DUAP) is the State's prime planning agency and is made up of several Divisions. These deal with planning policy and regulation, rural and urban management and the development of housing policies. Linkages between the Divisions are aimed at the delivery of integrated planning and environmental policy programs within sustainable management frameworks.

In relation to environmental matters, several Divisions of DUAP are of significance, covering State and regional planning, environmental policy and assessment, and resources and conservation. Some of the key activities of these (and other) Divisions include: reforms to planning practices to ensure more open and efficient planning and approval processes; rigorous environmental impact assessment of forestry, mining and infrastructure projects; protection of coastal, rural and other natural resources, and co-ordination of biological, social and economic data for land and resource allocation decisions.

A Department of State and Regional Development is the principal body for promoting metropolitan and regional development, emphasising economic growth through innovation, sustainability and employment. Regional Development Boards have operated since 1972 (a number still exist) and are ministerial appointments. Eighteen State offices act as points of entry to the various regional development programs.

Until the 1980s, and in common with the rest of Australia, planning and development in non-metropolitan areas of NSW had generally escaped conventional planning controls (with the exception of economic development regions such as Albury–Wodonga and Bathurst–Orange). Where they existed, the government used its powers through various natural resources policies and regulations pertaining to the management of soils, water, vegetation and land release. In a formal sense, environmental and ecological matters were largely ignored.

For all this, NSW was in some respects a trend setter in the development of natural resource and environmental management in non-metropolitan areas. Australia's first national park was established in NSW, which was also the first Australian State to set up soil conservation legislation and agencies in the late 1930s. The good record in soil conservation was no doubt enhanced by the introduction of another first, the *Catchment Management Act* in 1989, which in turn was supported by the more broadly based Landcare movement. But there is still no formal State-wide rural planning policy, although the State is involved where subdivisions fall below the minimum requirements of Local Environment Plans (LEPs—refer the following section). All subdivision matters are dealt with by local governments under LEPs, which are subject to State government review. The State also has powers over local government through soil, water, vegetation or conservation legislation; and the agricultural agency requires local governments to consider land capability classes in determining subdivision applications.

Regional environment plans or conservation strategies have been developed at local government levels, with some voluntary regional groupings of Councils adopting Agenda 21 and ESD principles. Amendments to the *Local Government Act* in 1997 also incorporated ESD into plans and policies. Local government boundary reform is expected to see Councils assuming more active and effective roles in natural resource management.

Regional and Environmental Planning

The *Environmental Planning and Assessment Act 1979* was particularly progressive, in that environmental, economic and social criteria were incorporated into its key objectives (Chapter 7). These objectives were achieved through State Planning Instruments (SPIs); namely, State Environmental Planning Policies (SEPPs), Regional (REPs) and Local (LEPs) Environment Plans, Development Control Plans (DCPs) and EIAs for significant developments. Australia's first regional environmental plans flowed from these SPIs. Although they significantly raised the profile of the biophysical environment as well as economic and social needs in planning, they were not always adequate; and they have been under constant attack since their inception. Stein (1998) believes that these instruments have now been stripped of their effectiveness through various amendments and exemptions (refer Chapter 14). On the other hand, the more recently released greater metropolitan region planning strategy (DUAP 1998), discussed below, may have partly compensated for some of the earlier deficiencies.

In its geographical, integrated approach, the *South Coast Study* (discussed in Chapter 9) was in some respects a forerunner of later regional planning in NSW; and the first of the regional environmental plans, the *Hunter Regional Plan* (DEP 1982), followed years of research in the Hunter Valley, notably involving the Hunter Valley Research Foundation and the Department of Geography at the University of Newcastle—whose Professor, Alan Tweedie, was also one of the four authors of CSIRO's earlier Hunter Valley land systems study (Story *et al.* 1963). However, according to Broadbent and Searle (1996), the Hunter Plan had little to say on ecological sustainability. Some land degradation issues were referred to and general measures outlined to minimise the conflict between development and the environment. These included pollution controls and the use of buffer zones. Development controls were unspecified. On the other hand, the *Illawarra Regional Plan* four years later (DEP 1986) was more specific with regard to the management of natural environments and rural lands. The subsequent *North Coastal Regional Plan* (DoP 1988), however, appeared to move back to a more generalised approach to ecological issues. The later *North Coast Strategy* (DoP 1993) showed a tendency to view natural habitats as impediments to development. This, and a South Coast Strategy, were responding to increasing pressures on coastal resources resulting from large inmigrations to the regions. Whatever the environmental weaknesses of these and other Strategies (including the Sydney–Canberra corridor and the *Murray Regional Plan No. 2*), however, proactive decision-making involving all relevant parties was pursued.

Planning Policy Reviews

Epps (pers. comm.) noted that a number of planning possibilities arose in New South Wales as a consequence of various national and State reviews of regional planning, and the increasing realisation that management of non-metropolitan environments is not within the domain of any single agency. National reviews included those by: the Taskforce on Regional Development (Taskforce 1994); the Department of Housing and Regional Development; the Bureau of Resource Science, and the Industry Commission. Several State reviews were produced under this 'umbrella' of reassessment of the potential for development at the national level. These reviews included the New South Wales Senate Standing Committee's reports entitled *Regional Business Development in New South Wales* and *Achieving Sustainable Growth*, the Sturgess review entitled *Thirty Different Governments*...(Epps comments that it became known as the

'Report of the Commission of Inquiry into Red Tape'), and *Regulation Review* (the Regulatory Impact Statement, Draft Best Practices and Draft Environmental Planning and Assessment Regulation) produced by the Department of Planning for public comment.

The Sturgess report was linked with the Department of Planning's review of regulations and recommended the delegation of further responsibilities to local Councils suitably staffed and organised to accept them. The report was critical of the 'present maze' of State environmental policies and recommended a comprehensive review (note: the *Environmental Planning and Assessment Act* was subsequently amended in 1997 to streamline approvals processes). However, it advised on the need to maintain the public inquiry process for controversial issues, and proposed that community concerns should be resolved prior to individual developments and that the government should arrive at a common policy position, providing a predictable context for development.

To illustrate the recommendation, Sturgess referred to the initiatives of the Wagga Wagga City Council with regard to its strategic planning for industrial development. This included environmental auditing of key attributes of suitable sites before development applications were received. Of particular relevance to rural areas, it involved appropriate preliminary public consultation and successful co-operation between the Department of Planning and the Environment Protection Authority.

Thus, by the early to mid-1990s, a number of changes were taking place in natural resource management and environmental planning and practice in NSW. Once again, these mirrored (or influenced) similar events in other States. During this period, several serious environmental disputes had entrenched public and political views over development (viewed as equating with the provision of employment) versus environmental protection. The disputes were related to particular forestry practices, mining developments, the very fast train proposal and the disposal of intractable waste. While the State appeared to espouse ESD principles in its public environmental policies, and acknowledged a need for community participation, its actions frequently indicated other priorities.

The Greater Metropolitan Region Planning Strategy

In 1998, DUAP released a planning document entitled *Shaping Our Cities— The Planning Strategy for the Greater Metropolitan Region of Sydney, Newcastle, Wollongong and the Central Coast*. Although focused on the metropolitan region, it is used as an example here to illustrate the way in which planners are incorporating both environmental and regional criteria into more integrated and strategic planning processes.

This Strategy is important in that it embraces over one-quarter of Australia's population in a region experiencing rapid population and economic development. There are strong challenges to planners, land managers and politicians to ensure that the future needs of the region, in relation to employment, infra-structure, housing and a clean environment, are provided for.

New sets of planning directions have been developed to guide local govern-ments and State agencies through detailed planning policies and guidelines. To reflect regional differences, a new series of regional planning strategies will also be released progressively. These will start with Western Sydney, Sydney South, and the Central Coast. Other initiatives are under way for the Hunter and Illawarra regions. Clear directions are also given to link planning issues and strategies more closely with government agencies, local governments and communities.

Attention is given to the pressures of population growth in the Strategy, and the need for sustainable management of natural resources to protect the envir-onment. Natural and healthy environments are accorded high priorities in the Strategy, while a regional approach to conserving biodiversity and Aboriginal heritage is adopted. Other environmental objectives cover: the creation of a comprehensive regional network of open spaces for recreation and public access; protection and rehabilitation of the coastline, lakes, rivers and water catchment areas; air quality improvements; more efficient use of water and energy resources, and minimisation of waste.

Initiatives of the past decade in relation to air and water quality, setting aside of more parks and reserves and improved urban design have contributed to a 'greening' of the region. Existing parks and revegetation programs, and new ini-tiatives in air, water and bushland management, are aimed at securing an envir-onmentally sustainable future. Biodiversity issues are addressed through comprehensive strategies. These are to ensure the adequate and representative setting aside of parks, reserves and conservation areas and to see that degraded areas are rehabilitated. Encouragement is given to private landowners to conserve biodiversity through voluntary agreements. A Lower Hunter and Central Coast Biodiversity Strategy is in preparation to provide a State model for the protection of threatened species (refer to Queensland equivalents earlier in this chapter).

Regional surveys and assessments are to assist in providing reliable informa-tion on which to base land-use decisions. A co-ordinated approach to the man-agement of water quality, supply and water environments is adopted through the State's catchment management, water design and water reform programs.

Perhaps the encouraging aspect of the Greater Metropolitan Region Plan-ning Strategy is not only its scale but its intent: that is, to pursue a newly tail-ored approach to planning through regional and local government frameworks. Its objectives are spelt out not just in terms of traditional town planning but also through a more integrated approach to planning which recognises the value of the natural environment both in its own right and as the living place of people. In this respect, we believe our desire to find an integrated, sensitive planning and environmental management model may have been realised, provided that political and economic imperatives do not override the hard-won environmental objectives of the strategy.

Several other regional programs for New South Wales are discussed in Dore and Woodhill (1999) and cover: Western Sydney (which incorporates the

Hawkesbury–Nepean Catchment); the Northern Rivers region; the western New South Wales pastoral region, and the Murray irrigation zone. While economic matters are once again paramount in these planning regions, attention is also given to social and environmental priorities through co-ordinated planning frameworks.

Losses and Gains

In his overviews of planning policies in non-metropolitan areas of NSW, Ian Bowie (contributions to *Progress in Rural Planning and Policy* from 1991 to 1994) noted shifts (in government policy) following the election of a new conservative government in 1988. As in Victoria, corporatisation of the government sector and pro-development policies saw a determined downgrading of environmental priorities in subsequent years. The National Parks and Wildlife Service, for example, lost 10% of its staff in 12 months. Ironically, the head offices of the Agriculture and Lands agencies were relocated to major regional towns in the name of efficiency and in support of decentralisation and regionalism policies—despite a) the dismantling of the Albury/Wodonga and Bathurst/Orange regional development corporations, and b) concerted government policies of community services withdrawal from rural areas. Furthermore, local governments fear greater centralisation of State planning powers and use of ministerial discretion; and amendments in 1998 to the *Environmental Planning and Assessment Act* appear to favour developer interests at the expense of public participation.

This said, there have been gains: in improved environmental protection legislation (Chapter 7); a Statewide coverage of ICM areas (Chapter 15); development of guidelines on rural planning for local governments (DoP 1988); streamlining of planning procedures (EP & A Act amendments), and a better definition between the different tiers of governments' roles and responsibilities in relation to environmental matters. A revision of the *Local Government Act* in 1997 subjects local governments to SoE reporting, with obligations on Councils to incorporate ESD in environmental management in the spirit of the IGAE (Chapter 7). For example, the Southern Sydney Regional Organisation of Councils has developed a draft list for member Councils' SoE assessments. The list covers biophysical and heritage matters as an aid to developing a regional SoE report. And, as the greater metropolitan region planning strategy (above) indicates, there has been a raised profile for the environment in planning matters, particularly at local level.

Nevertheless, and as Epps has commented, a major constraint mitigating against the strengthening of local government (not only in NSW) is not politics but the lack of training and appropriate staff resources to manage this increased responsibility effectively. Greater co-operation from all sectors and tiers of government, improved communications and networking and, for some smaller authorities, amalgamation, could minimise some of these possible shortcomings. But in some rural communities there is little scope for avoiding the responsibility for poor decision-making, and some local governments may choose not to accept additional powers.

VICTORIA

Regional Strategies

During the late 1990s, a rural and regional strategy was being developed for Victoria. The State government is working with regional and rural communities, building on a range of existing programs to develop a strategy to guide country development through a whole-of-government approach over the next 15 years.

In 1999, the State government was sponsoring five Regional Forums. The role of each is to work with local communities to develop short, practical, regional plans. The government will build on the work of the forums and a State Reference Group, with representation from peak bodies and organisations, to assist with the identification of State-wide issues and priorities. Draft regional action plans have been developed and are available on Websites at www.doi.vic.gov.au, through regional offices of the Department of Infrastructure.

The Action Plans are to develop goals towards the year 2011 and beyond; to identify the most important issues which need to be tackled, and the two or three actions a region can take towards achieving the goals; and to agree on priority tasks and partnership approaches in their implementation. The Regional Forums are encouraged to address and foster a sense of involvement and ownership, to retain young people, broaden the economic base and improve infrastructure needs. Although primarily economic in focus, it is clear that links are being made with sustainable management practices and other regional initiatives.

Thus the *Northern Region Action Plan* (draft November 1998) has as the first of its key objectives the promotion of sustainable environmental practices. Partnerships between local governments and industry are encouraged to enhance future economic performance. The Plan's three key action areas are: natural resource management (covering ICM, salinity, land capability, 'best practice' and new technologies); development of rural leadership (human-capacity building and youth leadership as aids to revitalising the region), and the development of infrastructure (transport and irrigation).

Strategic Planning

Where strategic planning has been carried out, whether by formal regional commissions or less formal associations of Councils, environmental assessment, land evaluation, land capability analysis and land suitability analysis have been integrated to the extent that the lead agency (and its consultants) and those parties which involved themselves in the consultation process have determined. Strategic planning has not been conducted in the context of a formal procedure, with pre-set parameters or consultative requirements. The environmental issues which have been addressed in specific cases have been those which were deemed important by the parties involved in the exercise.

As has occurred elsewhere in Australia, changes in planning legislation and policy during the 1990s have shown a marked shift towards a more open and flexible style of planning as distinct from traditional, prescriptive approaches. Environmental policies can be incorporated into planning schemes or through strategic plans, when assessing development applications, in negotiated planning, or by entering into agreements or partnerships with developers or land owners.

Moves towards a more integrated form of planning—that which combines the need for orderly development with environmental management imperatives—resulted in some important developments in Victoria during the period of its two Labor Governments in 1982 – 1992. To a considerable extent, this was reflected in national developments of the period, as discussed in previous chapters. Under the Victorian Labor governments, eleven environmental strategies were developed, covering areas such as salinity, wetlands, flora and fauna, energy, coasts, water management and native vegetation protection. The national park system was extended from 3.2% to 10% of the State's area. Strategic planning units were also established (Christoff 1998).

Major issues which were raised in submissions to the *Review of Rural Land Use in Victoria* that was released in February 1991 were: loss of agricultural land; problems associated with rural subdivision; environmental problems; lack of necessary data, and the need for a rural policy. The Labor Government's response included placing a Statewide Rural Planning Framework on exhibition as an Amendment to the State sections of all Victorian planning schemes (Leigh 1994). However, in November 1992 the Coalition Minister for Planning announced that he had decided to abandon the Amendment because it was not flexible enough to 'cope with individual circumstances' and would be a barrier to rural restructuring (Leigh 1994:292)—but, as noted above, in 1998–99 a rural and regional strategy was resurrected.

Various State policies protect agricultural and coastal land and are implemented through planning schemes. The State government also exerts an influence through regional Catchment Management Authorities and Landcare activities (see below). A more strategic approach is now taken for new rural/residential zoning. Rural Councils are required to address rural development as a key component of their Municipal Strategic Statements (WAPC 1997).

The prime agency now responsible for integrated management and protection of natural resources in Victoria is the Department of Natural Resources and Environment (DNRE). Although its focus appears to be essentially one of traditional 'resource' management, the Department's desire for more integrated approaches to NRM is reflected in the establishment of six sustainable development regions across the State (parallel to the Regional Forums referred to above). Furthermore, its stated 'vision' is to recognise the balance between development for quality of life and prosperity (wealth) and protection for sustainability and long-term benefit (care), with shared outcomes for the multiple stakeholders.

In broad terms, these objectives are also contained within the *Planning and Environment Act 1987*, a key piece of planning and environmental legislation. Conservation and ESD are explicit in many of the Act's objectives. The legislation establishes a framework to build both policies and environmental factors into planning decisions, which is achieved through various planning instruments at State, regional and local levels. The *Planning and Environment Act* is administered by the Division of Planning, Heritage and Market Information in the Department of Infrastructure for both metropolitan and non-metropolitan areas, and the Division also complies with other environmental legislation and policies (including those pertaining to coastal management, greenhouse issues, native vegetation protection, salinity and forestry).

In 1996, amendments to the planning legislation provided for the development of new planning schemes within a standard framework. A key aim was to set out statutory planning controls in strategic planning policies so that land use, development and protection of land could be dealt with in a co-ordinated framework (note: at this time, the number of planning schemes in Victoria were reduced from 206 to 80—Dore and Woodhill 1999:67). Significantly, the amendments also offered opportunities for ICM to be integrated into State planning strategies and to translate regional natural resources management to local level planning.

The nine Catchment Management Authorities (CMAs) set up under the 1994 *Catchment and Land Protection (CaLP) Act* (refer Chapter 15) operate with the support of other environmental and natural resource management legislation (Planning and Environment, Water, Environment Effects and Forest Acts). Effectively, they function at regional scales through ICM/NRM strategies. The CaLP Act sets out the general duties of landowners, addresses catchment planning, and contains prescriptions relating to land management, extractive activities, noxious weeds and pet animals.

The links between ICM and local authorities are well exemplified in the *Port Philip – Westernport Regional Catchment Strategy*, which covers strategic and statutory land-use planning, water quality, waterways management, flood protection, native vegetation, habitat protection and integrated land management (DNRE 1997). The Strategy sets out high priority actions for implementation. It involves 38 local government Councils in a diverse region, with a population of more than 3 million people, placing an emphasis on effective communication and networking, technical assistance and education of Councillors and contractors. A key objective is to incorporate catchment management objectives into municipal strategic statements and other planning controls under the Planning System Reform Program.

A further example of such integration may be found in Part E of Dore and Woodhill (1999). It discusses northern Victoria's Goulburn–Shepparton region in which an over-arching catchment strategy draws in various programs relating to salinity, water quality, floodplains, land care, biodiversity, farm forestry and sustainable regional development.

Downgrading Environment and Planning

As previous chapters have shown, restructuring of government agencies and legislative 'reform' (= change) were a feature of the 1980s and 1990s throughout Australia, driven in part by national economic and competition policies (Hilmer 1993) and the understandable desire for 'leaner' governments. The environment *per se* was not seen as a major income earner or employer. Environmental considerations increasingly took second place, with nominal attention being paid to policy statements: ESD, for example, may have been embraced as a policy but was not necessarily (or easily) implemented.

Many of the environmental management and planning gains of the 1980s in Victoria were demolished systematically during the 1990s, with a few exceptions. Significant portfolio changes made by the new Liberal–National Coalition Government included the transfer of responsibility for the Environmental Protection Authority from the Minister for Planning to the Minister for Conservation and Environment, reducing the prospects of meaningfully integrating environmental management and planning processes. Public involvement in decision-making and the rights of objection were increasingly denied, and planning and environmental decisions heavily weighted in favour of developers. Raff (1995a), for example, noted Premier Kennett's uncompromising attitude towards public protests over environmental and planning issues (Albert Park, Coode Island, Wilson's Promontory and freeway extensions). This attitude was reflected in 1993 and 1994 amendments to the planning and environment legislation, which limited the rights of objection either by restricting notification of development or planning proposals, or by granting exemptions from giving notice for some classes of development on the grounds of 'State's interest', even though adverse effects associated with the proposed development may have been considerable. Special industry agreements with developers and Ministerial discretion were used to side-track environmental legislation. But Victoria was not unique in this regard: similar modifications (some justified) to legislation have been made elsewhere around Australia, underpinning the simplification and 'fast tracking' of planning procedures for development and planning approvals (see also Chapter 14).

Christoff (1998) has provided an alarming overview of the downgrading of environment in government agendas—an overview which could have been written for other jurisdictions, in varying degrees. While the process commenced around Australia during the early 1990s, Premier Kennett's Coalition Government in Victoria was something of a trend-setter. For example, after the 1996 election, four major statutory authorities or advocacy bodies serving the Minister for Conservation and Land Management were dismantled in favour of smaller, corporate-style boards with no significant community representation. The Environment Council, established to oversee EPA activities, was replaced by a three-person Environment Protection Board, a corporate-style arrangement with no community representation—similar to that in NSW. In 1997, one of Victoria's oldest and most impressive environmental institutions, the Land Conservation Council (LCC—an independent, statutory authority with 12 community representatives), was abolished and replaced with an Environmental Conservation Council of three Ministerial appointees (although it has the power to appoint additional members for specific investigations). Although the LCC was in need of an overhaul, its replacement was clearly shaped by the government's desire for quick solutions. Christoff (1998) considered that such moves have reduced the Government's capacity to negotiate a path through complex environmental problems and has created a confrontational relationship with the community.

On the other hand, public consultation has been retained in some environmental initiatives. The *Catchment and Land Protection Act*, for example, was subjected to widespread public review, and catchment plans are also open to public input. The development of a coastal policy also involved the community, and the EPA has retained its public consultative powers (for example, in developing SEPPs) and remains attentive to the community consultative process. Nevertheless, there appears to be a greater willingness on the part of senior bureaucrats to override the outcomes of public consultation and to tighten executive rule. In addition, the Boards created under the catchment and coastal Councils are seen by environment groups as being:

> politically pliant and strongly development-oriented, dominated by representatives of resource-based industries and the government's rural or coastal development constituency (Christoff 1998:15).

In the strategic planning area, several planning units set up under Labor to develop and implement environmental policy have also been closed down—in particular the Department of Water Resources, the Salinity Bureau and the Office of the Environment (and its Commissioner), which was responsible for producing Victoria's SoE reports (arguably Australia's best at the time). Perhaps the most significant environmental management initiative under Premier Kennett was the establishment of the DNRE (discussed above). Structurally, 'integrating' is absent in a broad, environmental policy and implementation sense; and the primacy of *resource use* ('development') is quite evident. But there are signs of change, as in NRM regional policy development and promotion of the State's rural and regional strategy referred to above.

Perhaps even more important has been the loss of personnel and skills from the environmental management sector of government. For example, although staffing levels in the Department of Conservation had fluctuated considerably under Labor Governments, the first Kennett Government cut staff by almost 40% between 1992 and 1995. This constituted a critical loss of co-ordinating, advisory and administrative support, limiting the capacity to respond effectively to environmental challenges, to innovate or to implement programs. In summary, Christoff observed that the impact of the foregoing changes were profound and negative. In effect, they constituted a 'hollowing out' of the Victorian environmental public sector with the contraction of the environmental part of DNRE to its core executive and regional components. He continued:

> to an extent, this has involved privatisation as a sleight of hand, as quango-style contractors, often former bureaucrats and ex-public sector researchers, have established stable relations as sole suppliers to an increasingly deskilled and dependent executive core (Christoff 1998:31).

The consequences of the changes are that:

> major issues are now crowding in on Victoria, which is worse equipped to handle them than at the start of the 1990s...but the real results of the fragmentation and demolition of the institutions of environmental governance—worsening trends in environmental data and conditions, and increasing costs for Victorians as they seek to repair the damage—will not appear for decades, long after the Kennett government has passed into history (Christoff 1998:32).

CHAPTER 20

THE INTEGRATION OF ENVIRONMENTAL PROTECTION WITH REGIONAL PLANNING AND MANAGEMENT: MODELS FROM TASMANIA AND NEW ZEALAND

TASMANIA

Tasmania is a particularly interesting case which, despite a somewhat chequered history of regional planning, provides some important and constructive pointers for other parts of Australia. Although the Department of Primary Industries, Water and Environment (DPIWE) does not undertake regional planning of non-metropolitan regions, there is some history of such planning in the past, with three schemes abandoned for various reasons (discussed below). A Regional Forest Agreement in 1997 adopted an integrated regional approach but, as with similar agreements for other parts of Australia, it is surrounded by controversy over issues of management and sustainability.

According to John Pretty (pers. comm., Manager, Planning Services, DPIWE), there was an acrimonious period of attempted Council amalgamations under the minority Liberal Government, the disputes contributing to the Government's downfall. A first wave of reform between 1989 and 1994 reduced the number of Councils from 46 to 29; a second attempt to drop the number of Councils to 11, effectively creating regional governments, was blocked by a High Court decision in 1998. With a change of government the matter has been put on hold (Dore and Woodhill 1999:72). The wounds between State and local governments are still healing, and the Labor Government is developing 'partnership' agreements with local Councils but without any underpinning, formal regional policy. Even so, a State NRM steering committee has been set up, driven in part by the strategic regional requirements of the NHT. In the Launceston/Tamar region of northern Tasmania, local governments are working with a reference group to trial a co-ordinated regional approach to NRM which could be applied to the rest of Tasmania (Banks 1998). It is also hoped that this will clarify various State NRM policies covering water quality, ICM (under development), protection of agricultural land and coasts (refer to Dore and Woodhill 1999 for further details on the Launceston/Tamar region).

The Resource Management and Planning System (RMPS)

Tasmania's *Land Use Planning and Approvals Act 1993* and the *State Policies and Projects Act 1993* comprise a package of integrated, environmental management and planning policies and legislation under the umbrella of the *Resource Management Planning System* (RMPS), which came into effect in 1994. By 1996 it incorporated eleven 'core pieces of legislation' (State of the Environment Unit 1996:13.9). Although the *Land Use Planning and Approvals Act* does not specifically refer to regional planning, it requires consistency and co-ordination between planning schemes and land-use decisions in adjacent areas, and 'they must have regard for the use and development of the region as an entity in environmental, economic and social terms' (Section 21). The objectives of the RMPS introduced by this and the *State Policies and Projects* Act are to:

- promote the sustainable development of natural and physical resources and the maintenance of ecological processes and genetic diversity;
- provide for the fair, orderly and sustainable use and development of air, land and water;
- encourage public involvement in resource management and planning;
- facilitate economic development in accordance with the above objectives, and
- promote the sharing of responsibility for resource management and planning between the different spheres of government (State, regional and local), the community and industry.

As has occurred with environmental and planning legislation in other States, these aims reflect national ESD policies and IGAE objectives. Under other key objectives of the RMPS, consistent decisions are sought on matters such as State and local planning, SoE reporting, State policies, significant State projects, land and water use, development controls, environmental management and pollution control, public land allocation and appeals.

The concept of sustainable development underpins the whole system. It is defined in the legislation as managing the use, development and protection of natural and physical resources in a way or at a rate which enables people and communities to provide for their social, economic and cultural well-being and for their health and safety while:

1   sustaining the potential of natural and physical resources to meet the reasonably foreseeable needs of future generations;
2   safeguarding the life-supporting capacity of air, water, soil and ecosystems, and
3   avoiding, remedying or mitigating any adverse effects of activities on the environment.

It is therefore clear that any regional plans prepared in the context of the RMPS will explicitly incorporate environmental protection and management with land-use planning. Local governments play a key role in implementing the RMPS through the use of various planning instruments and preparation of integrated strategic plans; and there is the opportunity within the *Local Government Act 1993* for two or more Councils to form a Joint Authority to perform any function or duty of the Councils. ICM is also an important component of natural resource management, insofar as State policy seeks to rationalise municipal boundaries on a catchment basis. The *Environmental Management and Pollution Control Act 1994* provides for EIA and complements the above and other legislation in the RMPS.

State Policies are being developed in accordance with the *State Policies and Projects Act 1993*. Policies create standards and objectives for NRM and will address matters such as reducing land degradation, protecting and enhancing remnant natural vegetation and wildlife habitats, improving water quality, and reducing or preventing conflicts between competing land uses (see, for example, agricultural policy below). In the event of inconsistencies between Policies and planning schemes, the Policies will prevail. The policies are implemented through local planning schemes.

A *Planning Policy Framework* 1997 assists decision-makers in the RMPS to understand links amongst national, State, regional and municipal policies and the planning system. The Framework includes principles for economic development, resource management, environment, conservation and social policy areas. It also lists State and national policies and legislation relevant to planning and management in each of these areas. Some of these principles are reflected in existing State and national legislation, while others are to be pursued through the RMPS.

A *Resource Planning and Development Commission* set up under legislation in 1997 amalgamated the former *Sustainable Development Advisory Council*, land-use bodies and planning review panel. The Commission includes community representatives. It guides the RMPS, makes final approval of planning schemes, conducts hearings, reports to the Minister on sustainable development policies and projects of State significance, and prepares State of the Environment reports. The Commission is subject to ministerial direction but not in relation to outcomes required of the Commission.

Protection of Agricultural Land

Of particular note in non-metropolitan planning was the release of a *State Policy on the Protection of Agricultural Land*. This policy is to further RMPS objectives by fostering agricultural development while ensuring consistent approaches to planning decisions in relation to agricultural land, and promoting sustainable development of agricultural land. The main vehicles for implementation of the State policy will be local government strategic and operational plans, and programs undertaken by the DPIWE. Guidelines to assist planning authorities and other bodies to implement the policy include the use of the Tasmanian Land Capability Classification System to identify prime agricultural land and to promote the use of agricultural land within its capability (see Chapter 10). Thus issues such as the sustainable use of agricultural land, land degradation and subdivision of rural land are addressed.

Previous Non-metropolitan Regional Planning in Tasmania

The Tamar Region Plan

The Tamar Regional Master Planning Authority (TRMPA) was established in September 1969 in accordance with the *Local Government Act 1962*, following a petition to the State government by the City of Launceston and six municipalities. Two other municipalities and the Port of Launceston Authority joined in 1974. The first *Tamar Region Plan* was produced in 1975 and a second, updated edition in 1979 (Tamar Regional Master Planning Authority 1979). The *Tamar Region Plan* grew out of a grass roots determination to deal with regional problems of isolation and low growth rates, to manage resources, develop facilities, and to prepare statements of planning policies on land use and environment. The Tamar Planning Authority is now defunct in the wake of local government reforms during the 1980s. The Region Plan has no status but it is useful to review the development of the plan and to note something of a revival in new and evolving regional NRM arrangements referred to at the start of this chapter.

The TRMPA had statutory responsibility under Tasmania's *Local Government Act* to prepare a master regional plan to direct regional development and provide guidelines for local town planning schemes. The first planning action was taken in 1970. In the absence of a State strategy, the region was to develop its own plan. Between 1971 and 1975, various local area studies were undertaken in relation to resources development, tourism, engineering schemes, river management, conservation, employment, recreation and local area planning (Tamar Regional Master Planning Authority 1979:Appendix). While the development of the plan took place in the 1970s within various State policy frameworks (political, social and economic), it had the objective of promoting a balance between development and conservation and to produce a plan which:

> is not merely a tool to reduce development conflicts, but a scenario of a better, more fulfilled region... (p. 7).

The regional plan was seen as being more of a 'process' than a 'plan', with the 1979 plan setting out statements of policy on land use, environment and promotion. The environment statement of policy covered conservation, recreational needs, and the consequences of planning and development policies on the environment. A Tamar Estuary report and a Conservation/Landscape study guided strategic environmental policies for the region.

According to a draft paper provided by Elizabeth Fowler (Department of Environment and Planning Tasmania, n.d.a), the 1989–90 financial year was the final year of support by way of special funding for the TRMPA for a two-year period to enable the preparation of a draft Master Plan for the Tamar region. Prior to that, untied grants had been provided to the Authority every year since 1977–78. At the end of the two-year period, the Master Plan was not in a form suitable for approval by the Authority. The government advised that it was not prepared to fund further work on the plan. The Launceston City Council withdrew from the TRMPA in December 1990, followed by three municipalities and the Port of Launceston Authority (by then the Marine Board of Launceston) in August 1991. Prior to its abolition, the Authority resolved to endorse a draft Master Plan which excluded the City of Launceston, and also a regional strategy for the area. Currently, the primary regional organisation in northern Tasmania is an economic one (Tasmania North). Formerly Commonwealth funded, it is now supported by local governments and the Tasmanian Department of State Development.

Other Regional Plans

The North West Master Planning Authority (NWMPA) was established in 1971 with eight municipalities. There have been some withdrawals, joiners and rejoiners; and in 1994 the NWMPA was continuing to meet every month.

Over the years various organisations have been established to consider the issue of regional planning in southern Tasmania. The Southern Metropolitan Master Planning Authority (SMMPA) was established in 1958, ceased to function in 1973 and was formally disbanded in 1981. The Hobart Metropolitan Area Strategy Plan Study Team (HMASPST) was brought together in 1975 and disbanded in 1976, while the Southern Metropolitan Planning Authority (SMPA) was established in 1978 and disbanded in 1982.

The SMMPA concentrated on the production of a master plan for Hobart and the mapping of the metropolitan area, whilst the HMASPST considered growth strategies for the development of the region. The SMPA undertook to provide comprehensive data and information on the region. Documents produced by the SMPA were considered to be difficult to interpret and use in the daily management and operation of local government 'because of their complex and academic nature'. Furthermore, the documents failed to provide recommendations which were capable of short-term implementation by the Council or the local community.

Nolan (1981) reported that the SMMPA had the most chequered history with regard to continuance and membership. It came to an end after 15 years because the Hobart City Council withdrew for various reasons, including an inequitable contribution towards running costs (50% of the total) and the fact that the SMMPA had failed to produce a working plan for the future development of Hobart.

Constraints to Regional Planning

Partly arising from the above experience, the Department of Environment and Planning (n.d.a) summarised some of the constraints to regional planning as follows:

- a regional planning authority is not a formal level of government. Such authorities are electorally difficult to account for, are accountable to many different local governments, do not have their own direct revenue-generating capability, and do not have a direct mechanism for plan implementation;
- the regional level often operates in a policy vacuum;
- regional planning considerations tend not to have direct implications for residents;
- regional plans are almost invariably out of date before they are published;
- attempting to produce a comprehensive statutory plan for each area leads to a pattern of compromises and watered-down policies which are incapable of achieving government objectives relating to control of development, and
- strategic plans suffer because if they are too general they will be ineffective and a poor basis on which to formulate local policies: on the other hand, if regional policies are precise and definitive they are very difficult to alter when circumstances change.

However, a regional approach to planning may be relevant where there is conflict over land-use management, to assist the delivery of programs by State or federal agencies, or to promote economic development. Regional planning facilitates both horizontal and vertical co-ordination across governments.

The East Coast and Dorset Shire Projects

Given this somewhat negative history of regional planning in Tasmania, it is of particular interest that late in 1994 a non-metropolitan regional planning exercise was about to commence on the east coast of Tasmania (Department of Environment and Planning Tasmania n.d.b). The briefing paper noted that the history of coastal and marine planning and management in the area covered by the study had been marked by *ad hoc*, fragmented decision-making and actions by many agencies. The stimulus for an integrated coastal and marine strategy arose from several different sources at local, State and federal levels.

The briefing paper described this as a major project covering about 20% of the State's coastline. It involved the development of a single, comprehensive coastal and marine management strategy for a region controlled by three local authorities (including Dorset Shire), in close consultation with the coastal and marine units in Tasmania's former Department of Environment and Land Management, the Hobart Marine Board, and the former Commonwealth Government Departments of Environment, Sport and Territories, and Housing and Regional Development. The project was in two parts.

Part One concerned the development of a regional coastal and marine management strategy for the three municipalities. Objectives included developing a framework to: co-ordinate coastal and marine policies at national, State, regional and local levels; encourage interaction amongst government agencies, the community and private sectors; develop coastal and marine strategies, policies and programs in which the roles of various interest groups are clearly identified; ensure that the strategies, policies and programs provide 'best practice' in the sustainable development and management of natural resources and in the adoption of integrated approaches to resource use (including economic development and environmental management); ensure accountability, and develop performance-monitoring and review systems. The strategy developed in the first part covered:

- existing settlements and population profile;
- environmentally and economically sustainable development, including tourism, forestry, aquaculture and agriculture;
- infrastructure needs;
- water quality;
- waste management;
- protection of natural assets;
- Landcare and Coast Care, and
- Aboriginal and cultural heritage.

The briefing paper stated that by developing a comprehensive coastal and marine strategy in partnership with major stakeholders, Part One of the project would contribute to the achievement of Commonwealth government objectives in efficiency, equity and sustainable development. It would also contribute to State government objectives, especially in coastal policy, as well as respond in an integrated way to many legislative reforms in areas such as local government, planning, environment, heritage and public health.

Part Two focused on the application of the strategy to the rural and coastal towns surrounding Georges Bay in the Municipality of Break O'Day. It identified the main planning issues, leading to the development of new or amended land and marine planning documents, assessed service and infrastructure needs and community expectations, developed strategies for the delivery of services and facilities, and clarified the roles and responsibilities of the community at the local level. The strategy in Part Two covered:

- land and marine use;
- physical infrastructure (roads, water, power, wharfs);
- social services and infrastructure;
- impact of the seasonal influx of holiday makers and tourists;
- environmental management including Landcare and Coast Care;
- economic and employment development;
- cultural and leisure activities;
- Aboriginal and cultural heritage;
- clarification of roles and responsibilities of the community as a whole, and
- integration of actions 'in-house'.

Thus Part Two implemented the broader strategy in a particular locality.

The overall project was managed by the Tasmanian Integrated Local Area Planning (ILAP) Committee (now disbanded), which comprised two representatives from each sphere of government and the private and community sectors, and was chaired by the Municipal Association of Tasmania. A local Steering Committee managed Parts One and Two, and reported to the State ILAP Committee. It comprised a Councillor and officer from each Municipality, a representative from the State Department of Environment and Land Management, a representative from an appropriate Federal agency, and community and private sector representation. Funds allocated to the project in 1994 totalled about $215 000.

Although we have been unable to ascertain the status of the foregoing project, we are aware that at least one of the Councils, the Dorset Shire, has completed a Sustainable Development Strategy. It was developed after extensive community consultation. Key issues were identified and objectives and actions set out within ESD principles. The Strategy is responsive to State land-use planning and management policies. Action and performance targets cover the environment, community, planning and economic matters and have been incorporated into the Council's statutory strategic and operational plans (Ransom 1999).

NEW ZEALAND'S *RESOURCE MANAGEMENT ACT 1991*

This Act is discussed here not simply as a 'model' which might be adopted in Australia, although it has been referred to in glowing terms by some planners, but also to alert Australian policy makers and planners to some avoidable problems should similar legislation be framed for Australia or its separate jurisdictions. A number of agencies and governments around Australia have paid close attention to the New Zealand reforms but have yet to introduce legislative equivalents. Tasmania's RMPS comes closest; Queensland's draft NRM legislation also incorporates elements of the New Zealand law, as (it is anticipated) will the national NRM policy.

The Act:
- establishes a comprehensive and integrated system of natural resource and environmental management to be achieved in a sustainable manner, recognising social, economic and cultural needs;
- establishes duties and responsibilities for all persons to avoid, remedy or mitigate adverse environmental impacts;
- replaces 75 statutes and 22 regulations (including major Acts relating to town planning, soil, water and geothermal resources, air and noise pollution, and coasts) with one statute;
- reduces more than 700 local governments and quasi-government bodies to fewer than 100;
- creates 13 Regional Councils (generally based on river basin boundaries) which are directly elected;
- develops a hierarchical, closely linked and standardised policy and planning framework for all levels of government;
- empowers local and regional Councils with increased regulatory and co-operative responsibilities for environmental management and planning, consistent with national priorities;

- sets up central environment and conservation bodies, and a Parliamentary Commissioner, responsible for national policies, standard setting, monitoring and economic instruments, with intervention powers;
- requires resource consent to be sought where any activity contravenes the Act. Applicants are under an obligation to produce an EIA;
- ensures public consultation in the preparation of all policy systems and plans and permitted public submissions on consent applications;
- legislates for a standardised, integrated consent process for land and water use, subdivision of land, coastal waters, environmental discharges, heritage issues, appeals and monitoring, and
- strengthens Town Planning Tribunal enforcement powers.

(Compiled from Willis 1992, 1993; Buhrs and Bartlett 1993; Miller 1994; Industry Commission 1997:68).

Comments on the New Zealand Resource Management Act 1991

The New Zealand *Resource Management Act 1991* (also supported by the *Environment Act 1983* and the *Local Government Amendment Acts 1988, 1989*) was one of the world's first major pieces of integrated natural resource management, planning and environment legislation. It has been regarded as revolutionary in terms of its institutional arrangements which assigned many former central government responsibilities to newly created local and regional Councils. The Act had a lengthy gestation but its passage was facilitated by New Zealand's government structure—one central government and a single House of Parliament— and its small population. However, the legislation might also be viewed as a not particularly revolutionary departure from previous natural resource management functions. Institutional restructuring and legislative reform was as much a response to corporate views of 'new right' governments seeking greater efficiency through delegated and devolved activities, as a revolutionary response to perceived social and environmental challenges.

Various authors (some are cited above, plus Smith (1996) and Gow (1997)) have raised a number of concerns or issues in relation to the legislation; and acknowledge that more time is needed for some of the problems to be resolved:

- a number of separate Acts and Departments still remain in areas such as minerals, quarantine, conservation, forests and lands. Although better integrated, the Act is thus not as comprehensive as it appears;
- the legislation operates with a highly complex structure;
- planners are uneasy at having to view planning in an entirely integrated way rather than through the previous, more familiar, resource sectoral approaches;
- there are perceived incompatibilities between major objectives of sustainability and protection of natural resources, and fast-tracking of development processes;
- there are concerns that 'development' prevails in decision-making through fast-tracking procedures and the strong influence of key people at Council level;

- the concept of 'sustainable' management is too diffuse to put into practice;
- considerable tinkering with the original Act has added to concerns over the stability of the legislation and the level of government commitment;
- there is potential for inconsistent and inappropriate consent decisions;
- there is concern over loss of 'contact' between central agencies and regional Councils;
- increased powers of regional Councils and costs of operating another tier of government have given rise to political anxieties;
- it is believed that insufficient attention is given to social and economic aspects of the environment and natural resource management. Urban environments and other matters such as environmental pollution tend to be ignored and preference given to nature protection;
- the Ministry for Environment has no clear mandate to require co-operation from other government departments: institutional barriers persist;
- EIA provisions have weaknesses regarding consideration of alternatives and assessing cumulative impacts;
- regional and local Councils often struggle with a lack of skills and resources to meet their responsibilities, resulting in confusion, uncertainty and frustration;
- central government is less than happy with the slow pace of local authority responses to development proposals and to review planning schemes, and
- Councils experience difficulties in administering the Act. (For example, in relation to water pollution there is no guidance on non-point discharges, which are a major problem in New Zealand's rural areas with agricultural chemicals. Local authorities need to address the problem but are reluctant to enforce or seek 'alternatives'. In any event, the appeals process would mitigate against such measures. Neither do local authorities have funds to offer incentives to farmers to modify practices. If plans require actions to be taken over water quality, then there is considerable potential for conflict and litigation.)

Many of the above points could apply equally to environmental and planning reforms taking place in Australia, but it is clear that many refinements will be required; indeed the New Zealand legislation is under review.

CONCLUSIONS TO PART 5

Benefits of Regional Environmental Management

It is often simply taken for granted or assumed that regional planning and management is a 'good thing'. From the work that has been reviewed here, some of the benefits can be listed as follows:

- working within regions often brings together local communities and environment groups, producing a sense of 'ownership' of the plan and community pride;

- the bringing together process enhances opportunities to achieve targeted and integrated management of economic, social and biophysical environmental issues within regions;
- planning processes can be adapted to local needs, resulting in more efficient use of resources and better targeting of outcomes;
- giving clearer responsibilities to local administrations cuts through the red tape associated with central bureaucracies;
- potentially, regional groups can raise their own levies and generate their own business and management plans within federal and State guiding environmental management frameworks;
- pooling of local resources generates greater efficiencies—GIS technicians, resource atlases, information networking facilities—generating stronger bases for planning and management;
- the quality of planning and management is likely to improve as training and awareness raising of local government officers is improved, and new purpose-trained graduates are appointed;
- opportunities are created for local governments to take integrated environmental planning into core business and strategy plans.

Problems with Regional Environmental Planning and Management

Unfortunately, the same work also reveals a rather longer list of problems—integrated regional planning and management is not the panacea many hoped it would be. But if the problems are recognised and confronted, then they can be overcome or at least avoided.

- Regional approaches can still be very fragmented, *ad hoc* and duplicatory. The three tiers of government in Australia operate at different scales with actions often contradicting policies of co-operation and integration.
- The lengthy time taken to prepare plans may mean they are out of date upon completion.
- Regional plans are often too general (or too complex), unrealistic or poorly articulated, creating difficulties for implementation. They are generally not well correlated with project/development planning.
- Governments and agencies are often insufficiently supportive of regional plans, especially where there are no statutory requirements. 'Resource' development is often promoted at the expense of environmental management, facilitated by fast-tracking legislation.
- Decisions made at local or regional level following considerable work and extensive community participation can be overridden (or marginalised) by ministerial discretion in response to developers' appeals or advice given by government agencies. The consequential disillusionment with the democratic process can take a very long time to overcome.
- The sensitivity or obstructionist behaviour of line agencies and interests can be a substantial barrier to sharing and co-ordination. Existing agencies have little incentive to give up responsibility or authority.

- Regional plans often lack statutory backing, though they can be formalised through statements of planning policies. These, however, can be limited in approach.
- ESD principles may be included in policies but there is still very little guidance on how to implement them. They may be 'too vague' to interpret.
- Some plans have not been fully implemented—if at all (partly related to the issue of statutory backing).
- In effect, planning policies are still centralised, particularly with the advent of the 'super' or mega agencies.
- Public participation is often ineffective, and opportunities are limited or not fully used. More sophisticated methods of achieving meaningful community involvement in plan preparation and implementation need to be developed.
- 'Power games' in local politics are often linked to influence and Councillors' agendas, which in turn are limited by their period in office. These agendas may ignore environmental needs or allocate a secondary role to them.
- Many local governments are uncommitted to local Agenda 21, EMS or SoE reporting, due to low levels of awareness, lack of skills or inadequate enabling processes or resources to adopt them. Their concerns are traditionally more 'practical'.
- Few local governments have purpose-directed environment departments. Environment officers are located in planning, engineering, parks and gardens or personnel sections and often struggle to carve out a role or to secure legitimacy.
- There are big disparities between rich and poor regions and therefore in the resources available for regional environmental planning and plan implementation. This problem of equity has to be resolved at State or national level.
- Without direct grants, the preparation and implementation of environment plans struggle with poor levels of funding. There are important question marks over securing additional funds by imposing local or regional levies on top of existing local rates—particularly in rural areas where a significant proportion of farmers are in serious debt.
- Environment officers are often external to local structures and are dependent on external funding unrelated to the local governments. Their appointments are generally short term (for example, catchment co-ordinators are appointed for only three years), resulting in considerable inefficiencies through loss of continuity. A new appointee takes 12 months to settle in to the position and obtain the trust of the community, is effective for the second 12 months, and is then worrying about future employment for the final 12 months.
- Regional managers often find it difficult to separate economic or corporate objectives from land-use, environmental and strategic needs and issues. Short-term needs often take precedence over long-term imperatives.
- There is a general absence of co-ordinated, monitoring mechanisms for measuring local environmental impacts and reporting results (a few, national, pilot, environmental indicator trials have been undertaken).

Requirements for Integrated Planning to be Successful

For integrated planning to be successful, these requirements need to be included:

- short- and long-term proactive planning;
- appropriate planning regions;
- examination of the key issues, values and information *relevant* to the future of a particular region and formulation of achievable actions;
- the availability (accessibility and availability on time) of sound environmental data;
- incorporating environmental considerations in the planning process at an early stage;
- skilled leadership;
- involving an informed community throughout the decision-making process;
- formulating specific and objective recommendations and detailed action or strategy plans;
- flexibility;
- having one dedicated agency responsible for creating and implementing regional plans;
- having appropriately-trained staff and adequate financial and technical resources;
- using appropriate techniques to incorporate environmental protection into regional planning (as discussed in Parts 3 and 4).
- co-ordinating regional planning with government policies and programs at all scales;
- effective use of legislation and incentives to ensure plan implementation;
- regular reviews and audits, and progress reports to the communities, and
- patience—five- to ten-year periods are not unusual before results are seen.

A significant problem which we have encountered is a lack of communication and integration between regional or local planning and administrative bodies on the one hand, and line agencies on the other. The latter may often operate on a regional basis (water resources, agricultural, national parks, central planning and development agencies come to mind) but with little or no communication with each other or with the regional bodies. As a result, overlapping plans are developed—for water catchment management, reserves (sometimes by more than one agency), rural subdivisions and for various 'development' proposals (tourism and mining, for example)—in blissful ignorance of the work being carried out by the other agencies or groups.

One proposal to resolve this type of situation is for regional bodies (whether one or several local governments or a specially created regional management authority) to have agency personnel seconded to them for a period of (perhaps) two years (Conacher 1980). The objectives are to improve communication between agencies and the regional body, and amongst the agencies, and to pro-

vide essential expertise to the regional body: the latter problem has been identified repeatedly by commentators as a major constraint to the effective
involvement of local governments in regional planning and management. The
period of secondment should be long enough for the persons involved to make
a useful contribution to the regional body or authority, but not so long that they
lose their all-important connections with their 'home' agency.

Yet despite the considerable legislative and planning activities identified in
Part 5, there are still very few regional management bodies in Australia. This in
itself is a considerable constraint to the successful implementation of regional
environmental management around the nation. This and other matters are
considered further in the final Part.

Part 6

Conclusions

This book has been about environmental planning and management in Australia, with an emphasis on non-metropolitan regions. Following a brief synopsis, the main part of these conclusions highlights the more important problems relating to the extremely complex matter of managing the Australian environment in a sustainable manner, considers the causes of those problems and offers some solutions. As noted in the Introduction, a key issue has been the rapid and uneven rate of changes in legislation, policies, agencies and nomenclature across all jurisdictions, often associated with changes in government.

CHAPTER 21

SYNOPSIS

Underlying everything discussed have been two simple questions: how did Australia get to the present situation of extensive land degradation, other forms of environmental stress and numerous conflicts over environmental issues; and how can the problems be resolved?

In an attempt to provide answers to these two questions, the book has followed a sequence of ideas. In Part 1, the first chapter asked what in fact is meant by 'environment', 'environmental management', 'environmental problem' and 'environmental issue'; and also whether there is any difference between environmental management and resource management.

Two case studies were then used to demonstrate the need for environmental management and to illustrate recurring themes in later chapters. The first case study, the conflict over wood chipping, illustrated an environmental *issue*, and the second—secondary salinisation—an environmental *problem*.

Part 2 considered how Australia's legislators have responded to environmental problems and issues, commencing with concerns over public health and nuisance, then appending environmental protection clauses to resource development legislation, and finally bringing in specific environment protection Acts. Over the same period a plethora of other Acts relevant to environmental protection were introduced, dealing with matters such as soil conservation, vegetation clearance, national parks and reserves, and introduced pests and diseases. One chapter dealt with the important influence of international, environment-related agreements on Australian legislation and policies. Despite this abundance of legislation (and policies), environmental problems and issues persist: perhaps partly due to insufficient political will to implement the legislation and policies.

Various methods of measuring and evaluating land and environmental attributes were discussed in Part 3. These range from numerous environmental assessment techniques, including the search for environmental quality indicators, through several approaches to land evaluation, land capability and suitability assessment, to environmental impact assessment.

Public participation is crucial to the success of environmental planning and management and was considered in Part 4. As seen here and in Part 5, there has been a tendency for governments around Australia to fail to recognise adequately the crucial role of public participation in the democratic process, contributing to the persistence of environmental problems and, especially, issues.

Finally, Part 5 showed how natural resources and land planning and management have been or can be integrated on a regional basis with the more traditional planning approaches in Australia. Current methods seek sensible solutions to: environmental problems and issues; the allocation of natural resources; land-use conflicts, and land degradation. Again, insufficient recognition or understanding of the full range of approaches involved in 'regional environmental planning and management' has undoubtedly contributed to continuing land degradation and environmental conflicts.

The following section summarises legislative and institutional changes to environmental protection, planning and management over the 1980s and 1990s and the degree to which they have succeeded in meeting their objectives.

Legislative and Institutional Changes 1980s and 1990s: how much has changed?

Economic, social and biophysical environmental factors and ESD processes have been drawn into environmental law and policies in recognition of the need to manage the human environment in a more sustainable, integrated and comprehensive manner.

Numerous international agreements guide Australian environmental policy at Commonwealth, State and local levels—but remain weakly applied (or not at all) in some instances.

Major inter- or intra-state environmental management partnerships have been forged under co-operative arrangements—such as the Murray–Darling Basin, wet tropics, Lake Eyre and Great Artesian Basins and the Great Barrier Reef—but sometimes they suffer from the size of the task, inadequate resources or inappropriate program targeting.

Greater consistency in national and State environmental policies, objectives, standards and practices is being achieved—as in EIA, environmental indicators, NEPMs, industry codes, waste management, water and air quality standards.

Concerted efforts have been made to pursue legislative and institutional reform through repeal, revision or consolidation of State environmental protection laws which bring together noise, air, water and waste legislation; and some natural resource management agencies have amalgamated in mega-departments (notably in SA, Vic. and Qld). In other instances, organisational splits have centred around the separation of corporate or business functions (service delivery and cost recovery) from resource and environmental management functions (examples include water and agriculture sectors). Some arrangements see the central policy office separated from regional and local offices responsible for program delivery. Often, the separations have resulted in a sense of loss of contact and integration amongst personnel, although the Internet is a significant mitigating factor.

Environmental protection legislation has been narrowed down but also strengthened with a generally nationwide confluence of objectives and measures. These deal primarily with assessment of development proposals, regulatory matters (licensing and penalties, with an emphasis on voluntary compliance) and better industrial practices. But agency powers are not always upheld, or legislation is side-stepped via separate industrial agreements or by ministerial discretionary powers.

EIA has experienced some refinement, including incorporation in some planning legislation and some moves towards SEA (though this was explicitly provided for in the original Commonwealth Act), but it suffers from not being adequate for most environmental management tasks (regional or sectoral environmental problem solving, cumulative environmental impacts and conflict resolution in relation to environmental issues).

Most natural resources agencies now acknowledge the need to manage resources in a sustainable and integrated manner but, despite good intentions, this is not always achieved. For some, the focus remains dominantly utilitarian (forests, fisheries); for others, the importance of environmental considerations in their own right have gained in stature (agriculture, water). In general, natural resource management arrangements now stand as follows:

- *water*—once with a strong development focus, service delivery is now separated from management of the resource, with the latter recognising matters such as environmental flows and the habitat needs of aquatic fauna. ICM policies have facilitated the latter whilst also acknowledging the importance of socio-economic issues;

- *mining, energy*—more industries now recognise sustainable management and engage in self-regulation and environmental auditing practices; but more reporting is needed, and smaller operations tend to fall through the net;

- *forestry*—this sector has substantially revised its management policies, harmonising with national ESD and other objectives and extending the system of conservation reserves. However, serious public conflicts persist over the way in which native forests are managed through unsustainable practices. The long-term persistence of the conflicts is indicative of flawed policy and decision-making processes and suggests better approaches are required;

- *agriculture and pastoralism*—this industry faces a growing environmental degradation problem, aggravated by the sheer physical size of the sector. A range of biophysical, economic and social problems arises from a complex web of direct and indirect pressures on the land base. Several innovative, integrated management approaches (Landcare and ICM in particular) have distinctively characterised the sector and provided useful models in regional planning. The approaches have enjoyed substantial levels of government and community support since the 1980s and the momentum is expected to continue. However, some reviews of management performance indicate a need for a better targeting of program objectives and resource allocation, and there is a question mark over future funding.

Data collection and environmental monitoring have improved in terms of quantity, quality and access but have some distance to go. Interest turns to environmental indicators and their usefulness in tracking environmental change through SoE surveys and assessing the performance of environmental managers.

Public participation in environmental decision-making and management has at once thrived and suffered over several decades of change to environmental and planning laws. While community expectations of being heard have increased, their ability to challenge decisions appears to be diminishing, as are participation rights.

Traditional planning methods have undergone some transformation with a broadening out of objectives in comprehensive and strategic regional plans, which now embrace non-metropolitan areas. Planning is also giving greater recognition to a range of environmental protection laws and measures. But hundreds of local authorities around Australia still operate in a fragmented manner, constrained by a lack of resources, skills and statutory powers, and poor integration with sectoral government agencies.

Indigenous land rights have been strongly articulated in recent decades. But Aboriginal rights remain generally peripheral to environmental management processes despite legal acknowledgement of those rights—in a climate of shifting political priorities and equity devalued by legal wrangling and economic imperatives.

There is still a low level of environmental expenditure (about 1% of GDP). The $2.5 billion NHT funding disguises a real loss of environmental funding at Commonwealth and State levels.

CHAPTER 22

POINTERS

As shown in Table 4.0 (Introduction to Part 2), Australia has seen a progression in environmental planning and management; from the colonial period, to specific natural resource management legislation and agencies, environmental protection and environmental impact assessment, regional economic development, a consolidation of environmental protection with natural resource management and land-use planning, to comprehensive, integrated, local and regional planning and environmental management. Yet this final stage, discussed in Part 5, was found wanting. Each stage in the progression has useful, even essential elements; sometimes one approach builds on another, at other times a method may stand on its own. But it becomes apparent that there is no single, perfect approach to understanding and managing the highly complex systems associated with human interactions with their environments.

Over time, the changes incorporated social, economic and biophysical environmental factors in decision-making and planning in response to growing public pressures, increasing demands on the resource base and changing government policies. Essentially, however, they have been *incremental* changes, addressing situations as they arose and adding responses to the pool of experience. This 'muddling through', Lindblom (1959) believed, was characteristic of traditional planning. It is a model which appears to have adapted well to the more complex, contemporary world. Regional planning has surfaced at various times in different guises, in climates of changing priorities; from the economic foci of the 1960s and 1970s, to embrace the greater environmental and planning concerns of the 1980s and 1990s, yet increasingly constrained by politico-economic 'rationalism'. The process has latterly been formalised through the development of comprehensive and integrated State and regional planning strategies, the establishment of some regional planning bodies, and increased statutory powers supporting the regional planning processes.

A greater recognition of local needs has seen the preparation of detailed, local action plans which include environmental matters, and the devolution of greater responsibilities to local governments. In the face of diminishing resources, local governments have often grouped together in voluntary or other associations to strengthen their funding, skills, management and negotiating bases. These informal and shifting alliances increasingly characterised the mid-1990s. Yet they, too, have often struggled as resourcing, or commitment to regional processes, fail.

A particular question which needs to be asked, and for which answers need to be found, is: *Why have so many plans been abandoned or lain idle?* By the end of the 1990s, it had become evident that a string of planning initiatives had not worked well. Good strategies had been developed but not always implemented. What went wrong? Some possible answers:

- changes of government brought new priorities;
- globalisation, technological change, economic rationalism and competition policy replaced people and the tradition of 'service' (as in public service) with the economist's 'bottom line'. Unless something (national parks, reserves, repair of land degradation, control of pollution) earns money, it is placed low on the list of priorities—if indeed it makes the list at all;
- inefficiencies have been present due to overlapping laws, policies, regional structures and processes;
- there were inappropriate identification of issues and goals, inadequate mechanisms to translate plans into action, or insufficient attention given to outcomes;
- the transfer of responsibilities to local governments has perhaps been a sleight of hand shift of costs by national and State governments to local communities, which are expected to carry the burden but are poorly set up to do so in terms of resources, skills and power;
- difficulties in dealing with people (individuals, agencies) with varied attitudes—power, frustration, anger, suspicion or apathy—have been underestimated. Parochialism, uncertainty and fear of change obstruct 'progress';
- 'downsizing' and outsourcing—the loss of capable, key people from agencies—have resulted in a loss of corporate skills and memories;
- reduced education and training (to the point of inadequacy) with trimmed down and under-resourced universities, reduces the pool of skilled people; additionally, barriers to innovation arise from conservative disciplinary attitudes in the older (and more prestigious) universities. There is an urgent need for cross-disciplinary approaches to teaching and researching *integrated* environmental and resource planning and management; and (on the assumption that the future well-being of the planet is important) these need to be given priorities on a par with law, medicine and engineering.

(Re)shaping the Future

The following considers some of the processes or events which will influence environmental planning and management in Australia in coming decades, and also offers some suggestions. In particular, the *kaleidoscopic pattern* of the period since the 1970s is likely to continue, shaped by a wide range of dynamic influences. Proactive, responsive and strategic approaches will be critical for planners and managers to address change and uncertainty with a sense of purpose.

Global economic forces—trade agreements, and shifts in the locations of political and economic power, over which Australia has no control and little influence—will have unpredictable but probably significant effects on the country's economy.

The *corporatisation or industrialisation of agriculture* will accelerate, with fewer people in control of food production and distribution, and associated with contract farming and an increasingly intensive agriculture. While reducing the environmental pressures of broadscale farming and contributing to rural depopulation and land abandonment, there will be a concentration of larger problems (pollution and degradation) into smaller areas. Nevertheless, it is possible that such point-sources of pollution may be more manageable than the currently more diffuse, predominantly non-point sources of agricultural pollution.

In broader terms, the concentration of wealth in the hands of a few will have major consequences for political decision-making, making it more difficult for other members of the community to be heard, increasing demands for public participation and the likelihood of conflict.

Similarly, *rural readjustment* (including forestry and fishing) is likely to continue for some time as economies adjust to market and political forces, with profound consequences for the socio-economic fabric of rural Australia. Some areas will continue to lose people, leaving behind an ageing population and reduced key services. Other areas will be revitalised through value-adding, diversification and innovation in agriculture, forestry and other activities: wheat, oats and barley being partly replaced by canola and lupins; deer, kangaroos, emus and ostriches partly replacing sheep; developments in intensive horticulture, greenhouses and feedlots (aided by genetic engineering); tree farms, kenaf and hemp for paper; vineyard and olive products; marine and freshwater aquaculture; areas with specific attractions promoting tourism and outdoor recreation, alternative lifestyles and retirement. There will always be a demand for food and fibre, but new management frameworks will be required. However, managers and planners need to be alert to the danger of responding to market changes with eventually-unsustainable, short-term programs rather than developing more considered, long-term strategies.

There will be an increased push for 'clean' food, a generally 'cleaner environment' and 'green (sustainable) cities': organically produced food or eco-agriculture, ESD principles, alternative forms of energy and recycling will play a larger role in future lifestyles and world trade, depending on political considerations and the nature of future environmental, health and resource crises.

The *concentration of population in larger cities and coastal areas* appears to be a continuing trend which commenced a century ago, with various local and regional environmental consequences. These include problems associated with: increasing urban demands for land, water, building materials and energy; the continuing loss of prime agricultural land; pressure on diminishing landscapes for recreation and waste disposal; pollution of air, surface- and ground-water and soils; and the growing social and economic disparities between city and country. Careful, integrated, local and regional strategic plans and management appropriate to the regions will be needed to deal with these situations.

Other areas will also have special needs—inland mining, offshore hydrocarbons, ecotourism, irrigation agriculture and pressures on certain heritage areas will require particularly sensitive management, as will balancing Aboriginal culture against contemporary demands and values.

Some *northern areas* are potentially highly productive under farming systems such as those practised in parts of southeast Asia; but constraints of disease, climate and isolation remain and may limit Australia's carrying capacity.

The *notion of centre/periphery* (Rumley 1988) will continue to have a strong influence on Australia's politics, economy, society and environment. Some States, and country areas generally, are marginalised in relation to the southeastern metropolitan axis of the nation.

In broad terms, *public agencies often appear to have been unable to deal effectively with land degradation*. What needs to be done? Some mandated and targeted decisions (planning policies, penalties) will be required, but they will need to be balanced against voluntary measures and partnership arrangements, to secure and maintain goodwill and co-operation amongst land users. There will always be some conflict and indeed tension is sometimes healthy in the face of apathy. However, transparent negotiation, mediation, communication, education and resourcing will all need to be fully employed. Identifying barriers to innovation will be a particular challenge, as will identifying appropriate regional boundaries for achieving integrated planning/management goals.

Decisions need to be targeted with *achievable outcomes*, adequate resources and the time in which to achieve them.

Tensions must be addressed—between private interests (landholders, developers with private access to natural resources) and public responsibilities (common rights, equity and ethical issues, international responsibilities, agency responsibilities). The righteous, moral outrage which emanates from some rural areas and developers in response to the imposition of necessary environmental controls, needs to be tempered with the realisation that there are real problems which need to be resolved.

Conservationists similarly need to recognise that compromises may be sought in genuine ESD processes. Constant opposition to every proposal, and some tactics, alienate some members of the community; yet opposition is often required to prick a community's 'conscience' in the prevailing socio-economic and political paradigm of greed and power.

Some issues attract public attention or even notoriety—such as secondary salinity and clear felling for woodchips—and are important. But this kind of recognition may overlook other, more insidious (or irreversible) and *equally important problems* such as: loss of productivity due to soil decline (increasing acidity, reduced organic matter content, general loss of fertility, compaction); loss of biodiversity or system resilience; declining quality of food, air and water; the enhanced greenhouse effect, and cumulative environmental impacts. These also need proactive treatment through good management practices, resourcing and planning.

Water is likely to be an even more important issue than it is today. The issue will centre not so much on scarcity (which has regional and seasonal expressions) as on quality and cost. Water is available in the north and surrounding Australia (in the sea) in abundant quantities, and to a lesser extent in groundwater. The problem is the cost of purification and distribution to consumers. In Australia's mostly harsh environment the availability of affordable, potable water will be a major constraint on continued industrial developments and population growth. Efficiency in use will also be a key management priority.

Scale is important. Strategies and actions need to be matched appropriately to the different national, regional and local levels. Links need to be developed and harmonised amongst these levels in an improved, nested hierarchy of environmental planning and management.

Uncertainties. By definition and experience, the future is uncertain. Planning and management need to take unpredictability into account. Future unknowns relate to natural disasters, fire, disease, climate change scenarios, war, social disruption, global, regional and national economic disruptions, and unstable markets.

People. Australia has a small population in a large country. This means that there is a relatively small pool of people (compared with, say, the USA, Western Europe or the UK) from which to find sufficient people with appropriate skills to deal with such an extensive and mostly unpopulated area. In some cases the scale of environmental problems is huge. In relation to the country's size, the economic base is also too small to resource adequately some solutions. Management of some intractable problems may need to be downgraded, postponed or even abandoned.

Scientists and planners. There is a need to provide a bridge between scientists and planners. Environmental scientists are qualified in disciplines such as botany, zoology, hydrology or physical geography; planners have backgrounds in architecture, engineering and human geography. Political scientists and sociologists also play a role in planning. Thus there is little knowledge, experience or skills which are common to the two groups. Planners know little about the ecology of wetlands; environmental scientists have a wholly inadequate understanding of planning policies, methods and legislation. Yet they are asked to work together seamlessly in integrating environmental and land-use planning and management. This lack of a shared group of skills is an important reason for the failure of integration. It is suggested that if both groups are trained in the skills discussed in Part 3, this would go a long way towards bringing the two groups of specialists together.

Local expertise. At local government level, Council officers have found it difficult to respond adequately to the new demands of the 1990s. They often have insufficient training in environmental science or various aspects of environmental management. Yet local government planners are expected to undertake these additional, specialist tasks. Local government engineers, too, are often put in charge of environmental management; they may do a good job in controlling storm-water drainage but have little or no understanding of environmental or community dynamics, or the methods needed to resolve conflicts. Planners and engineers are trapped in their own training and experience—the 'prison of experience'—and may vigorously resist new ways of doing things or resent external expertise or impositions. In time, newly-qualified engineers and planners trained in environmental management will replace the old; but it will take time. In the interim, local staff retraining is required urgently (in preference to staff replacements) to meet the new challenges.

And finally, *there are no 'global' solutions*. Instead, at local and regional levels, there is a need to recognise appropriate planning models for the context in which they are to be used (Lane 1997; Stewart 1997). Traditional, centralised, 'top-down' autocratic models may be appropriate where last resort, regulatory approaches are needed. But where there are complex issues and a large number of players, then more flexible, decentralised planning models are more appropriate; in these situations, all players need to be heard in a process of negotiation to develop achievable (and satisfactory) outcomes. The latter model is preferred in regional settings for integrated environmental management.

APPENDIX 1

SOUTH AUSTRALIA—DEVELOPMENT OF PRINCIPAL ENVIRONMENTAL PLANNING LEGISLATION, NINETEENTH CENTURY TO 1990s

Adapted from Environment Protection Council of South Australia (1988:6–7), with additions from South Australian Year Books, Bates (1995), Environment Institute of Australia *Newsletters* and the *Environmental and Planning Law Journal.* Note there is some overlap: some earlier legislation has been replaced by later legislation.

Initial Response 19th c – 1969	Post–1970 Responses
National Parks Act 1881	Fisheries Act 1971
Woods and Forests Act 1882	Mining Act 1971–73
Real Property Act 1884	Industrial Safety and Welfare Act 1972
National Pleasure Resorts Act 1914	Environmental Protection Council Act 1972
Fisheries Act 1917	Coastal Protection Act 1972–75
Control of Subdivision Act 1917	National Parks and Wildlife Act 1972–81
Control of Waters Act 1919	Vertebrate Pests Act 1975
Animals and Birds Protection Act 1919	Pest Plants Act 1975
Mines and Works Inspection Act 1920 – 1978	Beverage Containers Act 1975–86
Town Planning and Development Act 1920	City of Adelaide Development Control Act 1976
Sand Drift Act 1923	Noise Control Act 1976–77
Lands Act 1929	Water Resources Act 1976–83
Sewage Act 1929	Industrial Noise Control Regulations 1978
Crown Lands Act 1929 (amended 1983)	South Australian Heritage Act 1978–85
Waterworks Act 1932	Dangerous Substances Act 1979–85
Local Government Act 1934	South Australian Waste Management Commission Act 1979–86
Boating Act 1934	Historic Shipwrecks Act 1981–83
Health Act 1935	Pitjantjatjara Land Rights Act 1981
River Murray Waters Act 1935	Radiation Protection and Control Act 1982
Pastoral Act 1936	Fisheries Act 1982
Water Conservation Act 1936	Petroleum (Submerged Lands) Act 1982
Harbours Act 1936	Planning Act (Development Control Regulations) 1982

Initial Response 19th c – 1969	*Post–1970 Responses*
Soil Conservation Act 1939	Roxby Downs (Indenture Ratification) Act 1982
Native Plants Protection Act 1939	Planning Act 1982–85
Stock Medicines Act 1939	River Murray Waters Act 1983
Marginal Lands Act 1940	Clean Air Act 1984 – 1990
Petroleum Act 1940	Maralinga Tjarutja Land Rights Act 1984
Noxious Trades Act 1943	Environment Protection (Sea Dumping) Act 1984
Forestry Act 1950	Adelaide Railway Station Development Act 1984
Commercial Fisheries Act 1952	Australia Formula One Grand Prix Act 1984
National Trust of Australia Act 1955	Golden Grove (Indenture Ratification) Act 1984
Agricultural Chemicals Act 1955–78	Native Vegetation Management Act 1985
Underground Water Preservation Act 1959	Unleaded Petrol Act 1985
Traffic Act 1961	Animal and Plant Control (Agricultural and Other Purposes) Act 1986
Prevention of Pollution of Waters by Oil Act 1961–82	Biological Control Act 1986
Aboriginal and Historical Relics Act 1965	Occupational Health, Safety and Welfare Act 1986
Planning and Development Act 1966	Pollution of Waters by Oil and Noxious Substances Act 1987
Continental Shelf (Living and Natural Resources Act) 1968	Murray–Darling Basin Agreement 1987
Fruit and Plant Protection Act 1968	Waste Management Act 1987
	Public and Environmental Health Act 1987 (1991)
	Aboriginal Heritage Act 1988
	Heritage Act 1988
	Pastoral Land Management and Conservation Act 1989
	Soil Conservation and Land Care Act 1989
	Water Resources Act 1990 – 1995
	Marine Environment Protection Act 1990
	Marine Pollution Act 1990
	Wilpena Station Tourist Facility Act 1990
	Native Vegetation Act 1991
	Public and Environmental Health Act 1991
	Wilderness Protection Act 1991
	Murray–Darling Basin Agreement 1992
	Environment Protection Act 1993 – 1996
	Development Act 1993
	Irrigation Act 1994
	South Australian Water Corporation Act 1994
	Development (Major Development Assessment) Act 1996
	Resource Management Plan (Draft) 1998

APPENDIX 2

INTERNATIONAL ENVIRONMENT, CONSERVATION AND HERITAGE TREATIES AND CONVENTIONS TO WHICH AUSTRALIA IS A SIGNATORY

Listed in chronological sequence according to the date of the instrument (not included). From State of the Environment Advisory Council 1996, Appendix 1. The first date shown is the year the treaty was signed by Australia; the second the year in which the treaty entered into force generally.

International Agreement for the Creation at Paris of an International office for dealing with Contagious Diseases of Animals (1924; 1925)

International Convention for the Regulation of Whaling (International Whaling Convention) (1946; 1948)

Convention of the World Meteorological Organisation (WMO) (1947; 1950)

Agreement for the Establishment of the Indo-Pacific Fisheries Council (under the auspices of FAO) (1949; 1948)

Convention on the Inter-Governmental Maritime Consultative Organisation (IMCO, later IMO) (1948; 1958)

International Plant Protection Convention (under the auspices of FAO) (1952; 1952)

Convention for the Protection of Cultural Property in the Event of Armed Conflict (under the auspices of UNESCO) (1954; 1956)

Plant Protection Agreement for the South-East Asia and Pacific Region (under the auspices of FAO) (1956; 1956)

The Antarctic Treaty (1959; 1961)

Treaty banning Nuclear Weapon Testing in the Atmosphere, in Outer Space and Under Water (Partial Test Ban Treaty) (1963; 1963)

Treaty on Principles Governing the Activities of States in the Exploration and Use of Outer Space, including the Moon and other Celestial Bodies (1967; 1967)

International Convention on Civil Liability for Oil Pollution Damage (under the auspices of IMO) (1970; 1975)

International Convention relating to Intervention on the High Seas in Cases of Oil Pollution Casualties, 1969 (under the auspices of IMO) (1970; 1975)

Convention on the Means of Prohibiting and Preventing the Illicit Import, Export and Transfer of Ownership of Cultural Property (1989; 1972)

Convention on Wetlands of International Importance especially as Waterfowl Habitat (the Ramsar Convention) (1974; 1975)

Treaty on the Prohibition of the Emplacement of Nuclear Weapons and other Weapons of Mass Destruction on the Sea-bed and the Ocean Floor and in the Subsoil Thereof (1971; 1972)

International Convention on the Establishment of an International Fund for Compensation for Oil Pollution Damage (under the auspices of IMO) (1994; 1978)

Convention on International Liability for Damage Caused by Space Objects (1975; 1972)

Convention on the Prohibition of the Development, Production and Stockpiling of Bacteriological (Biological) and Toxin Weapons and on their Destruction (1972; 1975)

Convention for the Conservation of Antarctic Seals (1972; 1978)

Convention for the Protection of the World Cultural and Natural Heritage (under the auspices of UNESCO) (1974; 1975)

International Convention on the Prevention of Marine Pollution by Dumping of Wastes and other Matter (London convention) (1973; 1975)

Convention on International Trade in Endangered Species of Wild Fauna and Flora (CITES) (1973; 1975)

Statutes of the International Centre for the Study of the Preservation and Restoration of Cultural Property of 5 December 1956, as amended 24 April 1963 and 12 April 1973 (ICCROM, under the auspices of UNESCO) (1975; 1958)

International Convention for the Prevention of Pollution from Ships (MARPOL 1973) and Protocols I and II (1973: the Convention did not enter into force but see Protocol of 1978 below)

Japan–Australia Agreement for the Protection of Migratory Birds and Birds in Danger of Extinction and their Environment (JAMBA) (1981; 1981)

Agreement on an International Energy Program, as amended 5 February 1975 (under the auspices of OECD) (1979; 1976)

Convention on the Conservation of Nature in the South Pacific (1990; 1990)

Convention on the Prohibition of Military or any other Hostile Use of Environmental Modification Techniques (ENMOD, under the auspices of the UN) (1978; 1978)

Protocol of 1978 relating to the International Convention for the Prevention of Pollution from Ships of 2 November 1973, as amended (MARPOL Protocol) (1979; 1983)

International Convention for the Protection of New Varieties of Plants of 2 December 1961, as amended 10 November 1972 and revised 23 October 1978 (1989; 1981)

Papua New Guinea–Australia Treaty concerning Sovereignty and Maritime Boundaries in the Area between the Two Countries, including the Area known as the Torres Strait, and Related Matters (Torres Strait Treaty) (1985; 1985)

Convention on the Conservation of Migratory Species of Wild Animals (1991; 1983)

South Pacific Forum Fisheries Agency Convention (1979; 1979)

Japan–Australia Agreement on Fisheries (1979; 1979—subsidiary Agreements were done yearly between 1979 and 1993)

Convention on the Physical Protection of Nuclear Material (under the auspices of IAEA) (1984; 1987)

Convention on the Conservation of Antarctic Marine Living Resources (1980; 1982)

United Nations Convention on the Law of the Sea (1982; 1994)

International Tropical Timber Agreement 1983 (1988—due to expire 1990 but extended to 1994; 1985 provisionally)

Republic of Korea–Australia Agreement on Fisheries, and Exchange of letters (1983; 1983)

Vienna Convention for the Protection of the Ozone Layer (1987; 1988)

South Pacific Nuclear Free Zone Treaty (Treaty of Raratonga) (1985; 1986)

Headquarters Agreement with the Commission for the Conservation of Antarctic Marine Living Resources (1986; 1986)

Convention on Assistance in the Case of a Nuclear Accident or Radiological Emergency (under the auspices of IAEA) (1986; 1987)

Convention on Early Notification of a Nuclear Accident (under the auspices of IAEA) (1986; 1986)

China–Australia Agreement for the Protection of Migratory Birds and their Environment (CAMBA) (1986; 1988)

Convention for the Protection of the Natural Resources and Environment of the South Pacific Region (SPREP) (1987; 1990)

Treaty on Fisheries between the Governments of certain Pacific Island States and the Government of the United States of America, and Agreed Statement of Observer Programme (1987; 1988)

Agreement among Pacific Island States concerning the Implementation and Administration of the Treaty on Fisheries between the Governments of Certain Pacific Island States and the Government of the United States of America (1987; 1988)

Montreal Protocol on Substances that Deplete the Ozone Layer (1988; 1989)

Basel Convention on the Control of Transboundary Movements of Hazardous Wastes and their Disposal (1992; 1992)

Convention for the Prohibition of Fishing with Long Driftnets in the South Pacific (1990; 1991)

Indonesia–Australia Treaty on the Zone of Cooperation in an Area between the Indonesian Province of East Timor and Northern Australia (Timor Gap Treaty) (1991; 1991)

Agreement (between Australia and the Union of Soviet Socialist Republics) on Cooperation in the Field of Protection and Enhancement of the Environment (1990; 1990)

Agreement concerning the Continuation of Marine Geoscientific Research and Mineral Resource Studies in the South Pacific Region (Tripartite Phase II Extended Agreement) (1984—due to expire on 9 September 1995; 1990)

International Convention on Oil Pollution Preparedness, Response and Co-operation (under the auspices of IMO) (1992; 1995)

Indonesia–Australia Agreement relating to Cooperation in Fisheries (1993; 1993)

United Nations Framework Convention on Climate Change (1992; 1994)

Convention on Biological Diversity (1992; 1993)

Niue Treaty on Cooperation in Fisheries Surveillance and Law Enforcement in the South Pacific Region (1992; 1993)

Convention on the Prohibition of the Development, Production, Stockpiling and Use of Chemical Weapons and on their Destruction (1993; not yet in force)

Convention for the Conservation of Southern Bluefin Tuna (1993; 1994)

United Nations Convention to Combat Desertification in those Countries experiencing Serious Drought and/or Desertification, particularly in Africa (1994; not yet in force)

Energy Charter Treaty (1994 subject to ratification; 1994 provisionally)

Japan–Australia Subsidiary Agreement (to Agreement on Fisheries of 27 October 1979) concerning Japanese Tuna Long-Line Fishing (1994; 1994—due to expire 31 October 1995)

Regional Convention on Hazardous Wastes (Waigani Convention) (1995, subject to ratification; not yet in force)

APPENDIX 3

SOME NATIONAL, STATE AND TERRITORY ENVIRONMENTAL PROTECTION POLICIES AND STRATEGIES AND OTHER MEASURES

Policies, Agreements etc.	Reference	Examples of State/Local Responses
Intergovernmental Agreement on the Environment	COAG 1992	
National Conservation Strategy	Department of Home Affairs and Environment 1984	Government of Victoria 1987; DCE 1987
National Soil Conservation Strategy	ASCC 1988	Natural Heritage Trust (NHT) www.nht.gov.au
Agenda 21	UNCED 1992	Local Agenda 21 (www.peg.apc.org/~councilnet/cln et/members/LA21.htm)
National Strategy for Ecologically Sustainable Development	Commonwealth of Australia 1992c	
Sustainable agriculture	ESD Working Group 1991	DARA 1990; RIRDC 1997
Australia's population capacity	House of Representatives Standing Committee 1994a	
National Greenhouse Response Strategy	Commonwealth of Australia 1992a DEST 1995	Government of Victoria 1990 DEP 1994
Strategy for Ozone Protection	ANZECC 1994	EPA 1990a
National Water Quality Management Strategy—fresh and marine waters	ANZECC 1995	EPA SA 1997
National Water Quality Management Strategy— groundwater	ARMCANZ/SCARM 1996	DWR 1991
Australian Drinking Water Guidelines	NHMRC/ARMCANZ 1995 (under revision)	

Policies, Agreements etc.	Reference	Examples of State/Local Responses
Water reform, water policies	COAG 1994, 1995; ARMCANZ 1994a	DENR 1995
Salinity		MDBC 1993a; State Salinity Council 1998
National Wetlands Policy	Environment Australia 1997	Brereton 1994 Government of WA 1997a
National Strategy for the Conservation of Australian Biodiversity	Commonwealth of Australia 1996a	DNRE 1998
National Strategy for Endangered Species	Endangered Species Advisory Council 1992	Threatened Species Steering Committee 1993
National Rangelands Strategy	ANZECC/ARMCANZ 1996	Government of WA 1997b
National Drought Policy	Taskforce 1991	
National Weeds Strategy	ARMCANZ/ANZECC 1997	Government of Tasmania 1994
National Strategy for Land Information Management	ALIC 1990	
Algae	ARMCANZ 1994b	MDBC 1993b
National Strategy for Aquaculture DPIE 1994a	DPIE 1994a	South Australian Development Council 1995
Vertebrate pests	Braysher 1993	
National Forest Policy	Commonwealth of Australia 1992b	Lands and Forests Commission 1994
Cleaner Australia	Keating 1992	
National Waste Minimisation & Recycling Strategy	CEPA 1992	Government of NSW 1994; Dept Commerce & Trade 1993
Contaminated Sites	ANZECC/NHMRC 1992 Streets & di Carlo 1999	
Beef feedlots	SCA 1992	
Energy Policy	Commonwealth of Australia 1996b	Office of Energy 1997
Green Cities/green jobs	AURDR 1995; House of Representatives Standing Committee 1994b	Greening Australia 1995; Bradby and Pearce 1997b
National Coastal Policy	Commonwealth of Australia 1995	QDEH 1991 EPA NSW 1993a
National Policy on Recreational Fishing	DPIE 1994b	PIRSA 1997

Policies, Agreements etc.	Reference	Examples of State/Local Responses
Recreational use of water	NHMRC 1990	
Ecotourism Strategy/Tourism	Department of Tourism 1994	SROC 1996
National Heritage	AHC 1996	AHC/CALM 1992
Decade of Landcare, Integrated Catchment Management	Hawke 1989; Commonwealth of Australia 1991	Working Group 1991 Government of Victoria 1992 NSW SCS (n.d.) ACT 1991
Environment and Health NHMRC 1992, 1993	NHMRC 1992, 1993	

CODES, STANDARDS, INVENTORIES

State of the Environment Reporting	Commonwealth of Australia 1994b Zann 1995	Table 8.3
Greenhouse Gases inventory	DEST 1994b	EPA NSW 1995b
Dangerous Goods	Department of Transport & Communication 1992	
National air quality standards National water quality standards	(see above, National Policies) CEPA et al. 1994	EPA NSW 1994, NHT (above) Streets & di Carlo 1999
Forests	National Forest Inventory 1998 JANIS 1997 BRS 1997	
Wetlands Directory	ANCA 1996	DENR 1996
Best Practice Manufacturing	AMC 1992	
Mining Code of Practice	Chamber of Minerals & Energy 1995 CEPA 1995	
Petroleum Industry	Australian Petroleum 1996	
Environmental Standards ISO 14000 Series	Standards Australia 1996	Maher and Associates 1996
National Environment Protection Council	www.nepc.gov.au (see Appendix 4)	
National Pollutant Inventory	CEPA 1994 (www.environment.gov.au/net/npi.html)	Streets & di Carlo 1999
Environmental Performance Audits	ANAO 1997	

Codes, Standards, Inventories	Reference	Examples of State/Local Responses
Contaminated sites, waste data bases		EPA Vic. 1991 AGC 1994
Wilderness	Lesslie and Masley 1995	
Land Information (soil, agriculture, geology, minerals etc.) (See also Table 9.1)	AUSLIG 1990 (www.auslig.gov.au) ALIC 1990 ANZLIC n.d. Graetz *et al.* 1992	
National Land and Water Resources Audit	www.nlwra.gov.au ACLEP Newsletters	
Pesticide residue surveys	ANZFA 1996 DPI 1980	Love 1991
National Estate	AHC 1998	
ABS – Facts and figures – Environmental expenditure – Environmental attitudes – Agriculture and environment	ABS 1992 ABS 1997b ABS 1997a ABS 1996a	
Environmental indicators	ANZECC 1998 (www.erin.gov.au) SCARM 1993 OECD 1994	
National environmental impact assessment	ANZECC 1997 www.environment.gov.au	

APPENDIX 4

AUSTRALIAN COMMONWEALTH ENVIRONMENTAL PROTECTION ARRANGEMENTS

DEH (formerly DEST) *Commonwealth Department of the Environment and Heritage,* responsible to the Federal Minister for the Environment. Has co-ordinating role to liaise and consult with State/Territory governments, develop international policy objectives. These are formalised through various ministerial Councils (such as COAG, ANZECC, NEPC (see below) and their various supporting committees and networks) addressing a wide range of national and international environmental issues. Some key policies and programs delivered by DEH include the NHT, ESD, Coasts and Oceans, Biodiversity, Extinct Species, Wetlands, National Forest Policy, Greenhouse and International Treaties (Appendix 3).

DEH incorporates Environment Australia, the Great Barrier Reef Marine Park Authority, the Australian Antarctic Division, Bureau of Meteorology and the Australian Heritage Commission. Other Federal government departments with environmental responsibilities include AFFA, Department of Industry, Transport and Science, and the Greenhouse office.

Environment Australia, created in 1997 following a Taskforce Report, has six functional Groups: Biodiversity (includes Parks Australia and Wildlife Australia); Marine; Environmental Protection; Science; Australian and World Heritage, and Portfolio Strategies. NEPA was replaced by NEPC (details below), CEPA became part of the Environment Protection Group within Environment Australia. Environment agencies were brought under one corporate plan without affecting their statutory independence. Projects include national parks, EIA, environmental technologies, waste dumping at sea, contaminated site guidelines, lead abatement, ambient air quality, gene protection, biotechnology management, managing hazardous chemicals, cleaner industrial production (www.environment.gov.au).

ANZECC (Australia and New Zealand Environmental Conservation Council) (formerly ANZEC—without Conservation—and CONCOM (Council of Nature Conservation Ministers). Established in 1991 as a non-statutory body, providing a forum for member governments to exchange information, co-ordinate policies and national and international issues relating to environmental conservation, including pollution control, flora and fauna management, land-use policies, EIA, climate, biotechnology and environment indicators. Comprised of Federal/State/Territory environment and conservation Ministers, supported by two permanent standing committees of officials nominated by members and which advise the ministerial Council. In turn, the committees are supported by specialist sub-committees in the form of working groups and task forces, as well as networks which provide advice on a wide variety of topics relevant to the Council's work. Environment Australia provides secretarial support to the Council and the standing committees. Specialist ministerial committees cover the Great Barrier Reef, Murray–Darling Basin, Tasmanian wilderness, wet tropics and World Heritage areas. Others encompass water resources, fisheries, agriculture, soil conservation and land use, minerals, energy and forestry, EIA and waste management; and include ARMCANZ (Agriculture and Natural Resource Management Council of Australia and New Zealand) and the Ministerial Council for Forestry, Fisheries and Aquaculture (MCFFA).

Other Environment-Related Councils. Several non-ministerial Councils, standing committees and other bodies play important roles in environmental protection, the development of standards and guidelines, and in supporting research and ministerial committees. They include the NHMRC (air pollution, pesticide assessment), ANZFA (food surveys), River Murray Commission, the Nuclear Codes Council, and Occupational Health and Safety (Worksafe Australia), which declare national codes of practice. Other relevant bodies include the Australian Agricultural Council, Australian Soil Conservation Council, Australian Water Resources Council, Australian Forestry Council, Australian Fisheries Council, and the Australian and New Zealand Minerals and Energy Council.

COAG: Council of Australian Governments: a peak body formed in 1992, comprising the Prime Minister and State and Territory leaders and the President of the Local Governments Association (representing more than 700 local governments). Its purpose is to oversee closer co-operation amongst governments and to seek clarification of roles and responsibilities and reform. One of its first tasks was to set up IGAE in 1992. COAG also endorsed the National ESD Strategy in the same year. In 1994 it set out a framework for national water reform.

ICESD: Intergovernmental Committee on Ecologically Sustainable Development: the primary forum for reporting to COAG on environmental and natural resource issues. Governments meet at an administrative level with senior official representatives through ICESD which co-ordinates and oversees implementation of the National Strategy for ESD, the IGAE and Greenhouse Strategy. It works with other ministerial Councils and reports to COAG. It is currently reviewing Commonwealth and State responsibilities and roles in environmental protection.

IGAE: Inter-Governmental Agreement on the Environment (COAG 1992). At a special heads of government and local government representatives meeting in Brisbane, October 1990, it was agreed that an inter-governmental agreement on the environment be developed, providing a mechanism which facilitated:

- a co-operative national approach to the environment;
- a better definition of the roles of governments;
- a reduction in the number of environmental disputes between the Commonwealth and the States and Territories;
- greater certainty in government and business decision-making, and
- better environmental protection.

Subsequently, the IGAE was signed in May 1992. Under the Agreement, it was recognised that consensual action was important in developing national and international policies for which the Commonwealth has responsibility, and that environmental impacts and concerns are becoming increasingly trans-border and trans-national in character, a factor which is not addressed in the Australian Constitution. Adoption of ESD principles and co-operative agreements is seen as necessary to the use and conservation of natural resources and essential to ecosystem function, and to the life needs of Australians.

With regard to the political and administrative processes necessary for the development of improved environmental policies and management, it was agreed that clearer definitions of government (Commonwealth, State and local) roles and responsibilities were needed—including simplification of processes and a reduction in duplication, and greater accountability and timeliness in decision-making. Where matters of mutual interest arise, each party may approve or 'accredit' practices, procedures or processes of the other party. To avoid duplication, expense or friction, parties will give 'credit' to outcomes, even if contradictions exist, to decision-making aspiring to conflict avoidance.

The Commonwealth's role was to relate principally to international agreements (but in consultation with State governments) and the management of its lands. States, however, remain primarily responsible for their own environmental matters, but are expected to take a greater role in the implementation of international environment agreements and the development and implementation of national environmental policies and standards. Local governments are also urged (but not required) to develop sound environmental policies.

National environmental policies are to be pursued in several ways. One is the adoption of the National ESD Strategy into State policies and economic development. Environmental considerations are to be a part of all levels of government decision-making. In valuing services and assets, environmental matters are required to be considered and are expressed through polluter- or user-pays policies in the use and development of natural resources. Acknowledging ESD principles, matters such as inter-generational equity, conservation of biodiversity and adoption of the precautionary principle are to be attended to in the formulation of policies and plans.

The Agreement spells out the means of achieving better environmental management through nine *Schedules*. These cover, in order: data collection and handling (co-ordinated by the ANZ Information Council); resource assessment; land-use decision and approval processes; EIA; national environment protection measures (NEPMs); climate change; biological diversity; National Estate and Heritage; World Heritage, and nature conservation.

In essence, the Schedules seek: the development of consistent, national databases for natural resource information and sharing of that information; ESD to be used at all levels of government in natural resources assessment and land-use decisions; a consistent approach to EIA avoiding duplication and delays and encouraging openness and public consultation; development of NEPMs through NEPC to establish measures and guidelines or standards for ambient air and water quality, noise, site contamination, hazardous waste, motor vehicle emissions, and recycling; adoption of targets for greenhouse gas emissions by implementing the Greenhouse Strategy; preparation of a national biodiversity strategy; conservation of the national estate; fulfilment of World Heritage obligations; and protection, conservation and management of flora and fauna.

NEPC: the *National Environment Protection Council* is a national, statutory body comprised of Ministers from each State/Territory and the Commonwealth, with powers to set uniform, national regulatory measures relating to environmental pollution. The Council is advised by an NEPC Committee, and both are assisted by an NEPC Service Corporation. The NEPC has two primary functions:

1   to make National Environment Protection Measures (NEPMs: details below), and
2   to assess and report on their implementation and effectiveness in participating jurisdictions.

Complementary NEPC legislation has been passed in all States and Territories which have agreed to implement and assess NEPMs within their jurisdictions.

NEPMs (National Environment Protection Measures)

NEPMs are broad, framework-setting instruments as defined in NEPC legislation and may consist of goals, standards, protocols or guidelines, which in turn contain goals, standards, reporting procedures and guidelines. An NEPM automatically becomes law once passed by the Council (unless disallowed by the Commonwealth Parliament). Implementation of NEPMs is the responsibility of each jurisdiction and they may be incorporated into State environment protection policies. NEPMs may relate to ambient air quality, water quality, noise, assessment of contaminated sites, transport of hazardous wastes, recycling, and the development of a National Pollutant Inventory. The Inventory will provide governments and the public access to local and regional information on diffuse and point source emissions and discharges. Industry will be required to provide point source data for a national database which will eventually cover 97 substances.

NEPMs are required to be consistent with the IGAE; to consider environmental, social and economic impacts; international agreements, and regional environmental differences. Draft NEPMs and impact statements are to be open to public comment. States will need to report on what actions are being taken to implement NEPMs (see further, www.nepc.gov.au).

Murray–Darling Basin Ministerial Council (MDBMC)—comprises Commonwealth and State (NSW, Vic., SA, Qld) water, land and environment Ministers. It addresses natural and cultural resource management problems in the Basin. The program is supported by major NHT funding (Chapter 4).

REFERENCES

Abbott, I., 1995: Prodomus of the occurrence and distribution of insect species in the forested part of south-west Western Australia, *CALM Science*, 1(4), 365–464.

ABS, 1992: *Australia's environment: issues and facts*, Australian Bureau of Statistics Catalogue No. 4140.0, Commonwealth of Australia, Canberra.

ABS, 1996a: *Australian agriculture and the environment*, Australian Bureau of Statistics Catalogue No. 4606.0, Australian Government Publishing Service, Canberra.

ABS, 1996b: *Australians and the environment*, Australian Government Publishing Service, Canberra.

ABS, 1997a: *Environmental issues: people's views and practices*, ABS Catalogue No. 4602.0, Australian Government Publishing Service, Canberra.

ABS, 1997b: *Environment protection expenditure Australia 1992–93 and 1993–94*, ABS Catalogue No. 4603.0, Commonwealth of Australia, Canberra.

ABS, 1998: *Australian social trends 1998*, ABS Catalogue No. 4102.0, Canberra.

ACT, 1991: *Decade of Landcare plan*, ACT, Parks and Conservation Service, Department of Environment, Land and Planning, Australian Capital Territory.

AGC (Woodward-Clyde Pty Ltd), 1994: *Castlereagh waste disposal depot: stage II environmental audit, final report*, Vol. 1, Sydney.

Agriculture WA, 1996: *Strategic plan 1996 – 2001*, Agriculture WA, Perth.

AHC, 1996: *Australian heritage charter. Standards and principles for the conservation of places of national heritage significance*, Australian Committee for the International Union for the Conservation of Nature, Australian Heritage Commission, Canberra.

AHC, 1998: Register of the national estate database, Australian Heritage Commission, Canberra.

AHC/CALM, 1992: *National estate values in the southern forest region of south western Australia, WA*, Australian Heritage Commission and Department of Conservation and Land Management, Perth.

Alexander, H., 1995: *A framework for change. The state of the community landcare movement in Australia*, National Landcare Facilitator's Annual Report, National Landcare Program, Canberra.

Alexandra, J., Hassenden, S. and White, T., 1996: *Listening to the land: a directory of community environmental monitoring groups in Australia*, Australian Conservation Foundation, Melbourne.

ALIC, 1990: *National strategy on land information management*, Australian Land Information Council, Canberra.

Allison, J. and Keane, J., 1998: Positioning planning in the new economic landscape, in B. Gleeson and P. Hanley (eds), *Renewing Australian planning? New challenges, new agendas*, Research School of Social Sciences, Australian National University, Canberra, 23–42.

AMC, 1992: *The environmental challenge: best practice environmental management*, Australian Manufacturing Council, Sydney.

ANAO, 1997: *Commonwealth natural resource management and environment programs: Australia's land, water and vegetation resources*, the Auditor-General Performance Audit Report No. 36 1996–97, Australian National Audit Office, Australian Government Publishing Service, Canberra.

ANCA, 1996: *A directory for important wetlands in Australia*, 2nd edn, Australian Nature Conservation Agency, Canberra.

Andrews, K., 1999: The Lake Eyre Basin regional initiative, in J. Dore and J. Woodhill (eds), *Sustainable regional development: final report*, Greening Australia, Canberra, 287–92.

Anon., 1997: *Proceedings of the national workshop on indicators of catchment health* held at the Waite Campus, Adelaide, 2–4 December 1996.

Anon., 1998: Farm monitoring package wins deserved acclaim, *Australian Landcare*, June 1998, 18–20.

Anon., 1999: Western Australian government framework to assist in achieving sustainable natural resource management, unpublished discussion paper.

ANZECC, 1994: *Revised strategy for ozone protection in Australia*, Australia and New Zealand Environment and Conservation Council, ANZECC Report No. 30, Canberra.

ANZECC, 1995: *Australian water quality guidelines for fresh and marine waters*, Australia and New Zealand Environment and Conservation Council, Australian Government Publishing Service, Canberra.

ANZECC, 1997: *Basis for a national agreement on environmental impact assessment*, Australia and New Zealand Environment and Conservation Council, Secretariat, Canberra.

ANZECC, 1998: *Core environmental indicators for reporting on the state of the environment: discussion paper for public comment*, ANZECC State of the Environment Reporting Task Force, Department of Environment, Canberra.

ANZECC/ARMCANZ, 1996: *Draft national strategy for rangeland management*, Australia and New Zealand Environment and Conservation Council, and Agriculture and Resource Management Council of Australia (ARMCANZ) Working Group, Department of Environment, Sport and Territories, Canberra.

ANZECC/ARMCANZ, 1999: *National water quality management strategy. Australian and New Zealand guidelines for fresh and marine water quality*, prepared under the auspices of Australia and New Zealand Environment and Conservation Council and Agriculture and Resource Management Council of Australia and New Zealand, public comment draft July 1999.

ANZECC/NHMRC, 1992: *Australian and New Zealand guidelines for assessment and management of contaminated sites*, Australia and New Zealand Environment and Conservation Council, and National Health and Medical Research Council, Australian Government Publishing Service, Canberra.

ANZFA, 1996: *The Australian market basket survey*, Australia and New Zealand Food Authority, Australian Government Publishing Service, Canberra (www.anzfa.gov.au—results at April 1999).

ANZLIC, n.d.: *Fundamental natural resource data sets. Lists of data sets in digital format*, Australia and New Zealand Land Information Council Secretariat, Canberra.

Aplin, G., 1998: *Australians and their environment: an introduction to environmental studies*, Oxford University Press, Melbourne.

ARMCANZ, 1994a: *Pricing systems for major water authorities in Australia*, Agriculture and Resource Management Council of Australia and New Zealand, FCMC No. 2, Canberra.

ARMCANZ, 1994b: *Algal bloom research in Australia: report on the current status of issues and the development of national research priorities*, Water Resources Management Council Occasional Paper No. 6, Agricultural and Resource Management Council of Australia and New Zealand, Canberra.

ARMCANZ/ANZECC, 1997: *The national weeds strategy. A strategic approach to weed problems of national significance*, Commonwealth of Australia, Canberra.

ARMCANZ/SCARM, 1996: *Allocation and use of groundwater. A national framework for improved groundwater management in Australia*, ARMCANZ and Standing Committee on Agriculture and Resource Management, Canberra.

Armour, A., 1986: Issue management in resource planning, in R. Lang (ed.), *Integrated approaches to resource planning and management*, The Banff Centre for Continuing Education, University of Calgary Press, Calgary, 51–65.

ASCC, 1988: *National soil conservation strategy*, Australian Soil Conservation Council, Australian Government Publishing Service, Canberra.

ASTEC, 1989: *Health, politics, trade: controlling residues in agricultural products*, report to the Prime Minister by the Australian Science and Technology Council, Australian Government Publishing Service, Canberra.

AURDR, 1995: *Green cities*, Strategy Paper No. 3, Australian Urban and Regional Development Review, Department of Housing and Regional Development, Canberra.

AUSLIG, 1990: *Atlases of Australian resources* (vegetation, agriculture, soils, mining, etc.), Australian Surveying and Land Information Group, Canberra.

Australian Academy of Science, 1994: *Environmental science*, Australian Academy of Science, Canberra.

Australian Department of Resources and Energy, 1983: *Water 2000: a perspective on Australia's water resources to the year 2000*, Australian Government Publishing Service, Canberra.

Australian Petroleum, 1996: *Code of environmental practice*, Australian Petroleum Production and Exploration Association Ltd, Perth.

AWRC, 1977: *A national approach to water resources management*, Australian Water Resources Council, Australian Government Publishing Service, Canberra.

AWRC, 1982: *Water research in Australia: new directions*, Australian Water Resources Council, Australian Government Publishing Service, Canberra.

AWRC, 1988: *Proceedings of national workshop on integrated catchment management*, Australian Water Resources Council, Australian Environment Council and Australian Soil Conservation Council, Conference Series No. 16, Victorian Department of Water Resources, Melbourne.

Bache, S.J., 1998: Are appeals an indicator of EIA effectiveness? Ten years of theory and practice in Western Australia, *Australian Journal of Environmental Management*, 5(3), 159–68.

Bache, S., Bailey, J. and Evans, N., 1996: Interpreting the Environmental Protection Act 1986 (WA): social impacts and the environment redefined, *Environmental and Planning Law Journal*, 13, 487–92.

Baird, J.A., 1983: SIRO-PLAN and its interpretation of the planning process, in I.A. Baird *et al.* (eds), *Environmental planning, SIRO-Plan and the Rural Land Evaluation Manual*, CSIRO Division of Water and Land Resources Technical Memorandum 83/17, Canberra.

Baker, R., 1997: Landcare: policy, practice and partnerships, *Australian Geographical Studies*, 35(1), 61–73.

Banks, A., 1998: *Using local government areas for regional approaches to natural resources management*, Discussion paper, Department of Premier and Cabinet, Government of Tasmania, Hobart.

Barossa Valley Review Steering Committee, 1993: *Barossa Valley Region Strategy Plan*, Department of Housing and Urban Development, Adelaide.

Barr, N.F. and Cary, J.W., 1992: *Greening a brown land. The Australian search for sustainable land use*, MacMillan Education Australia, Melbourne.

Basinski, J.J. (ed.), 1978: *Land use on the south coast of New South Wales: a study in methods of acquiring and using information to analyse regional land use options, Volume 1, General Report*, CSIRO, Melbourne.

Baston, R., 1999: An evaluation of the Land Administration Act 1997 (WA) as a tool for the environmental management of pastoral leases in Western Australia, *Environmental and Planning Law Journal*, 16(2), 164–74.

Bates, G.M., 1992: *Environmental law in Australia*, 3rd edn, Butterworths, Sydney.

Bates, G.M., 1995: *Environmental law in Australia*, 4th edn, Butterworths, Sydney.

Beale, B. and Fray, P., 1990: *The vanishing continent: Australia's degraded environment*, Hodder and Stoughton, Sydney.

Beanlands, G.E., 1990: Scoping methods and baseline studies in EIA, in P. Wathern (ed.), *Environmental impact assessment: theory and practice*, Routledge, London, 33–46.

Beanlands, G.E. and Duinker, P.N. 1983: *An ecological framework for environmental impact assessment in Canada*, Institute for Resource and Environmental Studies, Dalhousie University, and Federal Environmental Assessment Review Office.

Beder, S., 1996: *The nature of sustainable development*, 2nd edn, Scribe Publications, Newham, Victoria.

Beilin, R., 1997: The construction of women in landcare: does it make a difference? in S. Lockie and F. Vanclay (eds), *Critical Landcare*, Centre for Rural Social Research, Charles Sturt University, Wagga Wagga, NSW, 57–70.

Bell, D. and Williams, M., 1998: Becoming an expert witness, *Australian Journal of Environmental Management*, 5(3), 135–6.

Bennett, R.J. and Chorley, R.J., 1978: *Environmental systems: philosophy, analysis and control*, Methuen, London.

Berger, A.R. and Iams, W.J. (eds), 1996: *Geoindicators: assessing rapid environmental changes in earth systems*, A.A. Balkema, Rotterdam.

Berger, P.L. and Luckman, T., 1967: *The social construction of reality, a treatise in the sociology of knowledge*, Allen Lane, London.

Berry, C., 1992: The evolution of local planning in Western Australia, in D. Hedgcock and O. Yiftachel (eds), *Urban and regional planning in Western Australia: historical and critical perspectives*, Curtin University of Technology Paradigm Press, Western Australia, 17–29.

Best, E.P.H. and Haeck, J. (eds), 1982: *Ecological indicators for the assessment of the quality of air, water, soil and ecosystems*, D. Reidel Publ. Co., Dordrecht.

Bettenay, A., Blackmore, A.V. and Hingston, F.J., 1964: Aspects of the hydrologic cycle and related salinity in the Belka valley, Western Australia, *Australian Journal of Soil Research*, 2, 187–210.

Bisset, R., 1990: Developments in EIA methods, in P. Wathern (ed.), *Environmental impact assessment: theory and practice*, Routledge, London, 47–61.

Bisset, R. and Tomlinson, P., 1990: Monitoring and auditing of impacts, in P. Wathern (ed.), *Environmental impact assessment: theory and practice*, Routledge, London, 117–28.

Black, A.L., Brown, P.L., Halvorsen, A.D. and Siddoway, F.H., 1981: Dryland cropping strategies for efficient water use to control saline seeps in the northern Great Plains, USA, *Agricultural Water Management*, 4, 295–312.

Black, A.W., 1999: Impediments to the achievement of the commercial and conservation benefits of farm forestry, in A. Robertson and R. Watts (eds), *Preserving rural Australia—issues and solutions*, CSIRO Publishing, Melbourne, 82–91.

Bockheim, J.G., 1980: Solution and use of chronofunctions in studying soil development, *Geoderma*, 24, 71–85.

Booth, T.H. and Saunders, J.C., 1985: A trial application of the FAO 'Guidelines on Land Evaluation for Forestry', CSIRO Division of Water and Land Resources Divisional Report 85/3, Canberra.

Bossell, H., 1999: Indicators for sustainable development: theory, method and applications, International Institute for Sustainable Development, Winnipeg, Canada.

Bowman, M., 1979: Australian approaches to environmental management—the response of State planning, Environmental Law Reform Group, Hobart.

Braatz, S., 1997: State of the world's forests, 1997, Nature and Resources, 33 (3–4), 18–25.

Bradbury, A., 1995: The duty to observe procedural fairness in the New South Wales planning system, Environmental and Planning Law Journal, 12(6), 440–6.

Bradby, K. and Pearce, D., 1997a: South west environmental strategy, South West (WA) Local Government Association, Bunbury WA.

Bradby, K. and Pearce, D., 1997b: Green jobs for Australia's south west, Discussion Paper, Local Government Association, Bunbury.

Bradsen, J.R., 1988: Soil conservation legislation in Australia: report for the National Soil Conservation Programme, Faculty of Law, University of Adelaide.

Brandt, C.J. and Thornes, J.B. (eds), 1996: Mediterranean desertification and land use, Wiley, Chichester.

Braysher, M., 1993: Managing vertebrate pests: principles and strategies, Bureau of Resource Sciences, Canberra.

Brereton, G., 1994: Macquarie Marshes management strategy, Summary Report, Natural Resources Management Strategy, Canberra.

Briody, M. and Prenzler, T., 1998: The enforcement of environmental protection laws in Queensland: a case of regulatory capture?, Environmental and Planning Law Journal, 15(1), 54–71.

Brisbane City Council, 1998a: Brisbane City Council Biodiversity Strategy. Actions today for the 21st Century, Brisbane City Council, Brisbane.

Brisbane City Council, 1998b: State of the Environment Report, Brisbane 1998, Brisbane City Council, Brisbane.

Broadbent, J. and Searle, J., 1996: Planning for ecological sustainability in inland New South Wales, Sydney Vision Paper No. 5, Faculty of Design, Architecture and Building, University of Technology, Sydney.

Brown, G., n.d.: Environmental audit guidebook; environmental management systems guidebook; environmental awareness and obligations, Centre for Professional Development, Kew, Victoria.

BRS, 1997: National forests inventory, Bureau of Resource Sciences, Canberra.

Brunton, N., 1995: Directors, companies and pollution in Western Australia, Environmental and Planning Law Journal, 12(3), 159–62.

Buckley, R., 1988: Critical problems in environmental planning and management, Environmental and Planning Law Journal, 5(3), 206–25.

Buckley, R.C., 1989: Precision in environmental impact prediction: first national environmental audit, Australia, CRES/ANU Press, Canberra.

Buckley, R., 1997: Strategic environmental assessment, Environmental and Planning Law Journal, 14(3), 174–80.

Buhrs, T. and Bartlett, R., 1993: Environmental policy in New Zealand: the politics of clean and green, Oxford Readings in New Zealand Politics, Oxford University Press, Auckland.

Burvill, G.H., 1956: Salt land survey 1955, Journal of the Department of Agriculture Western Australia, 5 (3rd series), 113–19.

Busselli, G., Barber, C. and Williamson, D.R., 1986: The mapping of groundwater contamination and soil salinity by electromagnetic methods, Proceedings Hydrology and Water Resources Symposium 1986. National Conference Publication, 86/13, Institute of Engineers, Australia, 317–22.

CALM, 1989: *Fitzgerald River National Park draft management plan*, Department of Conservation and Land Management, Perth.

CALM, 1990: *Annual Report 1989/90*, Department of Conservation and Land Management, Perth.

CALM, 1994: *Annual Report July 1993 to June 1994*, Department of Conservation and Land Management, Perth.

CALM, 1996: *Annual Report July 1995 to June 1996*, Department of Conservation and Land Management, Perth.

Calver, M.C. and Dell, J., 1998a: Conservation status of mammals and birds in southwestern Australian forests. I. Is there evidence of direct links between forestry practices and species decline and extinction? *Pacific Conservation Biology*, 4, 296–314.

Calver, M.C. and Dell, J., 1998b: Conservation status of mammals and birds in southwestern Australian forests. II. Are there unstudied, indirect or long-term links between forestry practices and species decline and extinction? *Pacific Conservation Biology*, 4, 315–25.

Campaign to Save Native Forests, 1987: *Time for change: proposals for conservation and improved management in the forests of south-western Australia*, Campaign to Save Native Forests, Australian Conservation Foundation, Conservation Council of WA (Inc.), Coalition for Denmark's Environment, Environment Centre, Perth.

Campbell, A., 1994: *Landcare: communities shaping the land and the future*, Allen & Unwin, Sydney.

Canter, L.W., 1983: Methods for environmental impact assessment: theory and application (emphasis on weighting-scaling checklists and networks), in PADC Environmental Impact Assessment and Planning Unit, University of Aberdeen, UK (eds), *Environmental impact assessment*, Martinus Nijhoff Publishers, Boston / The Hague / Dordrecht / Lancaster, 165–233.

Carew-Hopkins, D., 1997: Opportunities for public involvement in the environmental impact assessment process, in *A guide to environmental law in Western Australia*, Environmental Defenders Office, Perth.

Carson, R., 1962: *Silent spring*, Houghton Mifflin, Boston.

Cave, S., 1993: The convention on biological diversity, *Our Planet*, 5(6), 8.

CEPA, 1992: *National waste minimisation and recycling strategy*, Commonwealth Environment Protection Agency, Canberra.

CEPA, 1994: *National pollutant inventory*, Public Discussion Paper, Commonwealth Environment Protection Agency, Canberra.

CEPA, 1995: *Best practice environmental management in mining*, Commonwealth Environment Protection Agency, Canberra.

CEPA *et al.*, 1994: *Monitoring rivers health initiative*, Commonwealth Environment Protection Agency, LWR-RDC, Department of Environment, Sport and Territories, Canberra.

Chamber of Minerals and Energy, 1995: *Code of practice for mineral exploration in environmentally sensitive areas*, Chamber of Minerals and Energy, Perth.

Cherrie, T., 1999: Assessing aesthetic landscape values: a community participatory methodology, unpublished MSc thesis in Geography, University of Western Australia, Nedlands.

Chorley, R.J. and Kennedy, B.A., 1971: *Physical geography: a systems approach*, Prentice Hall, London.

Choy, D.L. and McDonald, G., 1993: Queensland, in *Progress in rural policy and planning*, 3, 386–9.

Christian, C.S., 1958: The concept of land units and land systems, *Proceedings of the 9th Pacific Science Congress*, 20, 74–81.

Christian, C.S. and Stewart, G.A., 1953: *General report on survey of Katherine-Darwin region, 1946*, CSIRO Australia Land Research Series No. 1, Melbourne.

Christian, C.S. and Stewart, G.A., 1968: Methodology of integrated surveys, in *Aerial surveys and integrated studies*, UNESCO, Paris, 233–80.

Christoff, P., 1998: Degreening government in the Garden State: environment policy under the Kennett government, *Environmental and Planning Law Journal*, 15(1), 10–32.

Christoff, P. and Wishart, F., 1994: *Biodiversity strategy*, Discussion paper prepared for Upper Yarra Valley and Dandenong Ranges Authority, Lilydale, Vic.

Clark, B., Bisset, R. and Wathern, P., 1980: *Environmental impact assessment*, Mansell, London.

Clement, J.P. and Bennett, M., 1998: *The law of Landcare in Western Australia*, Environmental Defenders Office, Perth.

COAG, 1992: *Inter-governmental agreement on the environment*, Council of Australian Governments, Australian Government Publishing Service, Canberra.

COAG, 1994: *Water resource policy—communique and report of the Working Group on Water Resource Policy*, Council of Australian Governments, Canberra.

COAG, 1995: *Water reform agreement*, Council of Australian Governments, Canberra.

Cocks, K.D., 1992: *Use with care: managing Australia's resources in the 21st century*, University of New South Wales Press, Sydney.

Cocks, K.D., Baird, I.A., Anderson, J.R. and Craik, W., 1983: *Application of the SIRO-Plan planning method to the Cairns section of the Great Barrier Reef Marine Park, Australia*, CSIRO Division of Water and Land Resources Divisional Report 83/1, Canberra.

Cocks, K.D., Ive, J.R. and Clark, J.L. (eds), 1996: *Forest issues: processes and tools for inventory, evaluation, mediation and allocation*, CSIRO, Canberra.

Cocks, K.D., Ive, J.R., Davis, J.R. and Baird, I.A., 1983: SIRO-PLAN and LUPLAN: an Australian approach to land-use planning. I. The SIRO-PLAN land-use planning method, *Environment and Planning B*, 10, 331–45.

Collie Planning and Development Pty Ltd, 1998: *A comparison of planning systems in Australian States and Territories*, Canberra.

Comino, M.P., 1997: The Ramsar convention in Australia: improving the implementation framework, *Environmental and Planning Law Journal*, 14(2), 89–101.

Commission of Inquiry, 1973: *Commission of Inquiry into land tenures*, Commission of Inquiry first report (final 1976), Canberra.

Commonwealth Government and Western Australian Government, 1998a: *Towards a regional forest agreement for the south-west forest region of Western Australia: a paper to assist public consultation*, Commonwealth Government, WA Government, RFA Joint Steering Committee, Bentley WA and Barton, ACT.

Commonwealth Government and Western Australian Government, 1998b: *Comprehensive regional assessment: a regional forest agreement for Western Australia*, Commonwealth Government, WA Government, RFA Joint Steering Committee, Bentley WA and Barton, ACT.

Commonwealth of Australia, 1974: *National estate*, Report of Commission of Inquiry, Parliamentary Paper No. 195, Canberra.

Commonwealth of Australia, 1983: *National strategy for Australia: living resource conservation for sustainable development*, Australian Government Publishing Service, Canberra.

Commonwealth of Australia, 1991: *Decade of Landcare plan: Commonwealth component*, Australian Government Publishing Service, Canberra.

Commonwealth of Australia, 1992a: *National Greenhouse response strategy*, Australian Government Publishing Service, Canberra.

Commonwealth of Australia, 1992b: *National forest policy statement: a new focus for Australian forests*, Commonwealth of Australia, Canberra.

Commonwealth of Australia, 1992c: *National strategy for ecologically sustainable development*, Australian Government Publishing Service, Canberra.

Commonwealth of Australia, 1994a: *Working nation: policies and programs*, Australian Government Publishing Service, Canberra.

Commonwealth of Australia, 1994b: *State of the environment reporting. A framework for Australia*, Australian Government Publishing Service, Canberra.

Commonwealth of Australia, 1995: *Living on the coast: the Commonwealth coastal policy*, Australian Government Publishing Service, Canberra.

Commonwealth of Australia, 1996a: *The national strategy for the conservation of Australia's biological diversity*, Australian Government Publishing Service, Canberra.

Commonwealth of Australia, 1996b: *A sustainable energy policy for Australia: white paper*, Department of Primary Industries and Energy, Canberra.

Commonwealth of Australia, 1998: *Reform of Commonwealth environmental legislation*, Consultation Paper, Issued by Senator the Hon. Robert Hill, Commonwealth Minister for the Environment, Community Information Unit, Department of the Environment, Commonwealth of Australia, Australian Government Publishing Service, Canberra.

Community Education and Policy Development Group, 1993: *The state of the environment report for South Australia 1993*, Department of Environment and Land Management, Adelaide.

Conacher, A.J., 1975a: *Environment-industry conflict: the Manjimup woodchip industry proposal, southwestern Australia*, Geowest 4, Working Papers of the Department of Geography, University of Western Australia.

Conacher, A.J., 1975b: Throughflow as a mechanism responsible for excessive soil salinisation in non-irrigated, previously arable land in the Western Australian wheatbelt: a field study, *Catena*, 2, 31–68.

Conacher, A.J., 1978: Resources and environmental management: some fundamental concepts and definitions, *Search*, 9(12), 437–41.

Conacher, A.J., 1980: Environmental problem-solving and land-use management: a proposed structure for Australia, *Environmental Management*, 4(5), 391–405.

Conacher, A.J., 1982a: Salt scalds and subsurface water: a special case of badland development in south-western Australia, in R. Bryan and A. Yair (eds), *Badland geomorphology and piping*, Geo Books, Norwich, 195–219.

Conacher, A.J., 1982b: Dryland agriculture and secondary salinity, in W. Hanley and M. Cooper (eds), *Man and the Australian environment*, McGraw-Hill, Sydney, 113–25.

Conacher, A.J., 1983: Environmental management implications of intensive forestry practices in an indigenous forest ecosystem: a case study from south-western Australia, in T. O'Riordan and R.K. Turner (eds), *Progress in Resource Management and Environmental Planning*, Vol. 4, Wiley, Chichester, 117–51.

Conacher, A.J., 1988: Resource development and environmental stress: environmental impact assessment and beyond in Australia and Canada, *Geoforum*, 19, 339–52.

Conacher, A.J., 1994a: The integration of land use planning and management with environmental impact assessment: some Australian and Canadian perspectives, *Impact Assessment*, 12, 347–72.

Conacher, A.J., 1994b: Western Australia, in A.W. Gilg *et al.* (eds), *Progress in rural policy and planning*, Vol. 4, Wiley, Chichester, 300–7.

Conacher, A.J., 1998: Environmental quality indicators: where to from here?, *Australian Geographer*, 29(2), 175–89.

Conacher, A.J., Combes, P.L., Smith, P.A. and McLellan, R.C., 1983: Evaluation of throughflow interceptors for controlling secondary soil and water salinity in dryland agricultural areas of southwestern Australia: I. Questionnaire surveys, *Applied Geography*, 3, 29–44.

Conacher, A.J. and Conacher, J.L., 1982: *Organic farming in Australia*, Geowest No. 18, Occasional Papers of the Department of Geography, University of Western Australia, Nedlands.

Conacher, A.J. and Conacher, J.L., 1995: *Rural land degradation in Australia*, Oxford University Press, Melbourne.

Conacher, A.J. and Dalrymple, J.B., 1977: The nine unit landsurface model: an approach to pedogeomorphic research, *Geoderma*, 18, 1–154.

Conacher, A.J. and Murray, I.D., 1973: Implications and causes of salinity problems in the Western Australian wheatbelt: the York-Mawson area, *Australian Geographical Studies*, 11, 40–61.

Conacher, A.J., Neville, S.D. and King, P.D., 1983: Evaluation of throughflow interceptors for controlling secondary soil and water salinity in dryland agricultural areas of southwestern Australia: II. Hydrological study, *Applied Geography*, 3, 115–32.

Conacher, A.J. and Sala, M. (eds), 1998: *Land degradation in Mediterranean environments of the world: nature and extent, causes and solutions*, Wiley, Chichester.

Condon, R.W., 1968: Estimation of grazing capacity on arid grazing lands, in G.A. Stewart (ed.), *Land evaluation*, Macmillan, London, 112–24.

Connor, J., 1999: Australia's new environment laws—questions and answers, *Habitat*, 7(4), 8–9.

Conservation Through Reserves Committee, 1974: *Conservation reserves in Western Australia*, Report to the Environmental Protection Authority, Perth.

Cooke, R.U. and Doornkamp, J.C., 1990: *Geomorphology in environmental management: a new introduction*, 2nd edn, Clarendon Press, Oxford.

Coopers and Lybrand, 1994: *Environmental management practices. A survey of major Australian organisations*, Coopers and Lybrand Consultants, Sydney.

Costanza, R.E. (ed.), 1992: *Ecological economics*, Columbia University Press, New York.

Couch, W.J., 1985: *Environmental assessment in Canada, 1985 summary of current practice*, Canadian Council of Resource and Environment Ministers, Minister for Supply and Services Canada, Ottawa.

CSIRO, 1998: *Sustainable landscapes: a guide to the research of program H for students and science collaboration*, CSIRO Division of Wildlife and Ecology, Canberra.

Cunningham, N. and Sinclair, D., 1999: Environmental Management Systems, regulation and the pulp and paper industry: ISO14001 in practice, *Environmental and Planning Law Journal*, 16(1), 5–24.

Curtis, A., 1999: Landcare: beyond on ground work, *Natural Resource Management*, 2(2), 4–9.

Curtis, A., Robertson, A. and Race, D., 1998: Lessons from recent evaluations of natural resource management programs in Australia, *Australian Journal of Environmental Management*, 5(2), 109–19.

Daly, H. and Cobb, J., 1989: *For the common good: redirecting the economy toward community, the environment and a sustainable future*, Beacon Press, Boston.

DARA, 1990: *Sustainable development. A strategy for Victorian agriculture*, Department of Agriculture and Rural Affairs, Victorian Government Printing Office, Melbourne.

Dargavel, J., 1998a: Politics, policy and process in the forests, *Australian Journal of Environmental Management*, 5(1), 25–30.

Dargavel, J., 1998b: Regional Forest Agreements and the public interest, *Australian Journal of Environmental Management*, 5(3), 133–4.

Davidson, B.R., 1969: *Australia wet or dry? The physical and economic limits to the expansion of irrigation*, Melbourne University Press, Melbourne.

Davidson, D.A., 1992: *The evaluation of land resources*, Longman, London.

Davis, B., 1972: Waterpower and wilderness: political and administrative aspects of the Lake Pedder contro-versy, *Public Administration*, 31, 21–42.

Davis, J.R. and Ive, J.R., 1985: *Dungog Shire environmental plan: an application of SIRO-Plan to local govern-ment planning*, CSIRO Division of Water and Land Resources Divisional Report 85/2, Canberra.

DCE, 1987: *State conservation strategy for Western Australia*, Bulletin No. 270, Department of Conservation and Environment, Perth.

DCFL, 1988: *Forest cover changes in Victoria 1969–87*, Department of Conservation, Forests and Lands, Victoria.

De Broissia, M., 1986: *Selected mathematical models in environmental impact assessment in Canada*, a back-ground paper prepared for the Canadian Environmental Assessment Research Council, Minister for Supply and Services Canada, Ottawa.

Dee, N., Baker, J.K., Drobny, N.L., Duke, K.M., Whitman, I. and Fahringer, D.C., 1973: An environmental evaluation system for water resources planning, *Water Resources Research*, 3, 523–55.

Deegan, C., 1996: A review of mandated environmental reporting requirements for Australian corporations together with an analysis of contemporary Australian and overseas environmental reporting practices, *Environmental and Planning Law Journal*, 13(2), 120–32.

Deegan, C., 1998: Environmental reporting in Australia. We're moving along the road, but there's still a long way to go, *Environmental and Planning Law Journal*, 15(4), 246–63.

DENR, 1995: *Our water. Our future. Sustainable management of South Australia's water resources*, Department of Environment and Natural Resources, Adelaide.

DENR, 1996: *A directory of important wetlands in South Australia*, Department of Environment and Natural Resources, Adelaide.

Dent, D.L. and Ridgway, R.B., 1986: *A land use planning handbook for Sri Lanka*, FD 2, SRL 79/058, Colombo, Sri Lanka, Land Use Policy Planning Division, Ministry of Lands and Land Development.

DEP, 1982: *Hunter region regional environmental plan no. 1*, Department of Environment and Planning, Sydney.

DEP, 1986: *Illawarra regional environmental plan no. 1*, Department of Environment and Planning, Sydney.

DEP, 1994: *Revised Greenhouse strategy for WA*, Greenhouse Co-ordinating Council, Department of Envir-onmental Protection, Perth.

Department of Arts, Heritage and Environment, 1986a: *State of the environment in Australia*, Australian Gov-ernment Publishing Service, Canberra.

Department of Arts, Heritage and Environment, 1986b: *State of the environment in Australia: source book*, Australian Government Publishing Service, Canberra.

Department of Commerce and Trade, 1993: *State recycling blueprint: a plan to halve landfill by the year 2000*, Western Australian Department of Commerce and Trade, Perth.

Department of Commerce and Trade, 1999: *A regional development policy for WA: setting the direction for regional Western Australia*, a policy framework discussion paper, Synectics Creative Collaboration and Department of Commerce and Trade, Government of Western Australia, Perth.

Department of Environment and Conservation, 1974: *Land use in Australia*, Australian Advisory Committee on the Environment Report No. 4, Australian Government Publishing Service, Canberra.

Department of Environment and Planning, n.d.a: *Internal briefing paper: Non-metropolitan regional planning on the east coast of Tasmania*, provided by E. Fowler, Director of Planning, Department of Environment and Land Management, Hobart.

Department of Environment and Planning, n.d.b: *Survey of published literature on regional planning*, provided by F. Fowler, Director of Planning, Department of Environment and Land Management, Hobart.

Department of Environmental Protection, 1997: *State of the Environment Reference Group Report Draft Working Papers for Western Australia*, Department of Environmental Protection, Perth.

Department of Environmental Protection, 1998: *Environment Western Australia 1998: State of the Environment Report*, Department of Environmental Protection, Perth.

Department of Environment, Housing and Community Development, 1978: *A basis for soil conservation policy in Australia: Commonwealth and State Government collaborative soil conservation study 1975–77*, Report 1, Australian Government Publishing Service, Canberra.

Department of Home Affairs and Environment, 1984: *A national conservation strategy for Australia*, Australian Government Printing Service, Canberra.

Department of Lands, 1993: *Mulga Region. A study of interdependence of the environment, pastoral production and the economy*, Position Paper, Department of Lands, Queensland Government, Brisbane.

Department of Minerals and Energy, 1996: *Annual Report*, Western Australian Department of Minerals and Energy, Perth.

Department of National Development, 1979: *Towards a national approach to land resource appraisal. Commonwealth/States collaborative soil conservation study*, Report No. 2, Australian Government Publishing Service, Canberra.

Department of Natural Resources, 1976: *Review of Australia's water resources 1975*, Australian Water Resources Commission, Australian Government Publishing Service, Canberra.

Department of Planning and Development, 1995: *Guidelines for environmental impact assessment and the Environment Effects Act*, 6th fully revised edition, Government of Victoria, Melbourne.

Department of Science and the Environment, 1979: *Environmental economics*, Australian Government Publishing Service, Canberra.

Department of Tourism, 1994: *National ecotourism strategy*, Department of Tourism, Australian Government Publishing Service, Canberra.

Department of Transport and Communication, 1992: *Australian code for the transport of dangerous goods by road and rail (Australian dangerous goods code)*, Australian Government Publishing Service, Canberra.

DEST, 1994a: *Fire and biodiversity: the effects and effectiveness of fire management*, Proceedings Conference 8–9 October 1994, Melbourne, Biodiversity Series No. 8, Biodiversity Unit, Department of Environment, Sport and Territories, Canberra.

DEST, 1994b: *State and Territory Greenhouse gas inventory 1998, 1990*, National Greenhouse Gas Inventory Committee, Department of Environment, Sport and Territories, Canberra.

DEST, 1995: *Greenhouse 21: an action plan for a sustainable future*, Department of Environment, Sport and Territories, Canberra.

DEST, 1996: *The national strategy for the conservation of Australia's biodiversity*, Commonwealth Department of Environment, Sport and Territories, Canberra; and Devall, W. and Sessions, G., 1985: *Deep ecology: living as if nature mattered*, Gibbs Smith, Salt Lake City.

Devall, W. and Sessions, G., 1985: *Deep ecology: living as if nature mattered*, Gibbs Smith, Salt Lake City, Utah.

DHUD, 1993: *Mount Lofty Ranges Regional Strategy Plan*, Policy Branch, Planning Division, Department of Housing and Urban Development, Adelaide.

DHUD, 1995: *Southern Adelaide plan: forward planning for the south*, Department of Housing and Urban Development, Southern Region of Councils, Adelaide.

Dickinson, R.E., 1964: *City and region: a geographical interpretation*, Routledge and Keegan Paul Ltd, London.

Dixon, C.J. and Leach, Bidget, n.d. (1978): *Questionnaires and interviews in geographical research*, Concepts and Techniques in Modern Geography 18, Geo Abstracts, University of East Anglia, Norwich, England.

DLWC, 1995: *Window on water: the state of water in NSW 1993–94*, Department of Land and Water Conservation, Sydney.

DNRE, 1997: *Victorian regional catchment strategies: Port Philip*, Department of Natural Resources and Environment, Melbourne.

DNRE, 1998: *Victoria's biodiversity. Directions in management*, Department of Natural Resources and Environment, Melbourne.

DoP, 1988: *North coast regional environmental plan 1988*, Department of Planning, Sydney.

DoP, 1993: *North coast draft urban planning strategy*, Department of Planning, Sydney.

Dore, J. and Woodhill, J. (eds), 1999: *Sustainable regional development: final report*, Greening Australia, Canberra.

Doyle, T. and Kellow, A., 1995: *Environmental politics and policy making in Australia*, Macmillan, Melbourne.

DPI, 1980: *A manual of safe practice in the handling and use of pesticides*, Document PB377, Pesticides Section, Department of Primary Industry, Australian Government Publishing Service, Canberra.

DPIE, 1994a: *National strategy on aquaculture in Australia*, Department of Primary Industries and Energy, Canberra.

DPIE, 1994b: *Recreational fishing in Australia: a national policy*, Department of Primary Industries and Energy, Canberra.

DPIE/ASCC, 1988: *Report of Working Party on effects of drought assistance measures and policies on land degradation*, Department of Primary Industries and Energy/Australian Soil Conservation Council, Australian Government Publishing Service, Canberra.

DPUD, 1991: *Albany Regional Planning Study Regional Rural Strategy*, Perth.

DPUD, 1993a: *Bunbury–Wellington Region Plan*, Perth.

DPUD, 1993b: *Albany Regional Planning Study Albany Region Plan*, Perth. May 1993. For public comment.

DPUD, 1994: *Central Coast Regional Strategy*, draft public discussion paper prepared for the Central Coast Planning Study Steering Committee by DPUD, Perth.

Driscoll, D.A. and Roberts, J.D., 1997: Impact of fuel-reduction burning on the frog *Geocrinia lutea* in southwest Western Australia, *Australian Journal of Ecology*, 22, 334–9.

Druce, M., 1999: Hawkesbury–Nepean Catchment: a regional environment management experience, in J. Dore and J. Woodhill (eds), *Sustainable regional development: final report*, Greening Australia, Canberra, 227–39.

DUAP, 1998: *Shaping our cities: the planning strategy for the greater metropolitan region of Sydney, Newcastle, Wollongong and the Central Coast*, Department of Urban Affairs and Planning, Sydney.

Durack, M., 1959: *Kings in grass castles*, Constable, London (reprinted 1981 by Currey O'Neil, South Yarra, Victoria).

DWR, 1991: *Draft State ground water policy. Policy framework and actions paper. Discussion paper*, Department of Water Resources, Sydney.

Dyson, M., 1997: Water Resources Act 1997 (SA)—balancing flexibility, accountability, certainty, *Environmental and Planning Law Journal*, 14(4), 305–14.

EDO, 1997: *A guide to environmental law in Western Australia*, Environmental Defenders Office, Perth.

Edwardes, C., 1997: *Amendments to the Environmental Protection Act 1986. Amendments under consideration to update and improve the Act and correct some problems*, a Public Discussion Paper, Minister for the Environment, Government of Western Australia, Perth.

Ehrlich, P.R., Ehrlich, A.H. and Holdren, J.P., 1977: *Ecoscience: population, resources, environment*, W.H. Freeman & Co., San Francisco.

EMRC, 1997: *Regional environmental strategy*, Project Brief for NHT, Eastern Metropolitan Region Councils, Perth.

Endangered Species Advisory Council, 1992: *National strategy for the conservation of Australian species and ecological communities threatened with extinction*, Australian Nature Conservation Agency, Canberra.

Engel, R., McFarlane, D.J. and Street, G., 1989: Using geophysics to define recharge and discharge areas associated with saline seeps in south-western Australia, in M. Sharma (ed.), *Groundwater recharge*, Balkema, Rotterdam, 25–40.

England, P., 1996: Change of form or change of philosophy: a critical review of the Queensland PEDA Bill, *Environmental and Planning Law Journal*, 13(2), 133–45.

England, P., 1999: Toolbox or tightrope? The status of environmental protection in Queensland's Integrated Planning Act, *Environmental and Planning Law Journal*, 16(2), 124–39.

Environment Australia, 1997: *Wetlands policy of the Commonwealth Government of Australia*, Biodiversity Group, Environment Australia, Canberra.

Environment Canada, 1970: *The Canada land inventory*, Report No. 1, Lands Directorate, Environment Canada, Ottawa.

Environmental Assessment Board, 1986: *Environmental Assessment Board 10th Annual Report Fiscal Year ending March 31st 1986*, Environmental Assessment Board, Toronto, Canada.

Environmental Protection Council of South Australia, 1988: *The state of the environment report for South Australia*, Department of Environment and Planning, Government of South Australia, Adelaide.

EPA, 1973: *Interim report—woodchip industry—Manjimup*, Report to the Minister for Environmental Protection, 24 August 1973, Perth.

EPA, 1975: *Second interim report on the woodchips (Manjimup) project*, Report to the Minister for Conservation and Environment, September 1975, Perth.

EPA, 1976: *Conservation reserves for Western Australia as recommended by the Environmental Protection Authority 1976: Systems 1, 2, 3, 5*, Recommendation to the Minister for Conservation and the Environment, Perth.

EPA, 1990a: *Environmental Protection Policy (ozone-depleting substances) 1989 guidelines*, Environmental Protection Authority, Perth.

EPA, 1990b: *Protecting the ozone layer: taking action in Western Australia*, pamphlet, Environmental Protection Authority, Perth.

EPA NSW, 1993a: *Environmental guidelines. Discharge of wastes to ocean waters*, New South Wales Environmental Protection Authority, Sydney.

EPA NSW, 1993b: *New South Wales state of the environment 1993 highlights*, Environment Protection Authority New South Wales, Chatswood.

EPA NSW, 1994: *The air quality index*, Environmental Protection Authority, Sydney.

EPA NSW, 1995a: *New South Wales State of the Environment 1995*, Environment Protection Authority New South Wales, Chatswood.

EPA NSW, 1995b: *Greenhouse gas inventory final report*, Environment Protection Authority, Chatswood.

EPA SA, 1997: *Development of an environment protection (water quality) policy*, Discussion Paper, Environment Protection Authority, Department of Environment and Natural Resources, Adelaide.

EPA SA, 1998: *State of the environment report for South Australia 1998*, Environment Protection Authority in co-operation with the Department for Environment, Heritage and Aboriginal Affairs, Natural Resources Council, Adelaide.

EPA Vic, 1991: *Contaminated sites register*, Information Bulletin No. 90/03, Environmental Protection Authority, Melbourne.

ESD Working Group, 1991: *Final report. Agriculture*, Ecologically Sustainable Development Working Group, Australian Government Publishing Service, Canberra.

Everitt, R.R. and Colnett, D.L., 1987: *Methods for determining the scope of environmental assessments*, Final Report for the Federal Environmental Assessment Review Office, Vancouver, BC, Environmental and Social Systems Analysts, Vancouver BC, Canada.

FAO, 1976: *A framework for land evaluation*, FAO Soils Bulletin 32, Food and Agriculture Organisation of the United Nations, Rome.

FAO, 1983: *Guidelines: land evaluation for rainfed agriculture*, FAO Soils Bulletin 52, Rome.

FAO, 1984: *Land evaluation for forestry*, Forestry Paper 48, Rome.

FAO, 1985: *Guidelines: land evaluation for irrigated agriculture*, FAO Soils Bulletin 55, Rome.

FAO, 1991: *Guidelines: land evaluation for extensive grazing*, FAO Soils Bulletin 58, Rome.

FAO, 1993a: *Agroecological assessments for national planning. The example of Kenya*, FAO Soils Bulletin No. 67, Rome.

FAO, 1993b: *Guidelines for land-use planning*, FAO Development Series 1, Rome.

Ferdowsian, R., George, R., Lewis, R. and others, 1996: The extent of dryland salinity in Western Australia, *Proceedings 4th National Workshop on the Productive Use and Rehabilitation of Saline Lands*, Albany, March, Agriculture Western Australia (unpublished).

Fisher, D.E., 1980: *Environmental law in Australia: an introduction*, University of Queensland Press, St Lucia.

Fisher, D.E., 1993: *Environmental law: text and materials*, LBC Casebooks, The Law Book Co., Sydney.

Fisher, R., Ury, W. and Patton, B., 1991: *Getting to yes: negotiating without giving in*, Houghton Mifflin, Boston.

Flannery, T., 1995: *The future eaters: an ecological history of the Australasian land and people*, Reed Books, Melbourne.

Flannery, T., 1998: Overpopulate and we perish, edited text of the 1998 Templeton lecture, *The Australian*, 1/9/98:13.

Forests Department, n.d. (1973): *Marri woodchip project environmental impact statement*, Perth.

Forests Department, 1977: *General Working Plan* No. 86, Perth.

Forsyth, J., 1998: Anarchy in the forests: a plethora of rules, an absence of enforceability, *Environmental and Planning Law Journal*, 15(5), 338–49.

Fowler, R.J., 1982: *Environmental impact assessment, planning and pollution measures in Australia: an analysis of Australian environmental impact assessment procedures and their relationship with land-use planning and pollution controls*, Department of Home Affairs and Environment, Australian Government Publishing Service, Canberra.

Fowler, R.J., 1992: Environmental dispute resolution techniques—what role in Australia? *Environmental and Planning Law Journal*, 9(2), 122–31.

Fowler, R.J., 1996: Environmental impact assessment: what role for the Commonwealth?—an overview, *Environmental and Planning Law Journal*, 13(4), 246–59.

Fowler, R., 1999: Act of convenience threatens ecology, *The Australian*, 25 June, 15.

Fox, E., Jan, G. and Norris, V., 1998: Environmental auditing of mines in the Northern Territory, *Australian Journal of Environmental Management*, 5(1), 16–23.

George, P.R. and Lenane, L., 1982: Saltland drainage…can it pay? *Journal of Agriculture Western Australia*, 23, 136–8.

George, R.J., 1990a: The 1989 saltland survey, *Journal of Agriculture Western Australia*, 31, 159–66.

George, R.J., 1990b: Reclaiming sandplain seeps by intercepting perched groundwater with eucalypts, *Land Degradation and Rehabilitation*, 2, 13–25.

George, R.J., 1990c: PUMPS—a method of financially assessing groundwater pumping used to mitigate salinity in south-western Australia, Division of Resource Management *Technical Report* No. 87, Western Australian Department of Agriculture, Perth.

George, R.J., 1991a: Interactions between Perched and Deeper Groundwater Systems in Relation to Secondary, Dryland Salinity in the Western Australian Wheatbelt: Processes and Management Options, PhD Thesis in Geography, University of Western Australia.

George, R.J., 1991b: Management of sandplain seeps in the wheatbelt of Western Australia, *Agricultural Water Management*, 19, 85–104.

George, R.J., 1992: Hydraulic properties of groundwater systems in the saprolite and sediments of the wheatbelt, Western Australia, *Journal of Hydrology*, 130, 251–78.

George, R.J. and Conacher, A.J., 1993: The hydrology of shallow and deep aquifers in relation to secondary soil salinisation in southwestern Australia, *Geografia Fisica e Dinamica Quaternaria*, 16, 47–64.

Ghassemi, F., Jakeman, A.J. and Nix, H.A., 1995: *Salinisation of land and water resources: human causes, extent, management and case studies*, University of New South Wales Press and CAB International, Sydney and Wallingford.

Gifford, R.M., Kalma, J.D., Aston, A.R. and Millington, R.J., 1975: Biophysical constraints in Australian food production: implications for population policy, *Search*, 6, 212–22.

Gilpin, A., 1995: *Environmental impact assessment: cutting edge for the twenty-first century*, Cambridge University Press, Cambridge.

Glasson, J., Therivel, R. and Chadwick, A., 1994: *Introduction to environmental impact assessment*, University College London Press, London.

Gleeson, B. and Hanley, P. (eds), 1998: *Renewing Australian planning? New challenges, new agendas*, A discussion forum organised by the Urban Research Program held on 17–18 June 1998, Research School of Social Sciences, Australian National University, Canberra.

Gleick, J., 1987: *Chaos: the making of a new science*, Viking, New York.

Government of Australia, 1993: *Landcare information: land, water and vegetation programs 1994–95*, Department of Primary Industries and Energy, Canberra.

Government of NSW, 1994: *No time to waste: NSW Government policy statement on waste*, New South Wales Government, Sydney.

Government of SA, 1992: *2020 Vision. Planning strategy for metropolitan Adelaide*, Government of South Australia, Adelaide.

Government of Tasmania, 1994: *Towards a Tasmanian weed management strategy*, Ministerial Working Group Discussion Paper, Hobart.

Government of Victoria, 1987: *Protecting the environment: a conservation strategy for Victoria*, Victorian Government Printing Office, Melbourne.

Government of Victoria, 1990: *Greenhouse: meeting the challenge: the Victorian government's statement of action*, Victorian Government Printing Office, Melbourne.

Government of Victoria, 1992: *Victoria's decade of Landcare plan*, Decade of Landcare Plan Steering Committee, Victorian Government Printing Office, Melbourne.

Government of WA, 1987: *State conservation strategy*, Government of Western Australia, Perth.

Government of WA, 1990: *Working together: ICM policy for WA*, Integrated Catchment Management Policy Group, Perth.

Government of WA, 1992: *State of the environment report*, Government of Western Australia, Perth.

Government of WA, 1993: *Avon River system management strategy*, Avon River System Management Committee/Waterways Commission, Government of Western Australia, Perth.

Government of WA, 1995: *A review of Landcare in WA*, Report of Landcare Review Committee, Minister for Primary Industries, Government of WA, Perth.

Government of WA, 1995: *Western Australian Yearbook*, Australian Bureau of Statistics, Cat. No. 1301.5, WA Office, Perth.

Government of WA, 1996a: *Salinity: a situation statement for Western Australia*, Agriculture WA, CALM, Department of Environmental Protection, Water and Rivers Commission: a Report to the Minister for Lands and the Minister for the Environment, Government of Western Australia, Perth.

Government of WA, 1996b: *Western Australian salinity action plan*, Agriculture WA, CALM, Department of Environmental Protection, Water and Rivers Commission: a Report to the Minister for Lands and the Minister for the Environment, Government of Western Australia, Perth.

Government of WA, 1997a: *Wetlands conservation policy for WA*, Department of Conservation and Land Management, Perth.

Government of WA, 1997b: *Gascoyne–Murchison rangelands strategy*, Government of Western Australia, Perth.

Government of WA, 1997c: *Draft strategy for the management of green and solid organic waste*, Department of Environmental Protection, Perth.

Government of WA, 1999: *A regional development policy for Western Australia: setting the direction for regional Western Australia*, a Policy Framework Discussion Paper, Regional Development Division, Department of Commerce and Trade, Government of Western Australia, Perth.

Gow, L., 1997: The New Zealand Resource Management Act, *Australian Planner*, 34(3), 132–6.

Graetz, D., Fisher, R. and Wilson, M., 1992: *Looking back: the changing face of the Australian continent 1972–92*, CSIRO Office of Space Science and Applications, Canberra.

Graetz, D., Gentle, M.R., Pech, R.P. and O'Callaghan, J.F., 1982: *The development of a land image-based resource information system (LIBRIS) and its application to the assessment and monitoring of Australian arid rangelands*, Proceedings of the International Symposium on Remote Sensing of the Environment, Cairo.

Grant, W.E., Pedersen, E.K. and Marin, S.L., 1997: *Ecology and natural resource management: systems analyses and simulation*, John Wiley & Sons, New York.

Gray, V., 1998: Hot air at Kyoto, *New Scientist*, 157(2117), 51.

Green, R.H., 1979: *Sampling design and statistical methods for environmental biologists*, Wiley-Interscience, New York.

Greening Australia, 1995: *Local Greening Plans*, a guide for vegetation and biodiversity management, Greening Australia Ltd, Canberra.

Greenwood, E.A.N. and Beresford, J.D., 1979: Evaporation from vegetation in landscapes developing secondary salinity using the ventilated-chamber technique. I. Comparative transpiration from juvenile Eucalyptus above saline groundwater seeps, *Journal of Hydrology*, 42, 369–82.

Greenwood, E.A.N., Klein, J.L., Beresford, J.D. and Watson, G.D., 1985: Differences in annual evaporation between grazed pasture and Eucalyptus species in plantations on a saline farm catchment, *Journal of Hydrology*, 78, 261–78.

Grinlinton, D., 1990: The 'environmental era' and the emergence of 'environmental law' in Australia—a survey of environmental legislation and litigation 1967 – 1987, *Environmental and Planning Law Journal*, 7, 74–105.

Grinlinton, D.P., 1992: Integrated resource management—a model for the future, *Environmental and Planning Law Journal*, 9(1), 4–19.

Haggett, P., 1972: *Geography: a modern synthesis*, Harper & Row, New York.

Hall, A.D. and Fagen, R.E., 1956: Definition of system, *General Systems*, 1, 18–28.

Hallam, S.J., 1975: *Fire and hearth*, Australian Institute of Aboriginal Studies, Canberra.

Hallsworth, E.G., 1978: *Purposes and requirements of land resource survey and evaluation*, Commonwealth/States collaborative soil conservation study 1975–77, Department of Environment, Housing and Community Development, Australian Government Publishing Service, Canberra.

Hamilton, C. and Attwater, R., 1997: Measuring the environment: the availability and use of environmental statistics in Australia, *Australian Journal of Environmental Management*, 4(2), 72–87.

Hardin, G., 1968: Tragedy of the commons, *Science*, 162, 1243–8.

Harman, E., 1987: Regional economic strategies in Western Australia, in J.D. Country (ed.), *Australian regional developments*, Australian Government Publishing Service, Canberra.

Harvey, N., 1998: *Environmental impact assessment: procedures, practice, and prospects in Australia*, Oxford University Press, Melbourne.

Hauge, C.J., Furniss, M.J. and Euphrat, F.D., 1979: Soil erosion in California's Coast Forest District, *California Geology*, 32, 120–9.

Hawarth, R., 1997: Fine sentiments vs brute actions: the Landcare ethic and land clearing, in S. Locke and F. Vanclay (eds), *Critical landcare*, Centre for Rural Social Research, Charles Sturt University, Wagga Wagga, New South Wales, 165–74.

Hawke, R.J.L., 1989: *Our country, our future: PM's statement on the environment July 1989*, 2nd edn, Australian Government Publishing Service, Canberra.

Heathcote, R.L., 1965: *Back of Bourke: a study of land appraisal and settlement in semi-arid Australia*, Melbourne University Press, Melbourne.

Heathcote, R.L., 1994: *Australia*, 2nd edn, Longman Group UK Ltd, Essex.

Hellawell, J.M., 1986: *Biological indicators of freshwater pollution and environmental management*, Elsevier, Barking.

Heller, J., 1962: *Catch 22*, Cape, London.

Henschke, C.J., 1980: The extent of the dryland salinity problem, *Land and stream salinity seminar Western Australia*, Government Printer, Perth, 19.1–19.2.

Hepworth, A., 1998: Too early for celebration, Special Report on the Environment, *Australian Financial Review*, Tuesday 15 September, p. 37.

Hilmer, F., 1993: *National competition policy*, Australian Government Publishing Service, Canberra.

Hingston, F.J., Turton, A.G. and Dimmock, G.M., 1979: Nutrient distribution in karri (*Eucalyptus diversicolor* F. Muell.) ecosystems in southwest Western Australia, *Forest Ecology and Management*, 2, 133–58.

Hobbs, R.J., 1996: Ecosystem dynamics and management in relation to conservation in forest systems, *Journal of the Royal Society of Western Australia*, 79, 293–300.

Hodgson, J., 1996: Third party appeals in South Australia 1972 – 1993, *Environmental and Planning Law Journal*, 13(1), 8–28.

Hoey, W.A., 1994: Rangelands reform in south west Queensland, in *Outlook 94. Vol. 2. Natural Resources*, Australian Bureau of Agricultural and Resource Economics, Canberra, 137–47.

Holling, C.S. (ed.), 1978: *Adaptive environmental assessment and management*, Wiley, Chichester.

Holmes, J.H., 1986: *The pastoral lands of the Northern Territory Gulf District: resource appraisal and land use options*, Report to the Northern Territory Department of Lands, Uniquest.

Holmes, J.H., 1994: Changing rangeland resource values: implications for land tenure and rural settlement, in *Outlook 94. Vol. 2. Natural Resources*, Australian Bureau of Agricultural and Resource Economics, Canberra, 160–75.

Holmes, J.W., 1971: Salinity and the hydrologic cycle, in T. Talsma and J.R. Philip (eds), *Salinity and water use*, Macmillan, London, 25–40.

Horstman, M., 1998: One continent or seven countries? *Habitat*, 26(2), 21.

Horwitz, P. and Calver, M., 1998: Credible science? Evaluating the Regional Forest Agreement process in Western Australia, *Australian Journal of Environmental Management*, 5(4), 213–25.

House of Representatives Select Committee, 1972: *Wildlife conservation*, House of Representatives Select Committee, Canberra.

House of Representatives Standing Committee, 1976: *Land use pressures on areas of scenic amenity*, House of Representatives Standing Committee on Environment and Conservation, Canberra.

House of Representatives Standing Committee, 1980: *Management of the Australian coastal zone*, House of Representatives Standing Committee on Science and Environment, Australian Government Publishing Service, Canberra.

House of Representatives Standing Committee, 1982: *Hazardous Chemicals*, Report of the House of Representatives Standing Committee on Environment and Conservation. Second Report of the Inquiry into Hazardous Chemicals, Australian Government Publishing Service, Canberra.

House of Representatives Standing Committee, 1989: *The effectiveness of land degradation policies and programmes*, House of Representatives Standing Committee on Environment, Recreation and Arts, Australian Government Publishing Service, Canberra.

House of Representatives Standing Committee, 1994a: *Australia's population 'carrying capacity': one nation—two ecologies*, House of Representatives Standing Committee for Long Term Strategies, Australian Government Publishing Service, Canberra.

House of Representatives Standing Committee, 1994b: *Working with the environment: opportunities for job growth*, House of Representatives Standing Committee on Environment, Recreation and Arts, Australian Government Publishing Service, Canberra.

House of Representatives Standing Committee, 1999: *Review of the Department of the Environment's Annual Report for 1997–98*, House of Representatives Standing Committee on Environment and Heritage, Australian Government Publishing Service, Canberra.

Huggett, R.J., 1998: Soil chronosequences, soil development, and soil evolution: a critical review, *Catena*, 32, 155–72.

Hunt, N., and Gilkes, R.J., 1995: *The farm monitoring handbook*, Land Management Society, Perth.

Hunter, R., 1979: *Warriors of the Rainbow: a chronicle of the Greenpeace movement*, Holt, Rinehart and Winston, New York.

Iles, A.T., 1996: Adaptive management: making environmental law and policy more dynamic, experimentalist and learning, *Environmental and Planning Law Journal*, 13(4), 288–308.

Industry Commission, 1997: *A full repairing lease: inquiry into ecologically sustainable land management*, draft report, Commonwealth of Australia, Canberra.

Industry Commission, 1998: *A full repairing lease: inquiry into ecologically sustainable land management*, final report, Commonwealth of Australia, Canberra.

IUCN/WWF, 1980: *World conservation strategy*, International Union for the Conservation of Nature and the World Wildlife Fund, United Nations Environment Programme, Nairobi.

Ive, J.R., 1992: Lupis: computer assistance for land use allocation. Preprint of paper to Resource Technology 92 Taipei: Information Technology for Environmental Management.

Jackson, S. and Crough, G.J., 1995: International environmental treaties and the rights of indigenous peoples in Australia, *Australian Geographer*, 26(1), 44–52.

James, B., 1991: Public participation in Department of Conservation management planning, *New Zealand Geographer*, 47(2), 51–9.

JANIS, 1997: *National agreed criteria for the establishment of a comprehensive, adequate and representative reserve system for forests in Australia*, Joint ANZECC/MCFFA National Forests Policy Implementation Sub-Committee, Canberra.

Jeffrey, M.I., 1992: Environmental Bill of Rights about to be added to Ontario's environmental arsenal, *Enviroflash*, 8, Fraser and Beatty Barristers and Solicitors, Ontario.

Johnson, A., McDonald, G., Shrubsole, D. and Walker, D., 1998: Natural resource use and management in the Australian sugar industry: current practice and opportunities for improved policy, planning and management, *Australian Journal of Environmental Management*, 5(2), 97–108.

Johnson, D.L., Ambrose, S.H., Bassett, T.J. and others, 1997: Meanings of environmental terms, *Journal of Environmental Quality*, 26, 581–9.

Johnston, R.M. and Barson, M.M., 1990: *An assessment of the use of remote sensing techniques in land degradation studies*, Report No. 5, Bureau of Rural Resources, Department of Primary Industries and Energy, Australian Government Publishing Service, Canberra.

Joint Committee, 1979: *Ord River irrigation area review*, Australian Government Publishing Service, Canberra.

Jones, R. (ed.), 1975: *The vanishing forests? Woodchip production and the public interest in Tasmania*, 4th publication of the Environmental Law Reform Group, University of Tasmania, Hobart.

Joss, P.J., Lynch, P.W. and Williams, O.B. (eds), 1986: *Rangelands: a resource under siege*, Proceedings of the Second International Rangelands Congress, Australian Academy of Science, Canberra.

Keating, P., 1992: *Australia's environment: a national asset. Prime Minister's statement on the environment*, Australian Government Publishing Service, Canberra.

Keating, P., 1996: *On conserving natural Australia*, Statement by the Prime Minister, 24 January 1996, Australian Government Publishing Service, Canberra.

Kelleher, B., 1998: Major offences under the Environmental Protection Act 1994 (Qld), *Environmental and Planning Law Journal*, 15(4), 264–74.

Kelly, G., Hamblin, A. and Johnston, R.M., 1990: *Remote sensing for agricultural production and resource management in Australia*, Bureau of Rural Resources Bulletin No. 8, Rural Industries Research and Development Corporation, Department of Primary Industries and Energy, Australian Government Publishing Service, Canberra.

Klingebiel, A.A. and Montgomery, P.H., 1961: *Land capability classification*, United States Department of Agriculture, Agriculture Handbook 210, Washington D.C.

Kozlowski, J., 1990: Queensland strategic planning: pit-falls and prospects, *Queensland Planner*, 30(2), 3–7.

Lake Pedder Committee of Inquiry, 1974: *The flooding of Lake Pedder*, Final Report, Canberra.

Landcare Review Committee, 1995: *A review of Landcare in Western Australia*, an initiative of the Minister for Primary Industry, Report of the Landcare Review Committee, Perth.

Land Resources Development Centre, 1966–: *Land resource studies*, Nos. 1–, Land Resources Development Centre, Ministry of Overseas Development, Surbiton, Surrey, England.

Lands and Forests Commission, 1994: *Forest management plan 1994 – 2003*, Department of Conservation and Land Management, Perth.

Lane, M., 1997: The importance of planning context: the wet tropics case, *Environmental and Planning Law Journal*, 14(5), 368–77.

Lane, M.B., 1999: Regional Forest Agreements. Resolving resource conflicts or managing resource politics, *Australian Geographical Studies*, 37(2), 142–53.

Lane, M.B., Brown, A.L. and Chase, A., 1997: Land and resource planning under Native Title: towards an initial model, *Environmental and Planning Law Journal*, 14(4), 249–58.

Lane, M., McDonald, G. and Corbett, T., 1996: Not all World Heritage Areas are created equal: World Heritage Area management in Australia, *Environmental and Planning Law Journal*, 13(6), 461–76.

LCC, 1989: *Mallee area review. Final recommendations*, Land Conservation Council, Victoria.

Leadbetter, P., 1991: Environmental legislative update, *Environment Institute of Australia Newsletter*, 17, 35–8.

Lee, B., 1998: Greenpower, *Groundwork*, 4(1), 10–11.

Leigh, C., 1994: Victoria, in G. McDonald (ed.), Recent developments in rural policy and planning in Australia, *Progress in Rural Policy and Planning*, 4, John Wiley & Sons Ltd, 290–5.

Leopold, Aldo, 1949: *A Sand County almanac*, Oxford University Press, New York.

Leopold, L.B., 1969: Landscape aesthetics, *Natural History*, 78, 37–44.

Leopold, L.B., Clarke, F.E., Hanshaw, B.B. and Balsley, J.R., 1971: *A procedure for evaluating environmental impact*, United States Geological Survey Circular 645, Washington D.C.

Lesslie, R. and Masley, M., 1995: *National wilderness inventory handbook*, Australian Heritage Commission, Canberra.

Lightfoot, L.C., Smith, S.T. and Malcolm, C.V., 1964: Salt land survey 1962, *Journal of Agriculture Western Australia*, 5, 396–410.

Likens, G.E., Bormann, F.H., Pierce, R.S. and others, 1977: *Biogeochemistry of a forested ecosystem*, Springer-Verlag, New York (2nd edn, 1995).

Lindblom, C.E., 1959: The science of 'muddling through', *Public Administration Review*, 19, 79–88.

Linton, D., 1968: The assessment of scenery as a natural resource, *Scottish Geographical Magazine*, 84, 219–38.

Lipman, Z., 1997: The New South Wales Protection of the Environment Operations Bill 1996: exposure draft, *Environmental and Planning Law Journal*, 14(2), 83–8.

Lipman, Z., 1998: Polluter pays for environmental crime, *Environmental and Planning Law Journal*, 15(1), 3–9.

Litton, R.B., 1972: Aesthetic dimensions of the landscape, in J.V. Krutilla (ed.), *Natural environments: studies in theoretical and applied analysis*, Johns Hopkins, Baltimore, 262–91.

Lockie, S., 1997: Beyond a 'good thing': political interests and the meaning of Landcare, in Lockie, S. and Vanclay, F. (eds), 1997: *Critical landcare*, Centre for Rural Social Research, Charles Sturt University, Wagga Wagga, NSW, 29–43.

Lockie, S. and Vanclay, F. (eds), 1997: *Critical landcare*, Centre for Rural Social Research, Charles Sturt University, Wagga Wagga, NSW.

Lottkowitz, S.N., 1989: *The greenhouse scenario and Victorian agriculture*, Department of Agriculture and Rural Affairs, Report No. 170, Melbourne.

Love, K.J., 1991: *Victorian produce monitoring results of residue testing 1989–1990*, Research Report Series No. 105, Department of Agriculture, Victoria.

Lovelock, J.E., 1979: *The Gaia hypothesis: a new look at life on earth*, Oxford University Press, Oxford.

Lovelock, J.E., 1988: *The ages of Gaia: a biography of our living earth*, W.W. Norton & Co., New York.

Lutz, E., 1993: *Toward improved accounting for the environment*, World Bank, Washington D.C.

LWRRDC, 1995: *Data sheets on natural resource issues*, Occasional Paper No. 06/95, Land and Water Resources Research and Development Corporation, Canberra.

LWRRDC, 1998: *Listing of LWRRDC funded R & D: current projects and final reports November 1998*, Land and Water Resources Research and Development Corporation, Canberra.

Lyster, M., 1985: *International wildlife law*, Grotius, Cambridge.

Macarthur Agribusiness, 1998: *South West Strategy. Strategic Plan 1998 – 2003*, Department of Natural Resources, South West Strategy Office, Charleville, Queensland.

MacGregor, C. and Pilgrim, A., 1998: Is Landcare funding hitting the target?, *Natural Resources Management*, 1(1), 4–8.

Macklin, K.M., 1987: Evaluative criteria for the assessment of integrative planning attempts and their application to the Shark Bay Region Plan (March 1987), unpublished BSc Hons thesis, Department of Geography, University of Western Australia.

MacRae, I. and Brown, D., 1992: The evolution of regional planning in Western Australia, in D. Hedgcock and O. Yiftachel (eds), *Urban and regional planning in Western Australia*, Paradigm Press, Perth, 205–17.

Maher and Associates, 1996: *Environmental management systems for Australian local government*, Australian Local Government Association National Seminar, EMS for Local Governments, Hobart.

Malcolm, C.V., 1969: Saltland pastures, *Journal of Agriculture Western Australia*, 10, 3–6.

Malcolm, C.V., 1986: Production from salt-affected soils, *Reclamation and Revegetation Research*, 5, 343–61.

Malcolm, C.V. and Stoneman, T.C., 1976: Salt encroachment—the 1974 saltland survey, *Western Australian Department of Agriculture Bulletin*, 3999.

Malone, N., 1997: Environmental impact monitoring, *Environmental and Planning Law Journal*, 14(3), 222–38.

Marshall, D., Sadler, B., Sector, J. and Wiebe, J., 1985: *Environmental management and impact assessment: some lessons and guidance from Canadian and international experience*, Federal Environmental Assessment Review Office Occasional Paper, Hull, Canada.

Masood, E. and Garwin, L., 1998: Costing the earth: when ecology meets economics, *Nature*, 395, 426–30.

Masterton, S., 1999: The Blackwood catchment initiative. An adaptive regional approach, in J. Dore and J. Woodhill (eds), *Sustainable regional development: final report*, Greening Australia, Canberra, 251–4.

McAllister, D.M., 1980: *Evaluation in environmental planning: assessing environmental, social, economic and political trade-offs*, MIT Press, Cambridge, Massachusetts.

McAuliffe, P., 1998: Native title: implications for Aboriginals, mining and the environment, unpublished seminar paper presented in the Advanced Environmental Planning and Management Honours Option, Department of Geography, University of Western Australia.

McDonald, J., 1995: One step forward, two steps back. Environmental assessment reform proposals in Queensland, *Environmental and Planning Law Journal*, 12(6), 367–72.

McFarlane, D.J. and George, R.J., 1992: Factors affecting dryland salinity in two wheatbelt catchments in Western Australia, *Australian Journal of Soil Research*, 30, 85–100.

McGuiness, S., 1991: *Soil structure assessment kit: a guide to assessing the structure of Red Duplex soils*, Department of Conservation and Environment, Victoria.

McHarg, I.L., 1969: *Design with nature*, Natural History Press, Garden City, New York.

McKinney, M. and Schoch, R., 1998: *Environmental science: systems and solutions*, Jones and Bartlett Publishers, Sudbury, Massachusetts.

MDBC, 1990: *Murray–Darling Basin natural resources management strategy*, Murray–Darling Basin Ministerial Council, Canberra.

MDBC, 1993a: *Dryland salinity management in the Murray–Darling Basin*, Working Group report to MDB Ministerial Council, MDB Commission, Canberra.

MDBC, 1993b: *Algal management strategy. Draft*, Murray–Darling Basin Commission, Canberra.

MDB Ministerial Council, 1987: *River salinity, waterlogging and land salinisation: towards an integrated management strategy*, Murray–Darling Basin Ministerial Council Technical Report No. 87/1, Canberra.

Meadows, D.H., Meadows, D.L., Randers, J. and Behrens, W.W.III, 1972: *The limits to growth: a report for the Club of Rome's project on the predicament of mankind*, Universe Books, New York.

Mercer, D. 1995: *A question of balance: natural resources conflict issues in Australia*, 2nd edn, Federation Press, Sydney.

Messer, J. and Mosley, G. (eds), 1983: *What future for Australia's arid lands?* Proceedings of the National Arid Lands Conference, Broken Hill, New South Wales, 21–25 May 1982, Australian Conservation Foundation, Hawthorn, Victoria.

Migratory Birds Program, n.d.: Copies of CAMBA and JAMBA Agreements and full lists of species under each agreement, at URL: www.anca.gov.au/index.htm, Australian Nature Conservation Agency, Canberra.

Miller, C., 1994: Sustainable management legislation—some emerging issues, *Progress in Rural Policy and Planning*, 4, 314–16.

Ministry for Planning, 1998: *Consolidation of the planning legislation: discussion paper for public comment*, State of Western Australia, Perth.

Minnery, J.R., 1987: Queensland: planning on the fringe, in S. Hammett and R. Baker (eds), *Urban Australia. Planning issues and policies*, Mansell Publishing Ltd, London and New York, 140–57.

Minson, K., 1994: *The proposed National Environment Protection Council. Analysis and criticism of the concept by the Government of Western Australia*, Minister for Environment, Perth.

Mitchell, B., 1979: *Geography and resource analysis*, Longman, London.

Mitchell, B. (ed.), 1990: *Integrated water management*, Belhaven Press, London.

Mitchell, B., 1991: *Integrated catchment management in Western Australia. Progress and opportunities*, Centre for Water Research, University of Western Australia, Perth.

Mitchell, B., 1997: Melbourne + 9: Canadian approaches and progress regarding integrated catchment management, in National ICM Workshop: *Advancing integrated resource management: processes and policies*, 2nd National Workshop on Integrated Catchment Management, Australian National University, Canberra.

Mitchell, C.W., 1991: *Terrain evaluation*, 2nd edn, Longman, London.

Mobbs, C. and Woodhill, J., 1999: Lessons from CYPLUS in Cape York. An external perspective, in J. Dore and J. Woodhill (eds), *Sustainable regional development. Final Report*, Greening Australia, Canberra, 267–85.

Moldan, B. and Cerny, J. (eds), 1994: *Biogeochemistry of small catchments: a tool for environmental research*, SCOPE 51, John Wiley & Sons, Chichester.

Moore, G. (ed.), 1998: *Soilguide. A handbook for understanding and managing agricultural soils*, a Joint National Landcare and Agriculture Western Australia Project, Agriculture Western Australia Bulletin 4343, South Perth.

Morelli, J. and de Jong, M., 1996: *A directory of important wetlands in South Australia*, Department of Environment and Natural Resources, Adelaide.

Morgan, R.P.C., 1995: *Soil erosion and conservation*, 2nd edn, Longman, Harlow.

Morris, L. and Therivel, R., 1995: *Methods of environmental impact assessment*, University College London Press, London.

Morrisey, P. and Lawrence, G., 1997: A critical assessment of Landcare in a region of central Queensland, in S. Lockie and F. Vanclay (eds), 1997: *Critical landcare*, Centre for Rural Social Research, Charles Sturt University, Wagga Wagga, NSW, 217–26.

Mosley, G., 1988: The Australian conservation movement, in R.L. Heathcote (ed.), *The Australian experience*, Longman Cheshire, Melbourne, 178–88.

Mould, B.D. and Bari, M.A., 1995: *Wellington reservoir catchment streamflow and salinity review*, Report WS 151, Water Authority WA, Perth.

Mould, H., 1998: The proposed Environment Protection and Biodiversity Conservation Act—the role of the Commonwealth in environmental impact assessment, *Environmental and Planning Law Journal*, 15(4), 275–86.

MRPA, 1970: *The corridor plan for Perth*, Metropolitan Region Planning Authority, Perth.

Mues, C., Chapman, L. and van Hilst, R., 1998: *Promoting improved land management practices on Australian farms: a survey of Landcare and land management related programs*, Australian Bureau of Agricultural and Resource Economics, Canberra.

Mues, C., Roper, H. and Ockerby, J., 1994: *Survey of Landcare and land management practices 1992 – 1993*, ABARE Research Report 94.6, Australian Bureau of Agricultural and Resource Economics, Canberra.

Mulcahy, M.J. and others, 1975: *Report on some aspects of the environmental effects of woodchip fellings in south western Australia*, CSIRO Division of Land Resources Management, Perth.

Mullins, P., 1980: Australian urbanisation and Queensland's underdevelopment: a first empirical statement, *International Journal of Urban and Regional Research*, 4, 212–38.

Munchenberg, S., 1994: Judicial review and the Commonwealth Environment Protection (Impact of Proposals) Act 1974, *Environmental and Planning Law Journal*, 11(6), 461–78.

Munchenberg, S., 1998: OECD reviews environmental performance, *Groundwork*, 4(1), 3–4.

Munn, R.E. (ed.), 1979: *Environmental impact assessment*, SCOPE Report 5, 2nd edn, Wiley, Chichester.

Munro, D.A., Bryant, T.J. and Matte-Baker, A., 1986: *Learning from experience: a state-of-the-art review and evaluation of environmental impact assessment audits*, Canadian Environmental Assessment Research Council, Minister of Supply and Services Canada, Ottawa.

Nader, R., 1965: *Unsafe at any speed: the designed-in dangers of the American automobile*, Grossman, New York (expanded edition 1972).

Naess, A., 1989: *Ecology, community and lifestyle. Outline of an ecosophy*, Cambridge University Press, Cambridge.

National Forest Inventory, 1998: *Australia's state of the forests report 1998*, Bureau of Rural Sciences, Canberra.

National ICM Workshop, 1997: *Advancing integrated resource management: processes and policies*, 2nd National Workshop on Integrated Catchment Management, Australian National University, Canberra.

National Rangeland Management Working Group, 1996: *Draft national strategy for rangeland management*, Australian and New Zealand Environment and Conservation Council, Agriculture and Resource Management Council of Australia and New Zealand, Canberra.

Natural Resources Management Services Unit, 1996: *Profile of the catchment hydrology discipline group, group profile as at June 18, 1996*, Agriculture Western Australia, Perth.

Naughton, T.F.M., 1992: Mediation and the Land and Environment Court of New South Wales, *Environmental and Planning Law Journal*, 9(3), 219–24.

Naughton, T., 1995: Court-related alternative dispute resolutions in New South Wales, *Environmental and Planning Law Journal*, 12(6), 373–87.

Negus, T.R., 1981: Interceptors on trial, *Journal of Agriculture Western Australia*, 22, 36–8.

Negus, T.R., 1987: Soil seepage interceptor drains and topsoil salinity, *Journal of Agriculture Western Australia*, 28, 44–9.

Neville, S., 1981: A methodology to assess recreation suitability in the karri forest, unpublished BSc(Hons) thesis in Geography, University of Western Australia.

Newman, P., Bathgate, C., Bell, K. *et al.*, 1994: *Australia's population carrying capacity: an analysis of eight natural resources*, Institute for Science and Technology, Murdoch University, Perth.

Newman, P. and Cameron, I., 1982: *Attitudes to conservation and environment in Western Australia*, School of Environmental Sciences, Murdoch University, Perth.

Newman, P. and Ross, W., 1996: *WA 2029: Stage II. Environmental implications*, WA Department of Commerce and Trade, Perth.

NHMRC, 1990: *Australian guidelines for recreational use of water*, National Health and Medical Research Council, Canberra.

NHMRC, 1992: *Ecologically sustainable development: the health perspective*, National Health and Medical Research Council, Canberra.

NHMRC, 1993: *National framework for environmental and health impact assessment*, National Health and Medical Research Council, Australian Government Publishing Service, Canberra.

NHMRC/ARMCANZ, 1995: *Australian drinking water guidelines*, National Health and Medical Research Council and Agriculture and Resource Management Council of Australia and New Zealand, Australian Government Publishing Service, Canberra.

Nix, H., 1988: Australia's renewable resources, in L.H. Day and D.T. Rowland (eds), *How many more Australians? The resource and environmental conflicts*, Longman Cheshire, Melbourne, 65–76.

Nolan, R., 1981: The regional perspective, in R. Nolan and P.D. Wilde (eds), *Designing a Tasmanian planning system for the 1980s*, Proceedings of a one-day conference held in Hobart on 3 April 1981, Department of Geography, University of Tasmania, 51–5.

NOLG, 1998: *A comparison of planning systems in Australian States and Territories*, National Office of Local Government/Department of Transport and Regional Development, Canberra.

Northern Territory Department of Lands and Housing, 1991: *Gulf Region Land Use and Development Study 1991*, Department of Lands and Housing, Casuarina, NT 0811.

Northern Territory Economic Development Strategy, 1988: *The Territory on the move*, N.T. Government Printer.

Nossin, J.J. (ed.), 1977: *Proceedings ITC symposium. Surveys for development*, Elsevier, Amsterdam.

NPWS, 1993: *Macquarie Marshes nature reserve plan of management*, NSW National Parks and Wildlife Service, Sydney.

NPWS/DLWC, 1996: *Macquarie marshes water management plan*, NSW National Parks and Wildlife Service and Department of Land and Water Conservation, Sydney.

NSW SCS, 1989: *Land degradation survey: New South Wales, 1978 – 1988*, New South Wales Soil Conservation Service.

NSW SCS, n.d.: *Total catchment management: a state policy*, New South Wales Soil Conservation Service.

NSW State Pollution Control Commission, 1991: *Proceedings of the New South Wales Government summit on air quality*, State Pollution Control Commission, Sydney.

Nulsen, R.A., 1984: Evaporation of four major agricultural plant communities in the south-west of Western Australia measured with large ventilated chambers, *Agricultural Water Management*, 8, 191–200.

Oates, N.M., Greig, P.J., Hill, D.G., Langley, P.A. and Reid, A.J. (eds), n.d.: *Focus on farm trees: the decline of trees in the rural landscape*, Proceedings of a National Conference, University of Melbourne, 23–26 November 1980.

O'Connor, A., 1985: *Towards a management plan for the Woorooloo drainage basin*, Geowest 21, Occasional Papers of the Department of Geography, University of Western Australia.

OECD, 1991: *State of the environment*, OECD, Paris.

OECD, 1994: *Environmental indicators: OECD core set*, OECD, Paris.

OECD, 1998: *Environmental performance reviews. Australia*, Organisation for Economic Co-operation and Development, Paris.

Office of Energy, 1997: Meeting Current and Future Energy Needs. A Sustainable Energy Policy for South Australia, unpublished draft, cited in EPA (SA) 1998:228.

Office of the Commissioner for the Environment, 1989: *State of the environment report 1988: Victoria's inland waters*, Victorian Government Printing Office, Melbourne.

Office of the Commissioner for the Environment, 1991: *1991 state of the environment report: agriculture and Victoria's environment resource report*, Government of Victoria, Melbourne.

Office of the Environment, 1992: *Victorian environmental policy guide: a planner's guide to Victorian conservation and environment policies*, Office of the Environment, Melbourne, September 1992.

Olley, J. (ed.), 1995: *Sources of suspended sediment and phosphorus to the Murrumbidgee River*, CSIRO Division of Water Resources, Consultancy Report No. 95-32.

Olsen, G. and Skitmore, E., 1991: *The state of the rivers of the South West Drainage Division*, Western Australian Water Resources Council Publication No. 2/91, Leederville.

O'Riordan, J., 1985: Cumulative assessment of the freshwater environment, in G.E. Beanlands *et al.* (eds), *Proceedings of the workshop on cumulative environmental effects*, Minister of Supply and Services Canada, Ottawa, 55–6.

O'Riordan, T. (ed.), 1995: *Environmental science for environmental management*, Longman, Essex.

O'Riordan, T., 1976: *Environmentalism*, Pion Ltd, London.

O'Riordan, T. and Turner, R.K., 1983: Planning and environmental protection: introductory essay, in T. O'Riordan and R.K. Turner (eds), *An annotated reader in environmental planning and management*, Pergamon Press, Oxford, 135–68.

Ostrom, E., Burger, J., Fields, C.B. and others, 1999: Revisiting the commons. Local lessons, global challenges, *Science*, 284, 278–82.

Owen, L. and Unwin, T. (eds), 1997: *Environmental management: readings and case studies*, Blackwell, Oxford.

PADC Environmental Impact Assessment and Planning Unit (ed.), 1983: *Environmental impact assessment*, Martinus Nijhoff Publishers, The Hague.

Panizza, M., Fabbri, A.G., Marchetti, M. and Patrono, A. (eds), 1996: *Geomorphologic analysis and evaluation in environmental impact assessment*, ITC Publication No. 32, Enschede, The Netherlands.

Parker, J. and Hope, C., 1992: The state of the environment: a survey of reports from around the world, *Environment*, 34(1).

Payne, A.L., Van Vreeswyk, A.M.E., Pringle, H.J.R., Leighton, K.A. and Hennig, P., 1998: *An inventory and condition survey of the Sandstone-Yalgoo-Paynes Find area, Western Australia*, Technical Bulletin No. 90, Agriculture Western Australia, South Perth.

Pearse, W., 1996: The legislative framework for the control of chemicals in Australia, *Environmental and Planning Law Journal*, 13(1), 29–39.

Peck, A.J., 1978: Salinisation of non-irrigated soils and associated streams: a review, *Australian Journal of Soil Research*, 16, 157–68.

Peterson, E.B., Chan, Y.-H., Peterson, N.M., Constable, G.A., Caton, R.B., Davis, C.S., Wallace, R.R. and Yarranton, G.A., 1987: *Cumulative effects assessment in Canada: an agenda for action and research*, a background paper prepared for the Canadian Environmental Assessment Research Council, Minister for Supply and Services, Ottawa.

Pickup, G. and Chewings, V.H., 1996: Correlations between DEM-derived topographic indices and remotely-sensed vegetation cover in rangelands, *Earth Surface Processes and Landforms*, 21(6), 517–29.

Pieri, C., Dumanski, J., Hamblin, A. and Young, A., 1995: *Land quality indicators*, World Bank Discussion Papers 315, The World Bank, Washington D.C.

Pigram, J.J., 1986: *Issues in the management of Australia's water resources*, Longman Cheshire, Melbourne.

Pilbara Study Group, 1974: *The Pilbara study*, Government Printer, Perth.

Pimentel, D., Andow, D., Dyson-Hudson, R. and others, 1980: Environmental and social costs of pesticides: a preliminary assessment, *Oikos*, 34, 126–40.

PIRSA, 1997: *Management and development of recreational fishing in South Australia*, SA Fisheries Management Series, Paper No. 23, Primary Industries South Australia, Adelaide.

Polkinghorne, L., 1998: Shrugging off bad publicity, *Australian Landcare*, June 1998, 57.

Powell, J.M., 1976: *Environmental management in Australia, 1788 – 1914. Guardians, improvers and profit: an introductory survey*, Oxford University Press, Melbourne.

Powell, J.M., 1988: Patrimony of the people: the role of government in land settlement, in R.L. Heathcote (ed.), *The Australian experience: essays in Australian land settlement and resource management*, Longman Cheshire, Melbourne, 14–24.

Powell, J.M., 1991: *Plains of promise, rivers of destiny. Water management and the development of Queensland 1824 – 1990*, Queensland Boolarong Publications, Brisbane.

Powell, J.M., 1993a: *'MDB'. The emergence of bioregionalism in the Murray–Darling Basin*, Murray–Darling Basin Commission, Canberra.

Powell, J.M., 1993b: *Griffith Taylor and 'Australia unlimited'*, John Murtagh Macrossan Lecture, University of Queensland Press, Brisbane.

Powell, J.M., 1998: *Watering the western third: water, land and community in Western Australia, 1826 – 1998*, Water and Rivers Commission, Perth.

Pringle, H.J., 1991: *Rangeland survey in Western Australia: history, methodology and applications*, Department of Agriculture, Perth.

Pritchard, L., Colding, J., Berkes, F., Suedin, U. and Folke, K., 1998: *The problem of fit between ecosystems and institutions*, Working Paper No. 2, Report for International Human Dimensions Programme on Global Environmental Change, IHDP Secretariat, Bonn.

Productivity Commission, 1999: *Impact of competition policy reforms on rural and regional Australia: draft report*, Commonwealth of Australia, Canberra.

Public Affairs Branch, 1995: *Guidelines for EIA and the Environmental Effects Act*, 6th rev. edn, Public Affairs Branch of the Department of Planning and Development, Melbourne.

Public Works Department, 1979: *Clearing and stream salinity in the south-west of Western Australia*, Document No. MDS 1/79, Perth.

QDEH, 1991: *Moreton Bay strategic plan*, Queensland Department of Environment and Heritage, Brisbane.

QDHLGP, 1993: *New planning and development legislation: a discussion paper*, Queensland Department of Housing, Local Government and Planning, Brisbane.

Queensland Department of Lands, 1993: *Mulga region: a study of the interdependence of the environment, pastoral production and the economy*, Position Paper, Brisbane.

Rab, M., 1994: Changes in physical properties of a soil associated with logging of *Eucalyptus regnans* forest in southeastern Australia, *Forest Ecology and Management*, 70, 215–29.

Raff, M., 1995a: Pragmatic curtailment of participation in planning in Victoria, *Environmental and Planning Law Journal*, 12(2), 73–7.

Raff, M., 1995b: The renewed prominence of environmental impact assessment: 'a tale of two cities', *Environmental and Planning Law Journal*, 12(4), 241–63.

Raff, M., 1997: Ten principles of quality in environmental impact assessment, *Environmental and Planning Law Journal*, 14(3), 207–21.

Ransom, T., 1999: Dorset Sustainable Development Strategy. A local government good news story, in J. Dore and J. Woodhill (eds), *Sustainable regional development. Final report*, Greening Australia, Canberra, 223–6.

Raynor, L., 1997a: The role of community Landcare coordinators: central wheatbelt Western Australia, unpublished Honours project in Environmental Management, Department of Geography, University of Western Australia, Perth.

Raynor, 1997b: Shire Council involvement in Landcare: Avon River Basin Western Australia, unpublished BSc (Hons) Thesis in Geography, University of Western Australia, Perth.

Rees, W.E., 1992: Ecological footprints and appropriated carrying capacity. What urban economics leaves out, *Environment and Urbanisation*, 4, 121–30.

Resource Assessment Commission, 1992: *Forest and timber inquiry. Final report*, Australian Government Publishing Service, Canberra.

Review Committee, 1976: *A review of the recommendations for reserves in the southwest and south coastal areas of Western Australia*, for the Environmental Protection Authority, Perth.

RIRDC, 1996: *Program plans and guidelines for researchers 1997 – 1998*, Rural Industries Research and Development Corporation, Canberra.

RIRDC, 1997: *Organic agriculture in Australia*, Proceedings of the National Symposium, Rural Industries Research and Development Corporation, Canberra.

Roberts, B., 1995: *The quest for sustainable agriculture and land use*, University of New South Wales Press, Sydney.

Robinson, G.M., 1994: Fast track planning in Queensland: the Bjelke-Petersen era and beyond, in A.W. Gilg (ed.), *Progress in Rural Policy and Planning*, 4, John Wiley & Sons Ltd, Chichester, 275–86.

Robinson, S. and Humphries, R., 1997: Towards Best Practice: observations on western legal and institutional arrangements for ICM, 1987, 1997, in *Advancing integrated resource management: processes and policies*, 2nd National Workshop on ICM, Australian National University, 29 September to 1 October 1997, Canberra. Not paginated.

Rolls, J., 1997: Integrated natural resource management in South Australia, in *Advancing integrated resource management: processes and policies*, 2nd National Workshop on ICM, Australian National University, Canberra (not paginated).

Rothery, B., 1995: *ISO 14000 and ISO 9000*, Gower Publishing Ltd, Hampshire, England.

Routley, R. and Routley, V., 1973: *The fight for the forests: the takeover of Australia's forests for pines and woodchips*, Research School of Social Sciences, Australian National University.

Rowan, J.N., 1990: *Land systems of Victoria*, Department of Conservation and Environment, Land Conservation Council, Victoria.

Rowe, R.K., Howe, D.F. and Alley, N.F., 1981: *Guidelines for land capability assessment in Victoria*, Soil Conservation Authority, Victoria.

Rumley, D., 1988: The political geography of Australian federal–State relations, *Geoforum*, 19(3), 367–79.

Rundle, G.E., 1996: History of conservation reserves in the south-west of Western Australia, *Journal of the Royal Society of Western Australia*, 79, 225–40.

Russel, J., 1998: World review: business and the environment, *Greener Management International*, 21, 6–28.

Sadler, B., 1990: The evaluation of assessment: post-EIS research and process development, in P. Wathern (ed.), *Environmental impact assessment: theory and practice*, Routledge, London, 129–42.

Sand, P.H. (ed.), 1992: *The effectiveness of international environmental agreements: a survey of existing legal instruments*, UN Conference on Environment and Development, Grotius Publications, Cambridge.

SCA, 1991: *Sustainable agriculture*, Report of Working Group on Sustainable Agriculture, Standing Committee on Agriculture, Report No. 136, CSIRO, Melbourne.

SCA, 1992: *National guidelines for beef cattle feedlots in Australia*, Standing Committee on Agriculture, Report No. 47, CSIRO, Melbourne.

SCARM, 1993: *Sustainable agriculture: tracking the indicators for Australia and New Zealand*, Agricultural Council of Australia and New Zealand Standing Committee on Agriculture and Resource Management, Commonwealth of Australia.

SCARM, 1998: *Sustainable agriculture: assessing Australia's recent performance*, Report to Standing Committee on Agriculture and Resource Management, Technical Report No. 70, Commonwealth of Australia, Canberra.

Schaefer, C. and Dalrymple, J., 1999: The importance of geopedology in sustainable use of tropical catchments: sodic soils and land use scenarios in northern Amazonia, in D. Harper and T. Brown (eds), *The sustainable management of tropical catchments*, John Wiley & Sons, Chichester, 89–109.

Schilizzi, S. and White, G., 1999: Dryland salinity control: what do we usefully know? *SEA Working paper* 99/03 (www.general.uwa.edu.au/u/dpannell/dpap9904f.htm (16K).

Schmitt, R.J. and Osenberg, C.W. (eds), 1996: *Detecting ecological impacts: concepts and applications in coastal habitats*, Academic Press, San Diego.

Schofield, N.J., 1990: Determining reforestation area and distribution for salinity control, *Hydrological Sciences Journal*, 35, 1–19.

Schofield, N.J. and Bari, M.A., 1991: Valley reafforestation to lower saline groundwater tables: results from Stene's farm Western Australia, *Australian Journal of Soil Research*, 29, 635–50.

Schramm, G. and Warford, J. (eds), 1989: *Environmental management and economic development*, World Bank, Hopkins, Baltimore.

Scott, J.F., 1975: Relationship between land and population: a note on Canada's carrying capacity, *Geografiska Annaler*, 57B, 128–32.

Sedgley, R.H., Smith, R.E. and Tennant, D., 1981: Management of soil water budgets of recharge areas for control of salinity in south-western Australia, *Agricultural Water Management*, 4, 313–34.

Select Committee into Land Conservation, 1991: *Final Report*, Western Australia Legislative Assembly, Perth.

Self, P., 1998: Democratic planning, in B. Gleeson and P. Hanley (eds), *Renewing Australian planning? New challenges, new agendas*, Research School of Social Sciences, Australian National University, Canberra, 45–9.

Senate Select Committee, 1990: *Report on agricultural and veterinary chemicals in Australia*, Senate Select Committee on Agricultural and Veterinary Chemicals in Australia, Commonwealth of Australia, Australian Government Publishing Service, Canberra.

Senate Standing Committee on Natural Resources, 1978: *Australia's water resources: the Commonwealth's role*, Australian Government Publishing Service, Canberra.

Senate Standing Committee on Rural and Regional Affairs, 1994: Soil and Land Conservation Council: *Land care policies and programs in Australia*, Senate Standing Committee on Rural and Regional Affairs—unpublished report.

Senate Standing Committee on Science and the Environment, 1976: *Woodchips and the Environment*, Australian Government Publishing Service, Canberra.

Senate Standing Committee on Science and the Environment, 1978: *Woodchips and the environment: Supplementary Report from the Senate Standing Committee on Science and the Environment*, Australian Government Publishing Service, Canberra.

Senate Standing Committee on Science and the Environment, 1982: *Pesticides and the health of Australian Vietnam veterans*, first report, Parliament of Australia, Canberra.

Senate Standing Committee on Science, Technology and Environment, 1984: *Land use policy in Australia*, Australian Government Publishing Service, Canberra.

Sewell, W.R.D., Handmer, J.W. and Smith, D.I. (eds), 1984: *Water planning in Australia: from myths to reality*, Centre for Resource and Environmental Studies, Australian National University, Canberra.

Sewell, W.R.D. and Phillips, S.D., 1979: Models for the evaluation of public participation programmes, *Natural Resources Journal*, 19(2), 338–57.

SFDF, 1979: *Non-compliance with the provisions for environmental protection in the Marri Woodchip Project Environmental Impact Statement*, South-West Forests Defence Foundation, Nedlands.

Shire of Mundaring, 1996: *Environmental management strategy*, Mundaring, WA.

Simmons, I.G., 1980: Ecological-functional approaches to agriculture in geographical contexts, *Geography*, 65, 305–16.

Simmons, I.G., 1993: *Interpreting nature: cultural constructions of the environment*, Routledge, London.

Simpson, T. and Jackson, V., 1997: Human rights and the environment, *Environmental and Planning Law Journal*, 14(4), 268–81.

Smailes, P., 1993: Socio-economic change and rural morale in South Australia 1983–1993, paper presented to the Institute of Australian Geographers Conference, Monash University, 27–30 September 1993.

Smith, D.G., 1990: *Continent in crisis—a natural history of Australia*, Penguin, Ringwood.

Smith, D.I., 1998: *Water in Australia: resources and management*, Oxford University Press, Melbourne.

Smith, G., 1996: The role of assessment of environmental effects under the Resource Management Act, *Environmental and Planning Law Journal*, 13(2), 82–102.

Smith, S., 1995a: Changing corporate environmental behaviour. Criminal prosecution as a tool of environmental policy, in R. Eckersley (ed.), *Markets and the state of the environment*, Macmillan, Melbourne.

Smith, S.L., 1995b: Doing time for environmental crimes: the United States approach to criminal enforcement of environmental laws, *Environmental and Planning Law Journal*, 12(3), 168–82.

Smith, S.T., 1962: Some Aspects of Soil Salinity in Western Australia, unpublished MSc (Agric) Thesis, University of Western Australia, Perth.

Sonntag, N.C., Everitt, R.R., Rattie, L.P. and others, 1987: *Cumulative effects assessment: a context for further research and development*, a Background Paper prepared for the Canadian Environmental Assessment Research Council, Minister of Supply and Services Canada, Ottawa.

South Australian Development Council, 1995: *Review of the South Australian aquaculture industry*, SA Department of Premier and Cabinet, Adelaide.

South West Natural Resource Management Group, 1999: *The natural resource management strategy. A living plan for south west Queenslanders, Interim Strategy*, Department of Natural Resources, South West Strategy Office, Charleville, Queensland.

SPC, 1989: *Peel regional planning study*, State Planning Commission, Perth.

SPC, 1994: *Albany Regional Strategy*, Adopted by the State Planning Commission as a Basis for Coordination of Local Planning, Control of Subdivision and Advice to the Hon. Minister and other Government Agencies, Perth.

SPC and CALM, 1988: *Shark Bay Region Plan*, State Planning Commission and Department of Conservation and Land Management, Perth.

Speck, N.H., Bradley, J., Lazarides, M., Patterson, R.A., Slatyer, R.O., Stewart, G.A. and Twidale, C.R., 1960: *The lands and pastoral resources of the North Kimberley area, Western Australia*, CSIRO Australia Land Research Series No. 4, Melbourne.

Spellman, P., 1998: The Western Australian Environmental Protection Amendment Act 1998, *Environmental and Planning Law Journal*, 15(4), 287–93.

Spennemann, D. and Marcar, N., 1999: Urban and heritage landscapes, *Natural Resource Management*, 2(1), 14–17.

Springett, J.A., 1976: The effect of prescribed burning on the soil fauna and on litter decomposition in Western Australian forests, *Australian Journal of Ecology*, 1, 77–82.

SROC, 1996: *Strategy for a sustainable Southern Region*, Prepared by Angela Hale, Environmental Planning Project officer for the Australian Local Government Association and the Southern Region of Councils, South Australia. 2 vols.

Standards Australia, 1996: *Australia/New Zealand Standard AS/NZS 14001: 1996. Environmental management systems—specifications with guidance for use*, Standards Australia, New South Wales.

State of the Environment Advisory Council, 1996: *State of the environment Australia 1996*, CSIRO Publishing, Collingwood, Victoria.

State of the Environment Unit, 1996: *State of the environment Tasmania. Volume 1. conditions and trends 1996*; State of the Environment Unit, Land Information Services, Department of Environment and Land Management, Hobart.

State of the Environment Unit, 1997: *State of the Environment Tasmania. Volume 2. Recommendations, 1997*, State of the Environment Unit, Land Information Services, Department of Environment and Land Management, Hobart.

State Pollution Control Commission, 1980: *Namoi environmental study*, New South Wales Department of Agriculture, National Parks and Wildlife Service, and others, New South Wales.

State Salinity Council, 1998: *Western Australian salinity action plan draft update, 1998*, State Salinity Council in association with community groups and government agencies, for the Government of Western Australia, Water and Rivers Commission, East Perth.

Steering Committee, 1980: *Research into the effects of the woodchip industry on water resources in south western Australia*, Department of Conservation and Environment Bulletin No. 81, Perth.

Steering Committee/DRE, 1983: *Water 2000. A perspective on Australia's water resources to the year 2000*, Report of the Steering Committee in conjunction with the Department of Resources and Energy, Australian Government Publishing Service, Canberra (plus 13 volumes of consultants' reports).

Stein, P., 1998: 21st Century challenges for urban planning—the demise of environmental planning in New South Wales, in B. Gleeson and P. Hanley (eds), *Renewing Australian planning? New challenges, new agendas*, Research School of Social Sciences, Australian National University, Canberra, 71–81.

Stephenson, G. and Hepburn, J., 1955: *Plan for the metropolitan region, Perth and Fremantle*, Government Printers, Perth.

Stewart, G.A. (ed.), 1968: *Land evaluation*, Macmillan, Melbourne.

Stewart, J., 1997: Australian water management: towards the ecological bureaucracy? *Environmental and Planning Law Journal*, 14(4), 259–67.

Story, R., Galloway, R.W., van de Graaff, R.H.M. and Tweedie, A.D., 1963: *General report on the lands of the Hunter Valley*, CSIRO Australia Land Research Series No. 8, Melbourne.

Streets, S. and di Carlo, A., 1999: Australia's first environmental protection measures: are we advancing, retreating or simply marking time? *Environmental and Planning Law Journal*, 16(1), 25–52.

Study Group on Environmental Assessment Hearing Procedures, 1988: *Public review: neither judicial, nor political, but an essential forum for the future of the environment*, A Report concerning the Reform of Public Hearing Procedures for Federal Environmental Assessment Reviews, Minister of Supply and Services Canada, Ottawa.

Swan–Avon Integrated Catchment management Group, 1997: *Working together. A recovery action plan for the Swan/Avon Catchment 1997 – 2007*, Swan Catchment Centre, Perth.

Taberner, J., Brunton, N. and Mather, L., 1996: The development of public participation in environmental protection and planning law in Australia, *Environmental and Planning Law Journal*, 13(4), 260–8.

Tamar Regional Master Planning Authority, 1979: *Tamar Region Plan—1979*, Tamar Regional Master Planning Authority, Launceston.

Taskforce, 1991: *National drought policy: final report*, vols 1–3, Drought Policy Review Taskforce, Australian Government Publishing Service, Canberra.

Taskforce, 1993: *Developing Australia: a regional perspective*, Report to the Federal Government, 2 vols, Taskforce on Regional Development, Department of Housing and Regional Development, Canberra.

Taskforce, 1997: *Taskforce for the review of natural resource management and viability of agriculture in Western Australia, Draft, May 1997*, Minister for Primary Industries, Government of WA, Perth.

Tasmanian Forestry Commission, 1994: *A state of the forests report*, Forestry Commission, Hobart.

Thackway, R. and Brunckhorst, D.J., 1998: Alternative futures for indigenous cultural and natural areas in Australia's rangelands, *Australian Journal of Environmental Management*, 5(3), 169–81.

Thackway, R. and Cresswell, I.D. (eds), 1995: *An interim biogeographic regionalisation for Australia: a framework for setting priorities in the national reserves systems co-operative program*, Australian Nature Conservation Agency, Canberra.

Thomas, I., 1996: *Environmental impact assessment in Australia: theory and practice*, The Federation Press, Leichhardt, New South Wales.

Thomas, I., 1998: *Environmental impact assessment in Australia: theory and practice*, 2nd edn, The Federation Press, Leichhardt, New South Wales.

Thorburn, L.J. and Till, M.R., 1988: *Overview of agricultural remote sensing in Australia*, CSIRO Office for Space Science and Applications, Canberra.

Threatened Species Steering Committee, 1993: *Threatened species strategy for South Australia*, Threatened Species Strategy Steering Committee, Adelaide.

Throne, J., 1997: Implementation of the Albany Regional Strategy, unpublished report produced for the Honours Option in Environmental Management, Department of Geography, University of Western Australia.

Tibor, T. and Feldman, I., 1996: *ISO 14000: a guide to new environmental management standards*, Irwin Professional Publishing, Chicago.

Tietenberg, T.H., 1993: Economic instruments for environmental regulation, in L. Owen and T. Unwin (eds), *Environmental management: readings and case studies*, Blackwell, Oxford, 386–99.

Tobias, M. (ed.), 1985: *Deep ecology*, Avant Books, San Diego.

Trayler, K.M., Davis, J.A., Horwitz, P. and Morgan, D., 1996: Aquatic fauna of the Warren bioregion, south-west Western Australia: does reservation guarantee preservation? *Journal of the Royal Society of Western Australia*, 79, 281–91.

Tribe, J., 1998: The law of the jungles. Regional forest agreements, *Environmental and Planning Law Journal*, 15(2), 136–46.

Trotman, C.H. (ed.), 1974: *The influence of land use on stream salinity in the Manjimup area, Western Australia*, Department of Agriculture Western Australia Technical Bulletin No. 27, Perth.

UNCED, 1987: *Our common future: Brundtland Report*, United Nations Commission on Environment and Development, Oxford University Press, Oxford.

UNCED, 1992: *Agenda 21: UN conference on environment and development*, United Nations Commission on Environment and Development Secretariat, Geneva.

UNESCO, 1986: *International coordinating council of the programme on Man and the Biosphere (MAB)*, MAB Report Series No. 60, UNESCO, Paris.

Valentine, P.S., 1976: *A preliminary investigation into the effects of clear-cutting and burning on selected soil properties in the Pemberton area of Western Australia*, Geowest No. 8, Occasional Papers of the Department of Geography, University of Western Australia.

Vanclay, F., 1997: The social basis of environmental management in agriculture: a background for understanding Landcare, in S. Lockie and F. Vanclay (eds), 1997: *Critical landcare*, Centre for Rural Social Research, Charles Sturt University, Wagga Wagga, NSW, 9–27.

Von Bertalanffy, L., 1968: *General System Theory: foundations, development, applications*, George Braziller, New York.

WAGCC, 1989: *Addressing the Greenhouse Effect*, WA Greenhouse Coordination Council, Perth.

Walker, J. and Reuter, D.J., 1996: *Indicators of catchment health: a technical perspective*, CSIRO, Collingwood.

Wallace, W.R., 1970: *Forests Department, Western Australia: Annual Report 1970*, Perth.

Walter, T.S., 1976: Some cost–benefit aspects of wood chipping in Western Australia, *Economic Activity*, 19, 56–65.

WAPC, 1995a: *State Planning Strategy Discussion Papers*, (8), WA Planning Commission, Perth.

WAPC, 1995b: *Bunbury–Wellington Region Plan*, WA Planning Commission, Perth.

WAPC, 1996a: *State Planning Strategy*, WA Planning Commission, Perth.

WAPC, 1996b: *Central coast regional strategy*, WA Planning Commission, Perth.

WAPC, 1997: *Planning for agricultural and rural land use: discussion paper*, WA Planning Commission, Perth and Agriculture WA, South Perth.

WAPC, 1998: *Perth's bushplan. Keeping the bush in the city*, for public comments, WA Planning Commission, Perth.

WAPC, 1999: *The Peel region scheme*, for public discussion, WA Planning Commission, Perth.

Warnken, J. and Buckley, R., 1998: Scientific quality of tourism environmental impact assessments, *Journal of Applied Ecology*, 35, 1–8.

Wathern, P. (ed.), 1988: *Environmental impact assessment: theory and practice*, Unwin Hyman Ltd, London.

Watson, B., Morrisey, H. and Hall, N. (1997): Economic analysis of dryland salinity issues, Paper presented at 'Outlook 97', National Agricultural and Resources Outlook Conference, Canberra, 4–6 February.

WCED, 1987: *Our common future*, World Commission on Environment and Development, Oxford University Press, Oxford (Australian edition Oxford University Press, Melbourne, 1990).

Wells, M.R. and King, P.D., 1989: *Land capability assessment methodology for rural residential development and associated agricultural land uses*, Land Resources Series Bulletin 1, WA Department of Agriculture, Perth.

Westman, W.E., 1985: *Ecology, impact assessment and environmental planning*, Wiley-Interscience, New York.

White, B.J., 1977: Focus on southern recreation and conservation management priority areas, *Forest Focus*, 18, 3–23.

White, M., 1997: *Listen…our land is crying. Australia's environmental problems and solutions*, Kangaroo Press, Sydney.

Whittington, H.S., 1975: *A battle for survival against salt encroachment at 'Springhill', Brookton, Western Australia*, H.S. Whttington, PO Box 9, Brookton, Western Australia.

WHO, 1993: *Biomarkers and risk assessment*, World Health Organisation, Geneva.

Wilcher, R., 1995: Water resource management in New South Wales and the Murray–Darling: integrating law and policy, *Environmental and Planning Law Journal*, 12(3), 200–10.

Williams, J., 1997: Indicators of catchment health workshop overview, in *Proceedings of the national workshop on indicators of catchment health*, Waite Campus, Adelaide, December 1996, 19–29.

Williams, J., Hook, R. and Gascoigne, H., 1998: *Farming action: catchment reaction. The effect of dryland farming on the natural environment*, CSIRO Publishing, Collingwood, Victoria.

Williams, M., 1993: The climate change convention, *Our Planet*, 5(6), 7.

Williams, M., McCarthy, M. and Pickup, G., 1995: Desertification, drought and landcare: Australia's role in an international convention to combat desertification, *Australian Geographer*, 26(1), 23–32.

Williamson, D.R. and Bettenay, E., 1979: Agricultural land use and its effect on catchment output of salt and water—evidence from southern Australia, *Progress in Water Technology*, 11, 463–80.

Willis, R., 1992: New Zealand, *Progress in Rural Policy and Planning*, 2, 262–5.

Willis, R., 1993: New Zealand, *Progress in Rural Policy and Planning*, 3, 389–92.

Wischmeier, W.H. and Smith, D.D., 1978: *Predicting rainfall erosion losses—a guide to conservation planning*, Agricultural Handbook 537, United States Department of Agriculture, Washington D.C. (updated and revised from the 1965 publication).

Withers, P.C. and Horwitz, P. (eds), 1996: *Symposium on the design of reserves for nature conservation in south-western Western Australia*, special issue of the *Journal of the Royal Society of Western Australia*, 79(Part 4).

Wood, C., 1995: *Environmental impact assessment: a comparative review*, Longman Scientific and Technical, Harlow, England.

Wood, W.A., 1924: Increase in salt in soil and streams following the destruction of the native vegetation, *Journal of the Royal Society of Western Australia*, 10, 35–47.

Woods, L.E., 1983: *Land degradation in Australia*, Australian Government Publishing Service, Canberra.

Woolf, B., 1992: Environment Protection Authority legislation, *Environment Institute of Australia Newsletter*, 19, 37–38.

Working Group, 1991: *Decade of Landcare plan for South Australia*, Working Group of Natural Resources Management Standing Committee, Department of Agriculture, South Australia.

World Bank, 1997: *Five years after Rio: innovations in environmental policy*, Environmentally Sustainable Development Studies and Monographs Series No. 18, The International Bank for Reconstruction and Development/The World Bank, Washington D.C.

Worster, D., 1990: Ecology of order and chaos, *Environmental History Review*, 14, 1–18.

WRC, 1997: *Draft policy and principles. Protection of waters from pollution in Western Australia*, Water Resources Protection Series No. 27, Water and Rivers Commission, Perth.

Wright, I., 1995: Implementation of sustainable development by Australian local governments, *Environmental and Planning Law Journal*, 12(1), 54–61.

Young, A. and Goldsmith, P.F., 1977: Soil survey and land evaluation in developing countries, *Geographical Journal*, 143, 407–31.

Zann, L., 1995: *Our sea, our future: major findings of State of Marine Environment Report*, Great Barrier Marine Park Authority, Department of Environment, Sport and Territories, Canberra.

INDEX